COMPLEX SURVEY
DATA ANALYSIS
WITH SAS®

COMPLEX SURVEY DATA ANALYSIS WITH SAS®

Taylor H. Lewis

Department of Statistics
George Mason University
Fairfax, Virginia, USA

CRC Press
Taylor & Francis Group
Boca Raton London New York

CRC Press is an imprint of the
Taylor & Francis Group, an **informa** business

A CHAPMAN & HALL BOOK

CRC Press
Taylor & Francis Group
6000 Broken Sound Parkway NW, Suite 300
Boca Raton, FL 33487-2742

First issued in paperback 2021

© 2017 by Taylor & Francis Group, LLC
CRC Press is an imprint of Taylor & Francis Group, an Informa business

No claim to original U.S. Government works

Version Date: 20160427

ISBN 13: 978-1-03-224200-2 (pbk)
ISBN 13: 978-1-4987-7677-6 (hbk)

DOI: 10.1201/9781315366906

Library of Congress Cataloging-in-Publication Data

Names: Lewis, Taylor H.
Title: Complex survey data analysis with SAS / Taylor H. Lewis.
Description: Boca Raton : Taylor & Francis, 2016. | "A CRC title." | Includes bibliographical references and index.
Identifiers: LCCN 2016008468 | ISBN 9781498776776
Subjects: LCSH: Multivariate analysis--Data processing. | Regression analysis--Data processing. | Sampling (Statistics) | SAS (Computer file) | Surveys.
Classification: LCC QA278 .L49 2016 | DDC 519.5/30285555--dc23
LC record available at https://lccn.loc.gov/2016008468

Visit the Taylor & Francis Web site at
http://www.taylorandfrancis.com

and the CRC Press Web site at
http://www.crcpress.com

To my three beautiful girls: Katie, Tessa, and Willow

Contents

Preface

I wrote this book to serve as a handy desk-side reference for students, researchers, or any other kind of data analysts who have an intermediate to advanced grasp on fundamental statistical techniques, but who may not be familiar with the nuances introduced by complex survey data. As I embarked on this project, my intent was never to supplant or even compete with any of the established textbooks on the subject, such as Kish (1965), Cochran (1977), Lohr (2009), or Heeringa et al. (2010), for that would have surely been a futile endeavor. Rather, my aim was to demonstrate easy-to-follow applications of survey data analysis for researchers like myself who rely on SAS primarily, if not exclusively, for conducting statistical analyses.

All material in this book was developed using SAS Version 9.4 and SAS/STAT 13.1. It features the SURVEY family of SAS/STAT procedures, of which there are six at the time of this writing: PROC SURVEYSELECT, PROC SURVEYMEANS, PROC SURVEYFREQ, PROC SURVEYREG, PROC SURVEYLOGISTIC, and PROC SURVEYPHREG. The last five listed are companions to PROC MEANS, PROC FREQ, PROC REG, PROC LOGISTIC, and PROC PHREG, respectively. As will be explained in great detail over the course of this book, you should use one of the SURVEY procedures when analyzing data from a complex survey design. Using one of the non-SURVEY procedures opens the door to making false inferences due to biased point estimates, biased measures of variability, or both.

The book is structured as follows. Chapter 1 is a brief introduction to the practice of applied survey research. Key terminology and notation associated with the process of estimating finite population parameters from a sample are defined, as are the four features of complex survey data. Chapter 2 covers PROC SURVEYSELECT, which boasts numerous useful routines to facilitate the task of selecting probability samples. Descriptive, univariate analyses of continuous variables are covered in Chapter 3. Categorical variable analysis, both univariate and multivariate, is discussed in Chapter 4. Chapters 5 and 6 deal with the analytic task of fitting a linear model to survey data using PROC SURVEYREG or PROC SURVEYLOGISTIC, respectively. Both chapters begin with a section reviewing the assumptions, interpretations, and key formulas within the milieu of data collected via a simple random sample and then segue into the conceptual and formulaic differences within the realm of complex survey data. Chapter 7 is a foray into survival analysis, highlighting in large part the newest addition to the SURVEY family of SAS procedures, PROC SURVEYPHREG, which was developed for fitting Cox proportional hazard regression models. Chapter 8 delves into the concept of domain estimation, referring to any analysis targeting a subset of the target population. We will see why the DOMAIN statement should be used instead of subsetting a complex survey data set. Chapter 9 touches on the increasingly popular class of variance estimators known as replication techniques, which offer analysts a flexible alternative to Taylor series linearization, the default variance estimation method used in all SURVEY procedures. Because missing data is a ubiquitous problem in applied survey research, in my humble opinion, any book would be remiss without some material discussing approaches used to compensate for it. To that end, Chapter 10 demonstrates methods to adjust the weights of responding cases to better reflect known characteristics of the sample (or population), and Chapter 11 considers methods to impute, or fill in, plausible values for the missing data.

Each chapter was designed to be read on its own and not necessarily in sequence. As each chapter progresses, the material tends to build on itself and become more complex. It is my belief that oscillating between numerous survey data sets can only serve to detract from the reader's ability to compare and contrast the current syntax example with those preceding it in the chapter. As such, examples in most chapters are motivated within the context of a single complex survey data set.

Many people contributed to this effort and are owed my sincere thanks. First and foremost, I am especially grateful for the encouragement I received from Shelley Sessoms at SAS, who believed in me and the vision I had for this book and on several occasions helped me overcome my self-doubt about whether I could complete this colossal task. I thank Rob Calver for resuscitating this project, and for the professionalism he and his entire team at CRC Press exhibited. I am also very grateful for the encouragement I received on both professional and personal levels from Frauke Kreuter, Partha Lahiri, Brady West, Roberto Valverde, Abera Wouhib, Sidney Fisher, Mircea Marcu, Karl Hess, Kimya Lee, Chris Daman, Glenn White, Eugene Pangalos, Graham Evans, Ryan Fulcher, Matt Fuller, and Doug Drewry. Richard Valliant and Richard Sigman were kind enough to endure several lengthy conversations during which I solicited their advice on certain content issues that surfaced as the book took shape. I also thank Richard Sigman for agreeing to serve as a reviewer. He, along with Patricia Berglund, Peter Timusk, Donna Brogan, and several anonymous reviewers at SAS Institute, provided valuable feedback that greatly improved the overall quality of the book. The same can be said for Casey Copen and Kimberly Daniels at the National Center for Health Statistics, who carefully reviewed all discussion and examples pertaining to the National Survey of Family Growth. I claim full responsibility for any and all errors that remain. Lastly, and most importantly, I thank my wife, Katie, for being extraordinarily patient and understanding during this process, selflessly tolerating my extended absences from home without (too much) complaint. None of this would have been possible without her love and support.

Taylor Lewis
Arlington, Virginia

Author

Taylor H. Lewis is a PhD graduate of the Joint Program in Survey Methodology at the University of Maryland, College Park, and an adjunct professor in the George Mason University Department of Statistics. An avid SAS user for 15 years, Taylor is a SAS Certified Advanced programmer and a nationally recognized SAS educator who has produced dozens of papers and workshops illustrating how to efficiently and effectively conduct statistical analyses using SAS.

1

Features and Examples of Complex Surveys

1.1 Introduction

In the era of *Big Data*, the variety of statistics that can be generated is ostensibly limitless. Given the copious and ever-expanding types of data being collected, there are many questions that can be answered from analyzing a data set already in existence or perhaps one even updated in real time. For instance, a credit card issuer seeking to determine the total amount of charges made by its customers on gasoline during a particular year may have this information readily retrievable from one or more databases. If so, a straightforward query can produce the answer. On the other hand, determining the average amount the typical U.S. household spends on gasoline presents a much more complicated estimation problem. Collecting data from all households in the United States would obviously be exceedingly costly, if not an outright logistical impossibility. One could probably make some progress pooling the comprehensive set of credit card issuers' databases and trying to group data into distinct households via the primary account holder's address, but not all households own and use a credit card, and so this would exclude any non-credit-card payment such as one made by cash or check. A survey of the general U.S. population is needed to acquire this kind of information. One such survey is the Consumer Expenditure Survey, sponsored by the Bureau of Labor Statistics, which reported in September 2015 that the average household spent approximately $2500 on gasoline during calendar year 2014 (http://www.bls.gov/news.release/cesan.htm).

It is truly a marvel to consider the breadth of statistics such as these available to answer a myriad of questions posed by researchers, policymakers, and the general public. The legitimacy of these statistics is attributable to the fields of survey sampling and survey research methodology, which together have engendered a wide variety of techniques and approaches to practical data collection problems. Hansen (1987) and Converse (1987) present nice summaries of the two fields' overlapping histories. The practice of modern survey sampling began with the argument that a sample should be *representative* (Kiaer, 1895) and drawn using techniques of randomization

(Bowley, 1906). The seminal paper by Neyman (1934) provided much needed theoretical foundations for the concept. Nowadays, there are textbooks entirely devoted to considerations for designing an efficient sample (Hansen et al., 1953; Kish, 1965; Cochran, 1977; Lohr, 2009; Valliant et al., 2013). Best practices in the art of fielding a survey have emerged more recently, and they continue to evolve in response to changes in information technology, communication patterns of the general public, and other societal norms. There are certainly a number of excellent volumes summarizing the literature on survey methods (Couper, 2008; Dillman et al., 2009; Groves et al., 2009), but they tend to have a shorter shelf life—or at least warrant a new edition more frequently—than those on sampling.

Survey research is the quintessence of an interdisciplinary field. While the opening example was motivated by expenditure data that might be of interest to an economist, there are analogous survey efforts aimed at producing statistics related to agriculture and crop yields, scholastic achievement, and crime, just to name a few. Of course, collecting data comes at a cost. In the United States, for-profit businesses do not generally conduct surveys and release raw data files to the public free of charge. More commonly, surveys are funded by one or more government agencies. These agencies are ideally apolitical and charged solely with the task of impartially collecting and disseminating data. The set of Principal Statistical Agencies listed at http:// fedstats.sites.usa.gov/agencies/ more or less fits this description. Aside from preformatted tables and reports, data dissemination often takes the form of a raw or microdata file posted on the survey website for open access. Indeed, many of the examples in this book are drawn from three *real-world* survey data sets sponsored by two of these agencies, the National Center for Health Statistics (NCHS) and the Energy Information Administration. Namely, the National Ambulatory Medical Care Survey (NAMCS), the National Survey of Family Growth (NSFG), and the Commercial Buildings Energy Consumption Survey (CBECS) will be formally introduced in Section 1.5.

As authoritative or *official* as these statistics seem, it is important to bear in mind they are *estimates*. The term *estimate* can sometimes be confused with the very similar term *estimator*, but the two terms have different meanings. An estimate is the value computed from a sample, whereas an estimator is the method or technique used to produce the estimate (see Program 3.7 for a comparison of two unbiased estimators of a total). If the entire survey process were conducted anew, there are a variety of reasons one would expect an estimate to differ somewhat, but this book focuses primarily on quantifying the portion of this variability attributable to *sampling error* or the variability owing to the fact that we have collected data for only a portion of the population. Using formal statistical theory and a single sample data set in hand, however, there are established ways to calculate an unbiased estimate of the sampling error, which can be reported alongside the estimate or used to form a confidence interval or to conduct a hypothesis test. A distinctive aspect of *complex* survey data, the features of which will be detailed in

Section 1.4 and which are all too often overlooked by applied researchers, is that the techniques and formulas for estimating sampling error one learns in general statistics courses or from general statistics textbooks frequently do not carry over intact. The reason is that complex surveys often employ alternative sample designs, either for the purpose of statistical efficiency or out of necessity to control data collection costs. The implied data-generating mechanism in general statistics courses is simple random sampling with replacement (SRSWR) of the ultimate units for which data are measured. In applied survey research, that particular data-generating mechanism is the exception rather than the rule.

Section 1.2 establishes some of the terminology pertaining to applied survey research that will be used throughout this book. Section 1.3 previews the SAS/STAT procedures that have been developed to facilitate complex survey data analysis. These are all prefixed with the word SURVEY (e.g., PROC SURVEYMEANS is the companion procedure to PROC MEANS). Section 1.4 introduces the four features that may be present in a survey data set to justify the qualifier *complex*: (1) finite population corrections (FPCs), (2) stratification, (3) clustering, and (4) unequal weights. This chapter concludes with a discussion of the three real-world complex survey data sets from which many of the book's examples are drawn. There is some brief commentary on the motivation behind each survey effort, the type of sample design employed, the complex survey features present in the data set, and specific examples of estimates produced.

1.2 Definitions and Terminology

Groves et al. (2009, p. 2) define a survey as a "systematic method for gathering information from (a sample of) entities for constructing quantitative descriptors of the attributes of the larger population for which the entities are members." They use the term "entities" to stress the fact that, although the word "survey" often has the connotation of an opinion poll or a battery of questions directed at humans, this is not always the case. Other example entities are farms, businesses, or even events. Parenthetically including the phrase "a sample of" serves to remind us that not all surveys involve sampling. A *census* is the term describing a survey that aims to collect data on or enumerate an entire population.

One of the first stages of any survey sampling effort is defining the *target population*, the "larger population" alluded to in the Groves et al. definition about which inferences are desired. The target population often carries an ambitious, all-encompassing label, such as "the general U.S. population." The next step is to construct a list, or *sampling frame*, from which a random sample of *sampling units* can be drawn. The totality of entities covered by this list is called the *survey population,* which does not always coincide perfectly with the target population. For example, there is no population registry in the United States as there is in many European countries to serve as a sampling frame. There is an important distinction to be made between the sampling units and the *population elements,* or the ultimate analytic units for which measurements are taken and inferences drawn. The two are not always one and the same. Sometimes, the sampling frame consists of clusters of the population elements. Considering the goal of making inferences on the general U.S. population, even if a population registry existed, it might be oriented by household or address instead of by individual. This would present a cluster sampling situation, which is permissible but introduces changes to the more familiar formulas used for estimation. We will discuss cluster sampling more in Section 1.4.4 with the help of a simple example.

A sampling frame's makeup is often influenced by the survey's *mode* or method of data collection (e.g., in-person interview or self-administered paper questionnaire by mail). For example, a popular method for administering surveys by telephone is *random-digit dialing* (RDD), in which the sampling frame consists of a list of landline telephone numbers. A survey opting for this mode may still consider the target population "the general U.S. population," but the survey population is actually the subset of U.S. households with a landline telephone.

Figure 1.1 illustrates how the target population and survey population may not always be one and the same. The area within the target population that does not fall within the survey population area is of most concern. That represents *undercoverage,* meaning population elements that have no chance of being selected into the sample. Continuing with the RDD example, households without a landline telephone fall within this domain. Sometimes, it is possible to supplement one sampling frame with another to capture this group (e.g., by incorporating a sampling frame consisting of cellular telephone numbers), but that can introduce duplicated sampling units (i.e., households with landline *and* cellular numbers, therefore present in both frames), which can be a nuisance to deal with in its own right (Lohr and Rao, 2000). Another remedy often pursued is to conduct weighting adjustment techniques such as post-stratification or raking. These techniques will be discussed in Chapter 10.

There is also an area in Figure 1.1 delineating a portion of the survey population falling outside the bounds of the target population. This represents extraneous, or ineligible, sampling units on the sampling frame that may be selected as part of the sample. With respect to an RDD survey of U.S. households, a few examples are inoperable, unassigned, or business telephone numbers. These are represented by the area to the right of the vertical line in the oval labeled "Sample." An appreciable rate of ineligibility can cause inefficiencies in the sense that these units must be *screened* out where identified, but it is generally easier to handle than undercoverage.

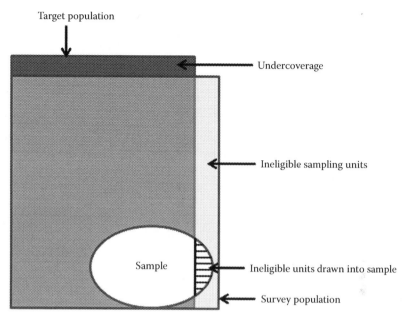

FIGURE 1.1
Visualization of a sample relative to the target population and survey population.

It is not unusual for those unfamiliar with survey sampling theory to be skeptical at how valid statistical inferences can be made using a moderate sample of size n from a comparatively large population of size N. This is a testament to the *central limit theorem*, which states that the distribution of a sequence of means computed from independent, and random samples of size n is always normally distributed as long as n is *sufficiently large*. Despite the vague descriptor *sufficiently large* fostering a certain amount of debate amongst statisticians—some might say 25, others 50, and still others 100—the pervading virtue of the theorem is that it is true regardless of the underlying variable's distribution. In other words, you do not need normally distributed data for this theorem to hold. This assertion is best illustrated by simulation.

Figure 1.2 shows the distribution of three variables from a fabricated finite population of $N=100,000$. The first variable is normally distributed with a population mean of $\bar{y}_1 = 5$. The second is right-skewed with a mean of $\bar{y}_2 = 2$, whereas the third variable has a bimodal distribution with a mean of $\bar{y}_3 = 3.75$. Figure 1.3 immediately following displays the result of a simulation that involved drawing 5000 samples of size $n = 15$, $n = 30$, and $n = 100$ from this population and computing the sample mean for each of y_1, y_2, and y_3. That is, the figure is comprised of histograms summarizing the distribution of the three variables' sample means with respect to each of the three sample sizes. As in Figure 1.2, the row factor distinguishes the three variables, while

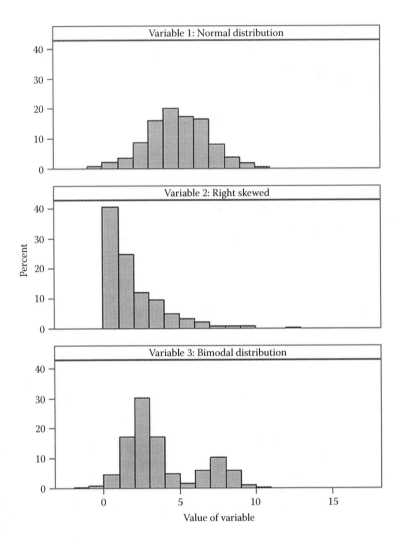

FIGURE 1.2
Distribution of three variables from a finite population of $N = 100,000$.

the column factor (moving left to right) distinguishes the increasing sample sizes. There are a few key takeaways from examining Figure 1.3:

- All sample mean distributions closely resemble a normal distribution, which has been superimposed on the histograms. Again, this is true regardless of the underlying distribution.

- The average or *expected* value of the 5000 sample mean values is the population mean, which is to say the sample mean is an unbiased estimate for the mean of the entire population.

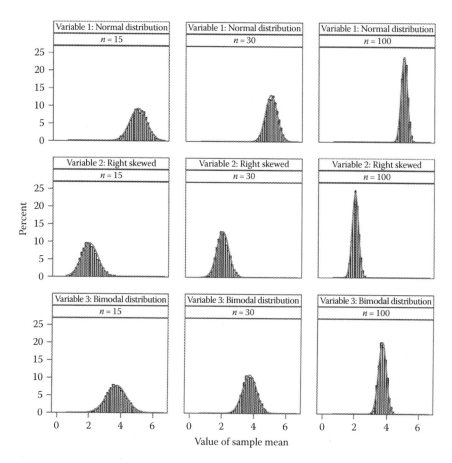

FIGURE 1.3
Sample mean distributions of 5000 samples of size 15, 30, and 100 drawn from the finite population in Figure 1.2.

- The distributions get "skinnier" as the sample size increases. This reflects increased precision or less deviation amongst the means from one sample to the next.

Knowing the sampling distribution of statistics such as the mean is what justifies our ability to make inferences about the larger population (e.g., to form confidence intervals, conduct significance tests, and calculate p values). In practice, we do not have a set of 5000 independent samples, but usually a single sample to work with. From this sample, a statistic is produced, which we typically top with a hat to distinguish it from the true population parameter it is estimating. Using general theta notation, we can denote $\hat{\theta}$ as the sample-based estimate or *point estimate* of θ. For example, $\hat{\bar{y}}$ refers to the point estimate of $\bar{y}$ from a particular sample. We acknowledge, however, that this estimate likely does not match the population parameter exactly. A fundamental quantification of the anticipated deviation is the *variance*, which can be expressed as $\mathrm{Var}(\hat{\theta}) = E[(\hat{\theta} - \theta)^2]$. The variance represents the average squared deviation of the sample-based estimates from the true population value over all possible samples. This quantity is rarely known, except perhaps in simulated settings such as the one discussed earlier. But drawing from formal statistical theory, there are established computational methods to formulate an unbiased estimate of it using only the single sample at hand. The sample-based estimate is referred to as the *estimated variance* or *approximated variance* and denoted $\mathrm{var}(\hat{\theta})$ (with a lowercase "v"). Over the course of the book, we will see expressions of variance estimates for a whole host of statistics, along with program examples demonstrating SURVEY procedure syntax to have SAS carry out the computational legwork for us.

Numerous related measures of uncertainty can be derived from the estimated variance. Since variability in squared units can be difficult to interpret, a common transformation is to take the square root of this quantity, which returns the *standard error* of the statistic denoted $\mathrm{se}(\hat{\theta}) = \sqrt{\mathrm{var}(\hat{\theta})}$. Along with the estimate itself, the standard error can be used to form a confidence interval $\hat{\theta} \pm t_{df,\alpha/2}\mathrm{se}(\hat{\theta})$, where $t_{df,\alpha/2}$ represents the $100(1 - \alpha/2)$th percentile of a t distribution with df degrees of freedom. Another popular measure is the relative standard error of the estimate, also known as the *coefficient of variation*, which is defined as $\mathrm{CV}(\hat{\theta}) = \mathrm{se}(\hat{\theta})/\hat{\theta}$. This has an appealing interpretation. A value of 0.1, for example, indicates the standard error of the statistic represents 10% of the magnitude of the statistic itself. A target CV can be used as a sample design criterion before the survey is administered or as a threshold for whether a survey estimate is deemed precise enough to be published. Unlike the variance and standard error, the CV is unitless, thereby permitting precision comparisons regardless of scale (e.g., dollars versus percentages) or type of estimate (e.g., means versus totals).

It should be emphasized that these assertions only apply if the sample is selected randomly. An impressive-sounding sample size of, say, 10,000 means little if it were drawn using a nonrandom process such as convenience sampling, quota sampling, or judgment sampling (see Lohr, 2009, p. 5). The essential requirement is that every sampling unit on the sampling frame has a known, nonzero chance of being selected as part of the sample. The selection probabilities need not be equal, but they should be fixed and nonzero.

1.3 Overview of SAS/STAT Procedures Available to Analyze Survey Data

Table 1.1 previews the SURVEY procedures to be covered in the coming chapters. PROC SURVEYSELECT was developed to facilitate the task of sample

TABLE 1.1

Summary of SURVEY Procedures Currently Available and Primary Chapter(s) in Which They Are Covered

Procedure	Analytic Tools	Chapter(s)
PROC SURVEYSELECT	Variety of built-in routines for selecting random samples; also contains a few methods for allocating a fixed sample size amongst strata and determining a necessary sample size given certain constraints	Chapter 2
PROC SURVEYMEANS	Descriptive statistics such as means, totals, ratios, quantiles, as well as their corresponding measures of uncertainty	Chapter 3
PROC SURVEYFREQ	Tabulations, tests of association, odds ratios, and risk statistics	Chapter 4
PROC SURVEYREG	Regression models where the outcome is a continuous variable	Chapter 5
PROC SURVEYLOGISTIC	Regression models where the outcome is a categorical variable	Chapters 6 and 7
PROC SURVEYPHREG	Cox proportional hazards models for time-to-event data (survival analyses)	Chapter 7

selection, so it is generally more useful at the survey design stage rather than the analysis stage. Once the data have been collected, there are five additional SURVEY procedures available to produce descriptive statistics and conduct more sophisticated analytic tasks such as multivariate modeling. Each procedure is typically covered in a chapter of its own, but later chapters covering cross-cutting topics such as domain estimation (Chapter 8) and replication (Chapter 9) demonstrate more than one specific SURVEY procedure. We will explore alternative (non-SURVEY) SAS/STAT procedures and user-defined macros in the final two chapters, which deal with techniques for handling missing data.

1.4 Four Features of Complex Surveys

1.4.1 A Hypothetical Expenditure Survey

To motivate exposition of the four features of complex survey data, suppose a market research firm has been hired to assess the spending habits of $N = 2000$ adults living in a small town. Two example estimates of interest are the average amount of money an adult spent in the previous year on over-the-counter (OTC) medications and the average amount spent on travel outside the town. The ensuing discussion centers around a few (hypothetically carried out) sample designs to collect this expenditure data for a sample of $n = 400$ adults and the particular complex survey features introduced.

1.4.2 Finite Population Corrections

Suppose the first sample design involved compiling the names and contact information for all $N = 2000$ people in the town onto a sampling frame and drawing a simple random sample (SRS) of $n = 400$ of them to follow up with to collect the expenditure information. From this sample, the estimated average for a given expenditure y would be calculated as $\hat{\bar{y}} = \sum_{i=1}^{n=400} y_i / n$. The research firm might then reference an introductory statistics textbook and calculate an estimated *element variance* of y as $s^2 = \sum_{i=1}^{n=400} (y_i - \bar{y})^2 / (n-1)$, an unbiased, sample-based estimate of the population element variance, or $S^2 = \sum_{i=1}^{N=2000} (y_i - \bar{y})^2 / (N-1)$, and use this quantity to estimate the variance of the sample mean by $\text{var}(\hat{\bar{y}}) = s^2/n$. They might then construct a 95% confidence interval as $\hat{\bar{y}} \pm 1.96 \sqrt{\text{var}(\hat{\bar{y}})}$, where $\sqrt{\text{var}(\hat{\bar{y}})} = \text{se}(\hat{\bar{y}})$ is the standard error of $\hat{\bar{y}}$. Note how the standard error differs conceptually and computationally from the *standard deviation* of y, which is $S = \sqrt{S^2}$ for the full population and estimated by $s = \sqrt{s^2}$ from the sample.

It turns out the market research firm's calculations would overestimate the variance of the sample mean because whenever the sampling fraction n/N is nonnegligible (as is the case with 400/2000), there is an additional term that enters into variance estimate calculations called the FPC. A more accurate estimate of the variance of the sample mean is $\text{var}(\hat{\bar{y}}) = (s^2/n)(1 - (n/N))$, where the term $(1 - (n/N))$ is the FPC. Notice how the FPC tends to 0 as n approaches N. Since the purpose of the estimated variance is to quantify the sample-to-sample variability of the estimate, an intuitive result is that it decreases as the portion of the population in each respective sample increases. In the extreme case when $n = N$, or when a census is undertaken, the variance accounting for the FPC would be 0. And despite the discussion in this section pertaining strictly to estimating the variance of a sample mean, a comparable variance formula modification occurs for other statistics such as totals and regression coefficients.

The difference between the two variance perspectives is that the traditional formula implicitly assumes data were collected under a SRSWR design, meaning each unit in the population could have been selected into the sample more than one time. Equivalently, the tacit assumption could be that data were collected using simple random sampling without replacement (SRSWOR) from an effectively infinite population, a corollary of which is that the sampling fraction is negligible and can be ignored. Sampling products from an assembly line or trees in a large forest might fit reasonably well within this paradigm. But in contrast, survey research frequently involves sampling from a finite population, such as employees of a company or residents of a municipality, in which case adopting the SRSWOR design formulas is more appropriate.

There are two options available within the SURVEY procedures to account for the FPC. The first is to specify the population total N in the TOTAL = option of the PROC statement. The second is to specify the sampling fraction n/N in the RATE = option of the PROC statement. With respect to the sample design presently considered, specifying TOTAL = 2000 or RATE = 0.20 has the same effect. The syntax to account for the FPC is identical across all SURVEY procedures, and the same is true for the other three features of complex survey data as well.

Suppose the SAS data set SAMPLE_SRSWOR contains the results of this survey of $n = 400$ adults in the town. Program 1.1 consists of two PROC SURVEYMEANS runs on this data set. We will explore the features and capabilities of PROC SURVEYMEANS in more detail in Chapter 3, but for the moment note that we are requesting the sample mean and its estimated variance for the OTC expenditures variable (EXP_OTCMEDS). The first run assumes the sample was selected with replacement. Since there are no complex survey features specified, it produces the same figures that would be generated from PROC MEANS. The second requests the same statistics but specifies TOTAL=2000 in the PROC statement, in effect alerting SAS that sampling was done without replacement and so an FPC should be incorporated. (The SURVEY procedure determines n from the input data set.)

Program 1.1: Illustration of the Effect of an FPC on Measures of Variability

```
title1 'Simple Random Sampling without Replacement';
title2 'Estimating a Sample Mean and its Variance Ignoring the
FPC';
proc surveymeans data=sample_SRSWOR mean var;
  var exp_OTCmeds;
run;

title2 'Estimating a Sample Mean and its Variance Accounting
for the FPC';
proc surveymeans data=sample_SRSWOR total=2000 mean var;
  var exp_OTCmeds;
run;
```

Simple Random Sampling without Replacement
Estimating a Sample Mean and Its Variance Ignoring the FPC

SURVEYMEANS Procedure

Data Summary	
Number of observations	400

Statistics			
Variable	**Mean**	**Std Error of Mean**	**Var of Mean**
exp_OTCmeds	17.645854	0.683045	0.466550

Simple Random Sampling without Replacement
Estimating a Sample Mean and Its Variance Accounting for the FPC

SURVEYMEANS Procedure

Data Summary	
Number of observations	400

Statistics			
Variable	**Mean**	**Std Error of Mean**	**Var of Mean**
exp_OTCmeds	17.645854	0.610934	0.373240

The sample mean is equivalent ($17.65) in both PROC SURVEYMEANS runs, but measures of variability are smaller with the FPC incorporated. Specifically, the estimated variance of the mean has been reduced by a factor of 20% as we can observe that $0.3732 = 0.4666 * (1-(400/2000))$. Since the standard error of the mean is just the square root of the variance, by comparison it has been reduced to $0.6109 = 0.6830* \sqrt{1-(400/2000)}$.

Where applicable, the FPC is beneficial to incorporate into measures of variability because doing so results in increased precision and, therefore, more statistical power. There are occasions, however, when the FPC is known to exist but intentionally ignored. This is often done when assuming a with-replacement sample design dramatically simplifies the variance estimation task (see discussion regarding the ultimate cluster assumption in Section 1.4.4), especially when there is only a marginal precision gain to be realized from adopting the without-replacement variance formula. Providing a few numbers to consider, with a sampling fraction of 10%, we would anticipate about a 5% reduction in the standard error; if the sampling fraction were 5%, the reduction would be around 3%. While the with-replacement assumption typically imposes an overestimation of variability, the rationale behind this practice is that the computational efficiencies outweigh the minor sacrifice in precision.

1.4.3 Stratification

The second feature of complex survey data is *stratification*, which involves partitioning the sampling frame into H mutually exclusive and exhaustive *strata* (singular: stratum), and then independently drawing a sample within each. There are numerous reasons the technique is used in practice, but a few examples are as follows:

- *Ensure representation of less prevalent subgroups in the population.* If there is a rare subgroup in the population that can be identified on the sampling frame, it can be sequestered into its own stratum to provide greater control over the number of units sampled. In practice, sometimes the given subgroup's stratum is so small that it makes more sense to simply conduct a census of those units rather than select a sample of them.

- *Administer multiple modes of data collection.* To increase representation of the target population, some survey sponsors utilize more than one mode of data collection (de Leeuw, 2005). When the separate modes are pursued via separate sampling frames, these frames can sometimes be treated as strata of a more comprehensive sampling frame.

- *Increase precision of overall estimates.* When strata are constructed homogeneously with respect to the key outcome variable(s), there can be substantial precision gains.

To illustrate how precision can be increased if the stratification scheme is carried out prudently, let us return to the expenditure survey example and consider an alternative sample design. Suppose there is a river evenly dividing the hypothetical town's population into an east and a west side, each with 1000 adults, and that adults living on the west side of the river tend to be more affluent. It is foreseeable that certain spending behaviors could differ markedly between adults on either side of the river. Since the two key outcome variables deal with expenditures, this would be a good candidate stratification variable.

For sake of an example, let us assume that the firm is able to stratify their sampling frame accordingly, allocating the overall sample size of $n = 400$ adults into $n_1 = 200$ adults sampled without replacement from the west side and $n_2 = 200$ from the east, and that the results have been stored in a data set called SAMPLE_STR_SRSWOR. To account for stratification in the sample design, we specify the stratum identifier variable(s) on the survey data set in the STRATA statement of the SURVEY procedure. For the present example, the variable CITYSIDE defines which of the $H = 2$ strata the observation belongs to, a character variable with two possible values: "West" or "East."

Like Program 1.1, Program 1.2 consists of two PROC SURVEYMEANS runs on the survey data set, except this time we are analyzing a measure of travel expenditures (EXP_TRAVEL) instead of OTC medications (EXP_OTCMEDS). The first run ignores the stratification and assumes a sample of size 400 was selected without replacement from the population of 2000. The second run properly accounts for the stratification by placing CITYSIDE in the STRATA statement. Observe how the FPC is supplied by way of a secondary data set called TOTALS in the second run. This is because the FPC is a stratum-specific quantity. When there is no stratification or the stratification is ignored (as in the first run), one number is sufficient, but you can specify stratum-specific population totals, or N_hs, via a supplementary data set containing a like-named and like-formatted stratum variable(s) and the key variable _TOTAL_ (or _RATE_, if you are opting to provide sampling fractions instead).

Program 1.2: Illustration of the Effect of Stratification on Measures of Variability

```
title1 'Stratified Simple Random Sampling without
Replacement';
title2 'Estimating a Sample Mean and its Variance Ignoring the
Stratification';
proc surveymeans data=sample_str_SRSWOR total=2000 mean var;
  var exp_travel;
run;

data totals;
  length cityside $4;
  input cityside _TOTAL_;
datalines;
East 1000
West 1000
;
run;

title2 'Estimating a Sample Mean and its Variance Accounting
for the Stratification';
proc surveymeans data=sample_str_SRSWOR total=totals mean var;
  strata cityside;
  var exp_travel;
run;
```

Stratified Simple Random Sampling without Replacement

Estimating a Sample Mean and Its Variance Ignoring the Stratification

SURVEYMEANS Procedure

Data Summary	
Number of observations	400

Statistics			
Variable	Mean	Std Error of Mean	Var of Mean
exp_travel	1363.179844	92.594306	8573.705490

Stratified Simple Random Sampling without Replacement

Estimating a Sample Mean and Its Variance Accounting for the Stratification

SURVEYMEANS Procedure

Data Summary	
Number of strata	2
Number of observations	400

Statistics			
Variable	Mean	Std Error of Mean	Var of Mean
exp_travel	1363.179844	77.564901	6016.313916

The sample mean reported by PROC SURVEYMEANS is the same ($1363.18) in either case, but accounting for the stratification reduced the variance by almost one-third. Aside from a few rare circumstances, stratification increases the precision of overall estimates. It should be acknowledged, however, that any gains achievable are variable-specific and less pronounced for dichotomous variables (Kish, 1965). For instance, expenditures on OTC medications are likely much less disparate across CITYSIDE as expenditures of this sort seem less influenced by personal wealth than those related to travel.

Because sampling is performed independently within each stratum, we are able to essentially eliminate the between-stratum variability and focus only on the within-stratum variability. To see this, consider how the estimated variance of the overall sample mean under this sample design is given by

$$\mathrm{var}(\hat{\bar{y}}) = \sum_{h=1}^{H=2} \left(\frac{N_h}{N}\right)^2 \frac{s_h^2}{n_h}\left(1 - \frac{n_h}{N_h}\right) = \sum_{h=1}^{H=2} \left(\frac{N_h}{N}\right)^2 \mathrm{var}(\hat{\bar{y}}_h) \qquad (1.1)$$

where
  N_h is the stratum-specific population size
  n_h is the stratum-specific sample size
  s_h^2 is the stratum-specific element variance

We can conceptualize this as the variance of a weighted sum of stratum-specific, SRSWOR sample means, where weights are determined by the proportion of the population covered by the given stratum, or N_h/N, where $\sum_{h=1}^{H=2} N_h/N = 1$.

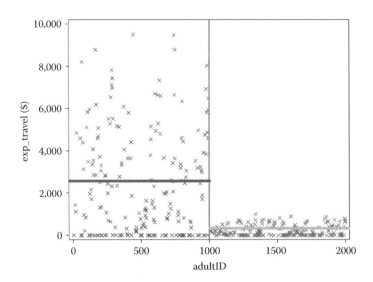

FIGURE 1.4
Visual depiction of the effect of stratification on variance computations.

Figure 1.4 is provided to help visualize what is happening in Equation 1.1 when the SAMPLE_STR_SRSWOR data set is analyzed by PROC SURVEYMEANS with CITYSIDE specified in the STRATA statement. The vertical reference line represents the boundary between the two strata. For the west side ($h=1$), the variance is a function of the sum of the squared distances between the points plotted and the horizontal reference line around $2500, the stratum-specific sample mean. For the east side ($h=2$), the same can be said for the horizontal reference line around $250. When the stratification is ignored, the vertical boundary disappears and a single horizontal reference line would replace the two stratum-specific lines around $1400, the mean expenditure for data pooled from both strata. The point is that the sum of the squared distances to this new reference line would be much greater, overall, which helps explains why measures of variability are larger in the second PROC SURVEYMEANS run when the stratification is ignored.

1.4.4 Clustering

The third feature of complex survey data is *clustering*. This occurs whenever there is not a one-to-one correspondence between sampling units and population elements; instead, the sampling unit is actually a cluster of two or more population elements. A few examples include sampling households in a survey measuring attitudes of individuals, sampling doctor's offices in a survey measuring characteristics of patient visits to doctors' offices, or sampling classrooms in an education survey measuring the scholastic aptitude of students. Clustering is rarely ideal as it generally decreases precision, but it is often a logistical necessity or used to control data collection costs. For instance, most nationally representative, face-to-face surveys in the United States sample geographically clustered units to limit interviewer travel expenses.

Whereas homogeneity within strata leads to precision gains, homogeneity within clusters has the opposite effect. Let us illustrate this phenomenon by considering the following alternative sample design for the expenditure survey example. Suppose the 2000 residents are evenly distributed across the town's $C = 100$ blocks—that is, exactly $N_c = 20$ adults reside on each unique block—and that the market research firm decides data collection would be easier to orchestrate if the sampling units were blocks themselves. Perhaps, they use a town map to enumerate all blocks and select an SRS of $c = 20$ of them, collecting expenditure data on all adults living therein. Note that this alternative design still maintains a sample size of 400. Suppose the survey is administered and the results are stored in the data set called SAMPLE_ CLUS. To isolate the effect of clustering, we will assume there was no stratification and, for simplicity, we will ignore the FPC.

Whenever the underlying sample design of the complex survey data set involves clustering, we should place the cluster identifier variable(s) in the CLUSTER statement of the given SURVEY procedure. In the present example, this identifier is the variable BLOCKID. Program 1.3 is comprised of two PROC SURVEYMEANS runs, one assuming simple random sampling and another properly accounting for the clustering. As before, we are requesting the sample mean and its estimated variance, only this time for both expenditure variables, EXP_OTCMEDS and EXP_TRAVEL.

Program 1.3: Illustration of the Effect of Clustering on Measures of Variability

```
title1 'Cluster Sampling';
title2 'Estimating a Sample Mean and its Variance Ignoring the
Clustering';
proc surveymeans data=sample_clus mean var;
  var exp_OTCmeds exp_travel;
run;

title2 'Estimating a Sample Mean and its Variance Accounting
for the Clustering';
proc surveymeans data=sample_clus mean var;
  cluster blockID;
  var exp_OTCmeds exp_travel;
run;
```

Cluster Sampling
Estimating a Sample Mean and Its Variance Ignoring the Clustering

SURVEYMEANS Procedure

Data Summary	
Number of observations	400

Statistics			
Variable	Mean	Std Error of Mean	Var of Mean
exp_OTCmeds	18.430203	0.709202	0.502968
exp_travel	1271.310549	101.074843	10,216

Cluster Sampling
Estimating a Sample Mean and Its Variance Accounting for the Clustering

SURVEYMEANS Procedure

Data Summary	
Number of clusters	20
Number of observations	400

Statistics			
Variable	Mean	Std Error of Mean	Var of Mean
exp_OTCmeds	18.430203	0.814593	0.663563
exp_travel	1271.310549	320.315188	102,602

This is yet another instance where ignoring a feature of the complex survey data set does not affect the point estimate since the sample means are identical in both PROC SURVEYMEANS runs, but the clustering does impact measures of variability. Failing to account for clustering is especially risky because clustering can prompt a significant increase in the estimated variances. One might notice the increase is far more dramatic for EXP_TRAVEL than EXP_OTCMEDS. The explanation has to do with the degree of homogeneity or how correlated adults' responses are within clusters with respect to the given outcome variable.

The reader might find a visualization of homogeneity useful prior to the presentation of an equation commonly used to quantify it. To this end, Figure 1.5 plots the distribution of the two expense variables within the sampled clusters. The cluster-specific means are represented by a dot and flanked by 95% confidence interval end points (not accounting for the any design features, only to illustrate the within-cluster variability). The idea is that the further away the dots are from one another or the more dissimilar the confidence intervals appear, the larger the expected increase in variance when factoring in the clustering in the sample design. Contrasting the right panel to the left helps explain why the variance increase for travel expenditures trumps that for OTC medications.

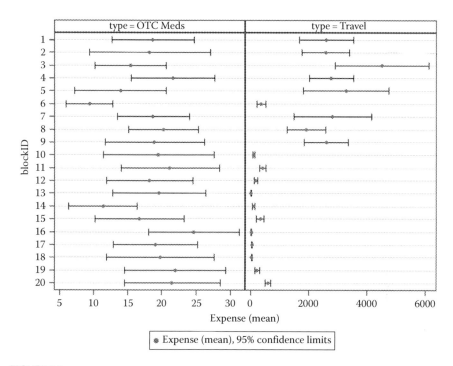

FIGURE 1.5
Expenditure distributions within blocks selected as part of a clustered sample design for the hypothetical expenditure survey.

For the special case of equally sized clusters, an alternative method to calculate the variance of a sample mean further elucidates this concept. Specifically, one can provide a summarized data set containing only the cluster-specific means to PROC SURVEYMEANS and treat as if it were an SRS. Program 1.4 demonstrates this method. It begins with a PROC MEANS step storing the cluster means of expenditure variables in a data set named CLUSTER_ MEANS. The resulting data set of 20 observations is then analyzed by PROC SURVEYMEANS without any statements identifying complex survey features.

Program 1.4: Illustration of the Reduced Effective Sample Size Attributable to Clustering

```
proc means data=sample_clus noprint nway;
  class blockID;
  var exp_OTCmeds exp_travel;
output out=cluster_means mean=;
run;

proc surveymeans data=cluster_means mean var;
  var exp_OTCmeds exp_travel;
run;
```

SURVEYMEANS Procedure

Data Summary	
Number of observations	20

Statistics			
Variable	Mean	Std Error of Mean	Var of Mean
exp_OTCmeds	18.430203	0.814593	0.665363
exp_travel	1271.310549	320.315188	102,602

Indeed, we can confirm the measures of variability output match those from the second run of Program 1.3 in which we provided a data set of the full sample (400 observations) to PROC SURVEYMEANS but specified BLOCKID in the CLUSTER statement. The point of this exercise is to illustrate how clustering reduces the effective sample size. Kish (1965) coined the phrase *design effect* to describe this phenomenon.

The design effect of an estimate $\hat{\theta}$ is defined as the ratio of the variance accounting for the complex survey features to the variance of an SRS of the same size or

$$Deff = \frac{\text{Var}_{complex}(\hat{\theta})}{\text{Var}_{SRS}(\hat{\theta})} \tag{1.2}$$

A *Deff* of 2 implies the complex survey variance is twice as large as that of a comparable SRS. Equivalently, this is to say that the effective sample size is one-half of the actual sample size. While it is possible for certain complex survey designs to render design effects less than 1, meaning designs that are more efficient than an SRS, clustering typically causes this ratio to be greater than 1.

In the special case of an SRS of equally sized clusters, an alternative expression for Equation 1.2 is

$$Deff = 1 + (N_c - 1)\rho \tag{1.3}$$

where
    N_c is the number of population elements in each cluster
    ρ is the *intraclass correlation coefficient* (ICC), a measure of the clusters' degree of homogeneity

The ICC is bounded by $-1/(N_c - 1)$ and 1. The extreme value on the lower end corresponds to all clusters sharing a common mean. At the other extreme, a value of 1 implies perfect homogeneity within clusters or all elements therein sharing a common value. In practice, negative values of ρ are rare. Most common are slightly positive values. Despite a seemingly small value, however, the variance increase can be substantial in the presence of large clusters. That is, all else equal, larger clusters produce larger design effects. Finally, note from Equation 1.3 how clusters of size 1 would yield a design effect of 1 and thus default to an SRS variance estimate. (This is why clusters of size 1 on a survey data set input to a SURVEY procedure result in variances equivalent to those generated if the CLUSTER statement was omitted.)

The true value of ρ is only known if we have knowledge about the entire population; however, a sample-based approximation is given by

$$\hat{\rho} = \frac{\textit{deff} - 1}{N_c - 1} \tag{1.4}$$

where *deff* is the sample-based estimate of *Deff*, or $\textit{deff} = \text{var}_{complex}(\hat{\theta}) / \text{var}_{SRS}(\hat{\theta})$.

There is no mandate to collect data on all population elements within a sampled cluster. For example, we could implement a *two-stage* sampling procedure in which a random sample of clusters is selected in the first stage, and then a subset of those elements therein is selected in the second stage. Returning to the current expenditure survey's clustered sample design, we could reduce the magnitude of the design effect while maintaining the same overall sample size if we altered the design such that we subsampled $n_c = 10$ adults within $c = 40$ sampled blocks as opposed to surveying all $N_c = 20$ adults within $c = 20$ sampled blocks, even though the net sample size is equivalent in either case. In the two-stage sample design, the blocks would be termed *primary sampling units* (PSUs) and adults *secondary sampling units* (SSUs). Although increasing the number of PSUs in the sample design is desirable from a variance perspective, it may be offset by an associated increase in data collection costs. This is especially likely when PSUs are defined geographically.

Indeed, there may be even more than two stages of sampling. For example, Figure 1.6 depicts the multistage sample design employed by the Residential Energy Consumption Survey, a nationally representative, interviewer-administered sample survey of U.S. households sponsored by the Energy Information Administration to collect data on energy usage and expenditures. The basic schema represented by Figure 1.6 is followed by countless other face-to-face surveys of U.S. households or the general U.S. population. The first step typically involves apportioning the land area of the United States into a set of PSUs consisting of single counties or groups of counties. A sample of these is then selected. Following that is a sequence of sampling stages for the various geographical units hierarchically nested within the PSUs, such as the census tracts, blocks, and households. For a practical discussion of area sampling, see Chapter 10 of Valliant et al. (2013).

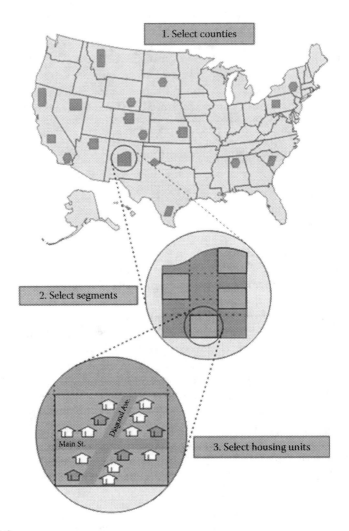

FIGURE 1.6
Visual representation of a multistage area probability sample. (From U.S. Energy Information Administration, 2009 Residential Energy Consumption Survey.)

It is natural to assume each distinct stage of clustering in the sample design should be accounted for within the SURVEY procedure. Some competing software offers this capability, but SAS does not. The documentation repeatedly states users should only identify the PSU in the CLUSTER statement. This implicitly invokes the *ultimate cluster* assumption discussed in Kalton (1983, p. 35) and Heeringa et al. (2010, p. 67). In a maneuver to achieve computational efficiencies, the notion is to assume with-replacement sampling of the PSUs, in which case one only needs to focus on the variability amongst PSUs. Within-PSU variability, or any variability attributable to subsequent stages of sampling, is ignored.

Those unfamiliar with this concept are quick to express their apprehension about not explicitly accounting for all stages of sampling. The knee-jerk reaction is to suppose variances would be underestimated since the terms associated with each stage are typically positive and additive. Most commonly, however, the with-replacement assumption leads to a slight *overestimation* of variability. For a few empirical studies demonstrating this tendency, see Hing et al. (2003) or West (2010).

To see why this occurs from an algebraic standpoint, consider Equation 1.5, the estimated variance for a sample mean expenditure for a sample design where c blocks are selected from the population of C blocks in the first stage, but only a sample of n_c of the N_c adults within each block are selected in the second stage (from Kalton, 1983, p. 34):

$$\text{var}(\hat{\bar{y}}) = \frac{s_c^2}{c}\left(1 - \frac{c}{C}\right) + \left(\frac{c}{C}\right)\frac{s_{n_c}^2}{cn_c}\left(1 - \frac{n_c}{N_c}\right) \tag{1.5}$$

Note that s_c^2 denotes the variance of PSU means among one another and $s_{n_c}^2$ is a term representing the average within-PSU variability. The with-replacement assumption is often employed when the first-stage sampling rate, c/C, is small, in which case the assumption is not far-fetched. Even if the without-replacement assumption were maintained, the first term of the summation tends to dominate since the second term has this *small* fraction as a leading factor. Adopting the with-replacement assumption of PSU selection eliminates all terms in Equation 1.5 except s_c^2/c, and the crux of the Hing et al. (2003) and West (2010) case studies is that the variance increase associated with dropping $\left(1 - (c/C)\right)$ from the first term tends to outweigh the decrease associated with eliminating the second term entirely.

1.4.5 Unequal Weights

The fourth feature of complex survey data is *unequal weights*. Unequal weights can arise for numerous reasons, but the reason we will focus on in this section is to compensate for unequal probabilities of selection (unequal weights resulting from nonresponse adjustment techniques are discussed in Chapter 10). Up to this point in the chapter, all example data sets have been derived from SRSWOR sample designs in which all sampling units have the same probability of being selected. Alternatively, variable selection probabilities can be assigned to sampling units prior to implementing the randomized sampling mechanism.

In spite of certain benefits, variable selection probabilities complicate matters because we must introduce a weight w_i, which is referred to synonymously as a *sample weight*, *base weight*, or *design weight*, to compensate for the relative degrees of over-/underrepresentation of units drawn into the sample. Specifically, Horvitz and Thompson (1952) showed that unbiased estimates of finite population quantities can be achieved by assigning each of the $i = 1, ..., n$ units in the sample a weight of $w_i = 1 / \Pr(i \in S)$, where $\Pr(i \in S)$ denotes the ith unit's selection probability into the given sample S. Because the selection probabilities are strictly greater than zero and bounded by 1, sample weights are greater than or equal to 1. These can be interpreted as the number of units in the survey population (i.e., sampling frame) the sampling unit represents. For instance, a weight of 4 means the unit represents itself and three other comparable units that were not sampled. To have the SURVEY procedure properly account for these weights during the estimation process, we simply specify the variable housing them in the WEIGHT statement.

Returning to the hypothetical expenditure survey, let us motivate a sample design rendering unequal weights. Discussed in greater detail in Section 2.2.3, when the sampling frame contains an auxiliary variable correlated with a key outcome variable to be captured by the survey effort, the technique of *probability proportional to size* (PPS) sampling can yield certain statistical efficiencies, particularly when estimating totals. Suppose the market research firm was able to obtain tax return information from the town's revenue office, information that included each of the $N = 2000$ adults' income during the prior year. Despite the acquisition of this information being highly unlikely, it would make for an ideal candidate variable to use in a PPS sample design for an expenditure survey.

The notion behind PPS sampling is to assign selection probabilities in proportion to each sampling frame unit's share of the total for some *measure of size* variable. In general, if we denote the measure of size for the ith unit MOS_i and the total measure of size for all units in the sampling frame $\sum_{i=1}^{N} MOS_i$, the selection probability for this unit would be

$$\Pr(i \in S) = n \times \left(MOS_i \Big/ \sum_{i=1}^{N} MOS_i \right),$$ and if the unit is selected into the sample, its sample weight would be assigned as the inverse of that quantity.

Suppose that the market research firm drew a PPS sample of $n = 400$ adults using income as the measure of size, carried out the survey, and the results have been stored in the data set SAMPLE_WEIGHTS. The base weights are maintained by the variable WEIGHT_BASE. The two PROC SURVEYMEANS runs in Program 1.5 both request the mean and variance of the two expenditure variables, but only the second run properly accounts for the weights (for simplicity, the FPC is ignored).

Program 1.5: Illustration of the Effect of Unequal Weights

```
title1 'Sampling with Weights';
title2 'Estimating Two Sample Means and Variances Ignoring the
Weights';
proc surveymeans data=sample_weights mean var;
  var exp_OTCmeds exp_travel;
run;

title2 'Estimating Two Sample Means and Variances Accounting
for the Weights';
proc surveymeans data=sample_weights mean var;
  var exp_OTCmeds exp_travel;
  weight weight;
run;
```

Sampling with Weights

Estimating Two Sample Means and Variances Ignoring the Weights

SURVEYMEANS Procedure

Data Summary	
Number of observations	400

Statistics			
Variable	Mean	Std Error of Mean	Var of Mean
exp_OTCmeds	18.042657	0.686190	0.470857
exp_travel	3771.900098	146.547307	21476

Sampling with Weights

Estimating Two Sample Means and Variances Accounting for the Weights

SURVEYMEANS Procedure

Data Summary	
Number of observations	400
Sum of weights	1715.42534

Statistics			
Variable	Mean	Std Error of Mean	Var of Mean
exp_OTCmeds	17.484794	0.857940	0.736061
exp_travel	1884.486868	124.739413	15,560

Notice how both the sample means *and* measures of variability differ between the two runs. Recall that in the three previous illustrations of failing to account for a particular complex survey feature, only the measures of variability differed. This is because the previous three employed equal probabilities of selection designs, a by-product of which is identical sample weights for all units in the sample. Although we could have computed and specified these weights in the WEIGHT statement, it would be technically unnecessary because when the weights are all equal, the weighted mean $\hat{\bar{y}}_w = \sum_{i=1}^{n} w_i y_i \Big/ \sum_{i=1}^{n} w_i$ is equivalent to the unweighted mean $\hat{\bar{y}}_{unw} = \sum_{i=1}^{n} y_i \Big/ n$ since $\hat{\bar{y}}_w = \sum_{i=1}^{n} w_i y_i \Big/ \sum_{i=1}^{n} w_i = w \sum_{i=1}^{n} y_i \Big/ wn = \sum_{i=1}^{n} y_i \Big/ n$.

In effect, omitting the WEIGHT statement is tantamount to instructing the SURVEY procedure to assign weights of 1 for all units in the sample.

Observe how the difference between the unweighted and weighted sample means for EXP_TRAVEL is much more pronounced than the difference for EXP_OTCMEDS. Kish (1992) showed that the difference in these two estimators can be expressed as

$$\hat{\bar{y}}_{unw} - \hat{\bar{y}}_w \approx \frac{-\mathrm{cov}(w_i, y_i)}{\bar{w}} \tag{1.6}$$

where
   $\mathrm{cov}(w_i, y_i)$ is the covariance between weight variable w_i and the outcome variable y_i
   $\bar{w}$ is the average weight in the sample data set

This equation is useful for explaining the potential discrepancies that can surface when failing to account for unequal weights when the estimate of interest is the mean. It implies there must be some kind of association between the weights and outcome variable for a difference to occur. In the current PPS sample design, there is an overrepresentation in the sample of higher-income adults, and it is evident that they spend more on travel; the converse is true for lower-income adults. Employing the weights rebalances these respective cohorts to better match their prevalence in the underlying population. On the other hand, the relationship between OTC medications and the weights is clearly much weaker as there is hardly any difference between two estimators. This serves to remind us how the impact is variable specific. It appears that $\mathrm{cov}(w_i, y_i) \approx 0$ for this particular outcome variable.

1.4.6 Brief Summary of the Features of Complex Survey Data

Table 1.2 summarizes the syntax required to account for the four features of complex survey data. The syntax is common across all SURVEY analysis procedures we will exploit over the course of this book. Although we discussed in this chapter each feature in isolation to focus attention on its impact on estimates and/or measures of variability, a given complex survey

TABLE 1.2

Summary of SURVEY Procedure Syntax to Account for the Four Complex Survey Data Features

Feature	Syntax	Comments/Warnings
Finite population correction	TOTAL=(*value(s)*) | *data-set-name* or RATE=(*value(s)*) | *data-set-name* in the PROC Statement	When sample design is stratified, you can specify a supplementary data set with stratum identifier variable(s) and their corresponding stratum-specific population total (_TOTAL_) or rate (_RATE_). Stratum variable(s) in supplementary data set must be equivalently named and formatted.
Stratification	STRATA statement	Stratum identifier variable(s).
Clustering	CLUSTER statement	Cluster identifier variable(s). Even if sampling involves multiple stages of clustering, specify only the PSU, or the first-stage identifier.
Unequal weights	WEIGHT statement	Only observations in the input data set with positive weights ($w_i > 0$) are maintained. Without a WEIGHT statement, an implicit weight of 1 is applied to all observations.

data set encountered in practice can be characterized by any combination of these four features.

Based on this author's experiences, the FPC is perhaps the least commonly encountered feature, particularly for surveys that involve clustering—whether in single or multiple stages—because these survey efforts are liable to adopt the ultimate cluster (with-replacement) assumption of PSUs, in which case the first-stage FPC would be ignored anyhow. Perhaps the most commonly encountered combination is stratification, clustering, and unequal weights. Indeed, each of the three example surveys discussed in the next section is characterized as such.

1.5 Examples of Complex Surveys

1.5.1 Introduction

The purpose of this section is to acquaint the reader with three real-world, complex surveys sponsored by two statistical agencies of the U.S. government. All survey efforts culminate with the release of one or more publically available data sets that we will make routine use of throughout the book to motivate examples. In addition to highlighting certain features of the data sets, some brief discourse is given regarding each survey's background and objectives.

1.5.2 The National Ambulatory Medical Care Survey

The NAMCS (http://www.cdc.gov/nchs/ahcd/about_ahcd.htm) is an annual survey of visits to office-based physicians sponsored by the NCHS, a branch of the Centers for Disease Control and Prevention (CDC). Note that the ultimate sampling unit—what constitutes an observation in the data set—is actually a visit to a doctor's office, not a patient, doctor, or doctor's office. The survey began in 1973 as a way to capture descriptive information about how patients utilize physician services as well as other characteristics of the visit, such as the primary reason for it, the amount of time spent with the physician, how the visit was paid for, whether any medications were prescribed, and whether any diagnostic procedures were ordered. Key stakeholders of this information include health-care policymakers, researchers, medical schools, and physician associations.

As with any reputable survey operation, details about the survey administration and definitions for what is considered in and out of scope are sufficiently documented on the website. Instructions on how to download and read the survey data into SAS can be found here: http://www.cdc.gov/nchs/ahcd/ahcd_questionnaires.htm. In this book, we will use the 2010 data set, which consists of information on 31,229 distinct visits that occurred during that calendar year. The sample design employed by NAMCS involves multiple stages of clustering. The first stage consists of a stratified sample of areas of the United States. That is, the PSU is a single county or block of adjacent counties. The second stage consists of a sample of physician practices in those areas selected in the first stage. The list (i.e., sampling frame) of physician practices is constructed using information obtained from the American Medical Association and American Osteopathic Association. Each sampled physician practice is then assigned at random a 1-week reporting period during the calendar year. During this week, a systematic sample of visits is selected.

The weight variable on the raw data file effectively magnifying the 31,229 visits to represent all such visits in the United States is PATWT. The stratification and clustering at the first stage can be accounted for by using the variables CSTRATM and CPSUM, respectively. Other analysis variables will be introduced in later chapters. For obvious reasons, maintaining respondent confidentiality is of utmost concern for surveys such as the NAMCS, and any variables containing geographical information pose a nontrivial risk of disclosure. To this end, the primary-stage clustering and stratification is innocuously differentiated on the raw data file via generic codes assigned by the survey sponsor to mask the true geographical information (Hing et al., 2003). (The "M" at the end of the variable names in the 2010 NAMCS data set stands for "masked.")

1.5.3 The National Survey of Family Growth

The NSFG (http://www.cdc.gov/nchs/nsfg.htm) is also sponsored by NCHS and also had its inaugural administration in 1973, but its mission is markedly different from NAMCS. The survey targets men and women in the United States of childbearing age, which they define as individuals aged 15–44. The objective is to collect data on marriage, divorce, pregnancy, contraceptive use, sexual behaviors, and attitudes about these topics. Historically, the survey has been periodically administered every 5–7 years. They term an administration a "cycle," of which there were six between 1973 and 2002. The NSFG went *continuous* in 2006 (Lepkowski et al., 2010), and the data set we will analyze in this book contains information obtained from in-person interviews of 12,279 women occurring between June 2006 and June 2010. All data sets used in this book were retrieved from http://www.cdc.gov/nchs/nsfg/nsfg_2006_2010_puf.htm.

In a similar manner to NAMCS, to limit interviewer travel costs associated with data collection, the NSFG sample design begins by randomly selecting a set of geographically clustered units from a comprehensive list of units covering the entire U.S. landmass. Subsequent stages of sampling involve finer geographical units, households, and ultimately individuals. But recall we only need to concern ourselves with clustering and stratification at the PSU stage. This information is identifiable by the distinct code combinations of the variables SECU and SEST. *SECU* is an acronym for *sampling error computation unit*, and a particular code for SEST reflects the stratum from which the SECU emanates. As with the NAMCS, the coding is assigned arbitrarily for disclosure avoidance purposes.

The NSFG oversamples blacks, Hispanics, and teenagers 15–19 years old. The weight variable WGTQ1Q16 on the raw data file compensates for the imbalances in the respondent pool this oversampling introduces, and so must be used to formulate nationally representative estimates. The suffix Q1Q16 differentiates this particular weight variable from other weight variables on the file (e.g., WGTQ1Q8) that were developed for estimation using a preliminary data release (i.e., containing data collected during the first 2 years of the survey's 4-year data collection period) or to account for questions added to the instrument while the survey was in the field. For more details, see page 14 of the data user's guide (http://www.cdc.gov/nchs/data/nsfg/NSFG_2006-2010_UserGuide_MainText.pdf#General).

1.5.4 Commercial Buildings Energy Consumption Survey

The CBECS (http://www.eia.gov/consumption/commercial/about.cfm) is sponsored by the Energy Information Administration, a statistical subagency of the U.S. Department of Energy. The ultimate sampling unit in this survey is a building. Eligible buildings include those at least 1000 ft² in size and having more than 50% of its floor space devoted to activities that are not residential, industrial, or agricultural in nature. Key statistics of interest include characteristics of the building such as square footage, year of construction, types and uses of heating/cooling equipment, and the volume and associated expenditures of energy consumed. One high-profile usage of these survey data is that it serves as a benchmark for EPA's ENERGY STAR rating system. Specifically, a given building's rating is based on a comparable set of buildings surveyed as part of CBECS with respect to size, location, number of occupants, and other factors. The survey was first administered in 1979 and is generally conducted every 4 years. Examples in this book analyze data from the 2003 CBECS, which is available for download at http://www.eia. gov/consumption/commercial/data/2003/index.cfm?view=microdata.

Like the NAMCS and the NSFG, the CBECS sample design begins with a stratified, clustered area probability sample. PSUs are counties or groups of counties, and subsequent stage sampling units are at finer levels of geographic detail. Ultimately, buildings in the smallest geographical unit sampled are listed and a random sample of them is selected. To limit costs associated with listing the eligible buildings and to improve coverage, the area sampling approach is augmented with several supplemental sample frames developed for certain classes of buildings, such as those occupied and managed by the U.S. government, colleges and universities, and hospitals. Data on the sampled building are collected on-site by interviews with the building owners, managers, or tenants, and a subsequent suppliers' survey is fielded to obtain additional information on energy consumption and expenditures.

The 2003 CBECS collected data on a total of 5215 buildings. As currently structured on the website, the data are segmented into a series of files housing thematically linked subsets of outcome variables, but any two or more files can be merged using the common building identification variable PUBID8. The stratification and clustering information is maintained by the variables STRATUM8 and PAIR8, and the weight variable is ADJWT8. (The data producers chose to end all variable names in the 2003 CBECS data set with an "8," because the 2003 administration was the eighth in the survey's history.)

1.6 Summary

This chapter began by laying out some of the terminology and issues involved when administering a survey. A hypothetical expenditure survey being conducted by a market research firm was used to motivate some of the decisions that must be made regarding the sample design and data collection protocol, as well as the particular complex survey features these decisions may introduce. The main takeaway message of this chapter—and the book—is that complex survey features alter virtually every estimator presented in an introductory statistics course. Some alterations lead to decreased measures of variability, some to increased measures of variability, while others lead to a different point estimate. When analyzing a data set derived from a complex survey, consult any resources available (e.g., subject matter experts, technical reports, or other forms of documentation) to determine whether any of the features discussed in this chapter are present. If so, you should use one of the SURVEY analysis procedures listed in Table 1.1 to formulate estimates from it.

2

Drawing Random Samples Using PROC SURVEYSELECT

2.1 Introduction

The purpose of this chapter is to briefly highlight some of the many ways PROC SURVEYSELECT can be used to select a random sample. As was stated in the Preface, the primary focus of this book is on the *analysis* of complex survey data, not *designing* a complex survey. The reader seeking a more in-depth treatment of survey sample design concepts and theory is referred to any of these excellent texts on the subject: Kish (1965), Cochran (1977), Kalton (1983), Scheaffer et al. (1996), Lohr (2009), and Valliant et al. (2013), to name a few.

This chapter is structured into three main sections. Section 2.2 touches on three fundamental randomized selection procedures: (1) simple random sampling, (2) systematic sampling, and (3) probability proportional to size (PPS) sampling. Section 2.3 discusses how the STRATA statement can be used to apply these techniques independently within two or more strata defined on the sampling frame. Section 2.4 demonstrates how the CLUSTER statement can be used to select a random sample of predefined groupings, or clusters, of observations on the input data set.

Continuing with the hypothetical expenditure survey introduced in Chapter 1, we will assume that the sampling frame is stored as the data set FRAME, which consists of one record for each of the $N = 2000$ adults living in the small city and the following variables:

- ADULTID—a numeric variable ranging from 1 to 2000 uniquely identifying each adult.
- BLOCKID—a numeric variable ranging from 1 to 100 denoting the block on which the adult lives. All $C = 100$ blocks in FRAME consist of exactly $N_c = 20$ adults.

- CITYSIDE—a character variable with two possible values—"East" or "West"—demarcating the side of the river bisecting the town in which the given block falls.
- INCOME—an aggregate measure of the each adult's income during the most recent year.

Over the course of the chapter, a variety of sample designs will be considered, each with the fixed sample size of $n = 400$. There is nothing particularly significant about this number, just a fixed value chosen to illustrate how a whole host of sample designs can be independently implemented to arrive at the same marginal count of individuals selected for a survey.

2.2 Fundamental Sampling Techniques

2.2.1 Simple Random Sampling

Program 2.1 shows syntax to conduct arguably the most basic method available in PROC SURVEYSELECT, simple random sampling without replacement (SRSWOR). Observe how all of the relevant options utilized appear in the PROC statement. The DATA= option points to the sampling frame, while the OUT= statement names the output data set. By default, PROC SURVEYSELECT outputs only the sampled records, but you can specify the OUTALL option in the PROC statement to have all records in the input data set output with the sampled records flagged by a 0/1 numeric variable named SELECTED. The SAMPSIZE= option is used to declare a sample size of $n = 400$, and METHOD=SRS invokes the SRSWOR method, which is the default. Assigning a random number in the SEED= option ensures an identical sample will be selected if the PROC SURVEYSELECT syntax is resubmitted at a later time, assuming the same input data set is provided in the same sort order. Although assigning a value to the SEED= option is technically optional, it is good practice to do so. To help reinforce this point, a different seed is assigned for each PROC SURVEYSELECT syntax example in this chapter.

Program 2.1: Drawing a Simple Random Sample without Replacement

```
proc surveyselect data=frame out=sample_SRSWOR sampsize=400
method=SRS seed=40029;
run;
```

SURVEYSELECT Procedure

Selection Method	Simple Random Sampling
Input data set	FRAME
Random number seed	40,029
Sample size	400
Selection probability	0.2
Sampling weight	5
Output data set	SAMPLE_SRSWOR
Input data set	FRAME

The real "output" from PROC SURVEYSELECT is the SAMPLE_SRSWOR data set, which consists of $n = 400$ observations drawn randomly from FRAME, but a brief rundown of what has occurred gets reported in the listing. For brevity, this is the only occasion in the chapter that we include that summary.

A variety of alternative sample selection procedures are available via the METHOD= option. For instance, to request simple random sampling with replacement (SRSWR), you can specify METHOD=URS. URS is an acronym for *unrestricted random sampling*. Program 2.2 illustrates the syntax to conduct this method for a sample size of 400.

Program 2.2: Drawing a Simple Random Sample with Replacement

```
proc surveyselect data=frame out=sample_SRSWR sampsize=400
seed=22207 method=URS outhits;
run;
```

Whenever a with-replacement design is specified, the default output data set structure consists of one row for each unique record sampled and a numeric variable called NUMBERHITS, indicating how many times the record was chosen. While there may be occasions when this summarized format is acceptable or preferable, the OUTHITS option in the PROC statement of Program 2.2 causes a separate record to be output for each selection such that the number of rows in the output data set matches the sample size—the NUMBERHITS summary variable is still output as well.

2.2.2 Systematic Sampling

Another widely used randomized selection procedure in applied survey research is *systematic sampling*. The basic idea is to select every *k*th unit into the sample. This method is particularly useful in scenarios in which it would be exorbitantly arduous or altogether impossible to construct a sampling frame. Consider a doctor's office that maintains each patient's information in a physical folder sorted alphabetically by surname. If the sampling unit is the patient, it would be much easier to sample, say, every 50th folder than to enumerate, sample, and retrieve folders piecemeal. Another example is a customer satisfaction survey for a grocery store. It is unlikely that an exhaustive list of patrons exists, but an effective random sampling method might be to try to engage every 20th customer leaving the store—of course, it would also be prudent to randomly assign the days and/or time(s) of day during which these attempts are made.

Systematic sampling can also be used when a well-defined sampling frame exists, as is the case with the hypothetical expenditure survey. Program 2.3 demonstrates syntax to select a systematic sample of $n = 400$ adults. Specifying METHOD=SYS in the PROC statement initiates the technique. PROC SURVEYSELECT calculates the sampling interval k as N/n, where N is determined by the number of observations in the input data set. If k is not an integer, a fractional interval is used in order to maintain the exact sample size requested (see the documentation for more details).

In the present example, however, we have an integer interval of $k = 2000/400 = 5$. PROC SURVEYSELECT begins by randomly choosing a starting point between the first and kth observation in the input data set. Let us denote this value by r. The sample will consist of the rth observation and the $(r + k)$th, $(r + 2k)$th, $(r + 3k)$th, and so on. For instance, if the first adult selected was the 4th, the sample in the output data set SAMPLE_SYS would consist of this individual followed by the 9th, 14th, 19th, ..., and 1999th. You can have the values r and k output to the listing if you specify METHOD=SYS(DETAILS) in the PROC SURVEYSELECT statement.

Program 2.3: Drawing a Systematic Random Sample

```
proc surveyselect data=frame out=sample_SYS sampsize=400
seed=65401 method=SYS;
run;
```

The advantages and disadvantages of systematic sampling are articulately delineated in Chapter 8 of Cochran (1977). One salient disadvantage is that there is no *standard* variance formula unbiased in all circumstances (Wolter, 1984). Cochran (1977) describes scenarios in which the sample can behave like a stratified *or* clustered sample. The former is much more welcomed than the latter. The real danger occurs when there is some periodicity in the sort order of the data set that coincides with the sampling interval. For example, suppose that the first stage of sampling involved randomly selecting a set of days within a three-month period for some data collection effort. Suppose further that the days in this span were enumerated sequentially (e.g., Monday, Tuesday,..., Sunday, Monday,...) and that a systematic sample of them was drawn using an interval of $k=7$. There would be no variation in the sample with regard to the day of the week, which would arguably defeat the inherent purpose of sampling these units of time in the first place.

To guard against these kinds of unfortunate scenarios, many practitioners advocate sorting the sampling frame beforehand using one or more *control variables*. You can certainly do this in a prior PROC SORT step, but a more syntactically efficient approach is to specify the control variable(s) in the CONTROL statement of PROC SURVEYSELECT. Program 2.4 illustrates syntax for doing so using the variable INCOME on the hypothetical expenditure survey's sampling frame.

Program 2.4: Drawing a Systematic Random Sample with a Control Variable

```
proc surveyselect data=frame out=sample_SYS_c sampsize=400
seed=94424 method=SYS;
   control income;
run;
```

2.2.3 Probability Proportional to Size Sampling

Another popular sampling approach is PPS sampling, a technique that was briefly introduced in Section 1.4.5. Among its many advantages is the ability to reign in sample-to-sample variability for certain common estimators. For instance, if the sampling frame contains an auxiliary variable strongly correlated with a key outcome variable to be measured in the survey, sampling with probability proportional to the size of this variable can dramatically reduce the variability of its estimated total.

To use more formal sampling terminology, the idea is to select each unit on the sampling frame with probability proportional to its share of the total with respect to some auxiliary variable's *measure of size* (MOS). A frequently used example by one of this author's professors is a survey of hospitals conducted to estimate the total profit of all hospitals in the population over a particular 1-year span. Substantial efficiencies could be achieved by sampling hospitals in proportion to their number of beds. If we denote the ith unit's measure of size MOS_i and the aggregate measure for all N units on the sampling frame $\sum_{i=1}^{N} MOS_i$, then the ith unit's probability of selection would be $n \times (MOS_i / \sum_{i=1}^{N} MOS_i)$.

Figure 2.1 is a pie chart portraying the concept of PPS sampling for a simple sampling frame with $N=8$ units of variable measures of size. We can think of PPS sampling as the event of throwing a dart onto the image, a dart that has an equal probability of landing anywhere inside the circle. Whichever pie slice the dart lands in is selected into the sample. We might

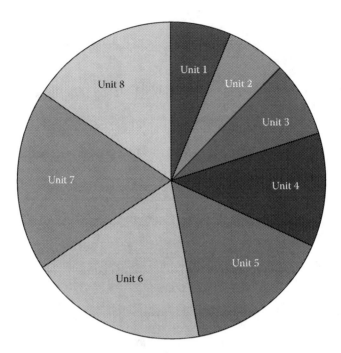

FIGURE 2.1
Visualization of PPS sampling: variable probabilities of selection attributable to variable measures of size.

expect the slice corresponding to Unit 8 being "hit" with higher probability than the slice corresponding to Unit 1.

There is marked difference, both conceptually and formulaically, between without- and with-replacement PPS sampling. We can think of without-replacement sampling as follows. Suppose the first dart fell within the boundaries of Unit 4. Prior to the second draw, we would remove that slice and redistribute the remaining slices' boundaries such that their respective areas (i.e., probabilities) sum to 1. Subsequent selections would be handled similarly. On the other hand, with-replacement sampling would involve throwing multiple darts independently at the same pie chart. Of course, this could lead to a particular unit being selected more than once, which is rarely desirable in single-stage sampling but may be acceptable in multistage cluster sampling.

The other complexity associated with without-replacement sampling is the need to account for the joint selection probabilities into measures of variability (see Equation 3.30 of Valliant et al., 2013). PROC SURVEYSELECT will output these onto the output data set if you specify the option JTPROBS in the PROC statement, but there is no easy way to account for them in the variances computed by the suite of SURVEY procedures. Because of these complexities, PPS sampling is another setting in which the with-replacement assumption is often adopted, even if sampling was actually without replacement.

As was suggested in Section 1.4.5, a prudent auxiliary variable for PPS sampling in the hypothetical expenditure survey is the INCOME variable on the sampling frame. This is because it seems plausible that certain expenditures may be a function of one's income. If true, PPS sampling with respect to this auxiliary variable may improve precision relative to an equal probability design. Program 2.5 shows the syntax to carry out this method for the same target sample size of $n = 400$ on the FRAME data set. Specifying METHOD=PPS invokes PROC SURVEYSELECT's PPS without-replacement (PPSWOR) sampling algorithm, which will then look for a numeric MOS variable to appear in the SIZE statement.

Program 2.5: Drawing a PPS Sample without Replacement

```
proc surveyselect data=frame out=sample_PPSWOR sampsize=400
seed=32124 method=PPS;
  size income;
run;
```

For complex survey samples, PROC SURVEYSELECT will calculate and store selection probabilities and corresponding sample weights in the output data set. These variables are not automatically appended for certain basic designs such as METHOD=SRS, but you can force them onto the output data set by specifying the STATS option in the PROC statement. Figure 2.2 shows these two variables in the data set SAMPLE_PPSWOR created by Program 2.5. The default labeling masks the underlying variable names, which are SELECTIONPROB and SAMPLINGWEIGHT, respectively. Note how the weights are larger for units with smaller values of INCOME and vice versa, which is to be expected.

PROC SURVEYSELECT uses an iterative procedure to select a PPSWOR sample based on an algorithm proposed by Vijayan (1968). In fact, there are numerous variations of PPS sampling available, including a with-replacement algorithm (METHOD=PPS_WR), a systematic selection algorithm (METHOD=PPS_SYS), and a few additional methods for PPS sampling of clusters proposed by Murthy (1957), Brewer (1963), Sampford (1967), and Chromy (1979). It is beyond the scope of this chapter to discuss all of these techniques, but you can consult the PROC SURVEYSELECT documentation for more details.

2.3 Stratified Sampling

Discussed in more detail in Section 1.4.3, the notion behind stratified sampling is to partition the sampling frame into H mutually exclusive and exhaustive

	Cityside	BlockID	AdultID	Income	Probability of Selection	Sampling Weight
1	East	83	1879	10242.66	0.0558853146	17.893788506
2	East	94	1163	11854.57	0.0646801098	15.46070349
3	West	6	98	12082.28	0.0659225259	15.169321666
4	East	58	1597	12463.11	0.068000385	14.705799096
5	East	83	1871	12608.34	0.0687927792	14.536409375
6	East	54	1102	12925.01	0.0705205728	14.180259185
7	East	53	1740	13288.86	0.0725057868	13.792002608
8	East	96	1604	13322.77	0.0726908042	13.756898286
9	East	56	1345	13514.92	0.0737392002	13.56130793
10	East	69	1499	13549.75	0.0739292373	13.52644822
11	East	100	1298	13872.73	0.0756914591	13.211530231
12	East	54	1108	14003.7	0.0764060488	13.087969021
13	East	88	1460	14157.06	0.0772428014	12.946190224
14	West	36	647	14264.82	0.0778307543	12.848391482
15	East	63	1014	14268.42	0.0778503963	12.845149763
16	East	71	1266	14450.01	0.0788411755	12.683727677
17	East	69	1495	14635.21	0.0798516513	12.523222542
18	West	21	678	14870.74	0.0811367343	12.324873663
19	West	11	122	14969.27	0.0816743271	12.24374948
20	East	90	1511	14972.49	0.0816918959	12.241116326

FIGURE 2.2
Partial data set view of PROC SURVEYSELECT output data set following the PPS sample selected in Program 2.5.

groups called *strata* and draw an independent sample within each. Note that there is no mandate to conduct the same randomized selection procedure within all strata, but implementing different methods would require multiple PROC SURVEYSELECT runs, each with a different METHOD= option specified.

Suppose a stratified design was desired in which $n_1 = 100$ adults were chosen from the east side of the city and $n_2 = 300$ adults from the west. Program 2.6 shows the syntax necessary to carry out this sample design. The two strata are defined in FRAME by the two distinct values of the CITYSIDE variable, specified in the STRATA statement. Be advised that the input data set must first be sorted by the stratification variable(s). The other requirement is to inform PROC SURVEYSELECT of the stratum-specific sample sizes, or n_hs. Although we could use syntax SAMPSIZE=(100 300) in the PROC statement, it is better practice (i.e., less error prone) to store this information in a supplemental data set. The supplemental data set need only consist of a like-named, like-formatted, and like-sorted stratification variable(s) and a variable named _NSIZE_. You then point PROC SURVEYSELECT to this supplemental data set using the SAMPSIZE=*data-set-name* option.

Program 2.6: Drawing a Stratified Simple Random Sample

```
proc sort data=frame;
  by cityside;
run;

data sampsizes;
  length cityside $4;
  input cityside _NSIZE_;
datalines;
East 100
West 300
;
run;

proc surveyselect data=frame out=sample_STR_SRS
sampsize=sampsizes seed=89045;
  strata cityside;
run;
```

Recall there are $N_1 = N_2 = 1000$ adults living on either side of the city. Therefore, the probabilities of selection are not the same for adults in either stratum. Accordingly, the output data set SAMPLE_STR_SRS is appended with the variables SELECTIONPROB and SAMPLINGWEIGHT.

Although we will not work through any examples here, be advised that PROC SURVEYSELECT is equipped with several sample allocation strategies you can use to allot a fixed sample size (e.g., one dictated by budgetary constraints) into a set of predefined strata. These can be called using the ALLOC= option after the slash in the STRATA statement. Three examples available at the time of this writing are proportional, Neyman, and optimal allocation (Lohr, 2009). Lewis (2013a) provides a few syntax examples illustrating these techniques. Another handy feature is the MARGIN= option after the slash in the STRATA statement, which can be used to determine the sample size necessary to achieve a targeted *margin of error*, or confidence interval half-width. One of the practical limitations to utilizing these techniques is that they are univariate in nature. That is, they are applicable to a singlepoint estimate, whereas the typical survey effort involves estimating a wide range of population quantities. Multicriteria optimization can be tackled, however, using Microsoft Excel's Solver add-in, or PROC NLP or PROC OPTMODEL in SAS/OR. See Chapter 5 of Valliant et al. (2013) for a demonstration of these tools.

2.4 Cluster Sampling

In Section 1.4.4, we discussed how the primary sampling units (PSUs) are sometimes clusters of the ultimate units of analysis. This may occur naturally or intentionally to facilitate survey data collection. All sample designs presented thus far in this chapter have treated the adult as the PSU. Another choice for PSUs is the $C = 100$ city blocks identifiable on the data set FRAME by the variable BLOCKID. If the expenditure survey were to be conducted in person by interviewers, it might be operationally more convenient to sample blocks instead of adults directly.

Recall exactly $N_c = 20$ adults live on each block. Alternatively to what was done in Program 2.1, a sample size of 400 could be achieved by selecting all adults living in a sample of $c = 20$ blocks. Program 2.7 shows how to do so using the CLUSTER statement in PROC SURVEYSELECT. When the CLUSTER statement is used, the value specified in the SAMPSIZE= option refers to the number of clusters to be chosen, not observations. Hence, the output data set SAMPLE_CLUSTER consists of 400 observations: 20 adults from 20 sampled blocks.

Program 2.7: Drawing a Simple Random Sample of Clusters

```
proc surveyselect data=frame out=sample_cluster sampsize=20
    seed=82908;
  cluster blockID;
run;
```

PPS sampling is frequently used in cluster sampling settings. To illustrate an example, Program 2.8 welds concepts of Programs 2.5 and 2.7 by conducting a PPSWOR sample of $c = 20$ blocks using the variable INCOME as the MOS. When a SIZE statement appears in combination with a CLUSTER statement, PROC SURVEYSELECT computes a PSU-level aggregate MOS equal to the sum of the SIZE variable for all units in the PSU. This aggregate measure is then utilized in whichever PPS algorithm is specified in the METHOD= option of the PROC statement. It is also added to the output data set along with the variables SELECTIONPROB and SAMPLINGWEIGHT.

Program 2.8: Drawing a PPS Sample of Clusters

```
proc surveyselect data=frame out=sample_PPS_cluster
sampsize=20 seed=49721 method=PPS;
  cluster blockID;
  size income;
run;
```

PPS sampling is sometimes executed with the MOS being the number of units comprising the PSU. This is particularly common in the context of area sampling, where geographical units might be selected in proportion to the population of individuals therein. If the sampling frame is oriented such that there is one observation for each underlying unit in the respective PSUs, you can carry out this variant by specifying the PPS option after the slash in the CLUSTER statement—the SIZE statement is no longer necessary.

Clusters can also be stratified prior to sample selection. This might be desirable as it could counteract some of the precision loss anticipated from sampling blocks at the first stage rather than adults. Given the greater expenditure variability anticipated for adults living on the west side, suppose the market research firm opted to oversample blocks in that stratum at a rate of 3 to 1 relative to the east side stratum. The syntax shown in Program 2.9 selects a simple random sample of blocks according to these criteria. With stratified cluster sampling, the supplemental data set should consist of stratum-specific PSU sample sizes.

Program 2.9: Drawing a Stratified Simple Random Sample of Clusters

```
proc sort data=frame;
  by cityside;
run;

data sampsizes_clusters;
  length cityside $4;
  input cityside _NSIZE_;
datalines;
East 5
West 15
;
run;

proc surveyselect data=frame out=sample_str_cluster
  sampsize=sampsizes_clusters seed=77472;
  strata cityside;
  cluster blockID;
run;
```

In applied survey research, cluster sampling rarely concludes after a single step. Indeed, all three surveys discussed in Section 1.5 involve *multistage cluster sampling*. As yet another example, an education survey's sample design might begin by selecting schools in the first stage (i.e., schools as PSUs), and classrooms in the second stage (i.e., classrooms as SSUs). That is, within each sampled school, a sample of classrooms is drawn. These types of designs are perfectly legitimate, but you must keep track of the associated selection probabilities at each step because the ultimate weight assigned should be inversely proportional to the product of them. With a little care, a sequence of PROC SURVEYSELECT runs can be employed to carry out these types of designs. A simple example involving the hypothetical expenditure survey data set is given next to help edify the concept.

Suppose that instead of surveying all 20 adults living in each block of the stratified sample, the market research firm decided to double the number of sampled blocks within each stratum, but survey only one-half of the adults therein. Even with the same number of adults interviewed, this design would probably improve precision since the number of PSUs has increased; however, if cluster sampling was chosen to limit on-site interviewer expenses, this alternative design would likely increase data collection costs.

Program 2.10 demonstrates syntax to select this two-stage sample. As with any stratified design, the sampling frame must first be sorted by the stratification variable(s). A new supplemental data set SAMPSIZES_CLUSTERS2 informs PROC SURVEYSELECT of the PSU sample sizes that have been doubled relative to those in Program 2.9. The first step is to sample blocks, which is output to a data set named SAMPLE_PRELIM. The variable renaming is done to distinguish the sampling probabilities and weights between this and the second stage, for which comparable suffixing nomenclature is used.

The data set SAMPLE_PRELIM consists of all adults within blocks sampled at the first stage. In the second sampling step, we randomly select one-half of them. Note that we no longer specify a CLUSTER statement in the second SURVEYSELECT run—if we did, we would be sampling whole clusters from the set of clusters already sampled. Instead, we add the BLOCKID variable to the STRATA statement. No separate PROC SORT step is needed because SAMPLE_PRELIM is already sorted by CITYSIDE and BLOCKID from the first PROC SURVEYSELECT step. An alternative method for defining the sample size is to specify a sampling rate using the RATE=*value* option. Since the (sub)sampling rate of adults is equivalent for all sampled blocks, providing one number is sufficient in this case. A little more work would be needed if the rates or sample sizes in the second stage were to vary, but a supplemental data set could be created and supplied via the SAMPRATE=*data-set-name* option or the SAMPSIZE=*data-set-name* option.

The DATA step after the second PROC SURVEYSELECT step computes the overall selection probabilities and sample weights for the adults ultimately chosen and output to the data set SAMPLE_MULTISTAGE. Adults in the east stratum are each assigned a weight of 10; adults in the west stratum are each assigned a weight of 3.33.

Program 2.10: Drawing a Stratified, Two-Stage Cluster Sample

```
proc sort data=frame;
  by cityside;
run;
```

```
data sampsizes_clusters2;
  length cityside $4;
  input cityside _NSIZE_;
datalines;
East 10
West 30
;
run;

* Step 1) selection of PSUs (blocks);
proc surveyselect data=frame seed=78410
    sampsize=sampsizes_clusters2
  out=sample_prelim (rename=(SelectionProb  = SelectProb1
                            SamplingWeight = SamplingWeight1));
  strata cityside;
  cluster blockID;
run;

* Step 2) selection of SSUs (adults within blocks);
proc surveyselect data=sample_prelim seed=64107 rate=.5
  out=sample_multistage (rename=(SelectionProb = SelectProb2
                            SamplingWeight = SamplingWeight2));
  strata cityside blockID;
run;

* assign variables SelectionProb and SamplingWeight to be the
  product of the corresponding variables output from the two
  SURVEYSELECT runs above;
data sample_multistage;
  set sample_multistage;
SelectionProb=SelectProb1*SelectProb2;
SamplingWeight=SamplingWeight1*SamplingWeight2;
run;
```

2.5 Summary

The purpose of this chapter was to highlight the versatility of PROC SURVEYSELECT for selecting random samples. After defining a few fundamental approaches for randomized selection, we demonstrated how the STRATA and CLUSTER statements could be utilized to draw stratified and/or cluster samples. Although the examples were rudimentary, the concepts can be extended to more complex problems in a straightforward manner. Admittedly, however, the material in this chapter merely scratched the surface regarding issues and considerations for designing samples in practice. For additional guidance, consult one or more of the textbooks referenced in the introduction.

One feature of PROC SURVEYSELECT that did not make its way into any examples but is worth mentioning is the SELECTALL option in the PROC statement. This option can be efficient in certain circumstances because it sidesteps the need to explicitly calculate a sample size. All records in the input data set are sampled. A second noteworthy feature not illustrated is the GROUPS= option in the PROC statement, which can prove useful in experimental design tasks, such as randomly allocating experimental units into a treatment group(s) or a control group. Yet another utile feature, we will see used in later chapters is the REPS= option in the PROC statement. This option will replicate the entire sample selection process as many times as specified. Each replicate is identifiable by a numeric variable REPLICATE appended to the output data set. This is especially handy for simulations and other iterative tasks. For example, you will see it embedded in Programs 9.10 and 11.5. Cassell (2007) asserts that it can be dramatically more efficient than using, say, a %DO loop in a macro.

3

Analyzing Continuous Variables Using PROC SURVEYMEANS

3.1 Introduction

This chapter explores the capabilities of PROC SURVEYMEANS for analyzing variables that can effectively be treated as continuous. Discussion and syntax examples exploit publically available data from the 2003 Commercial Buildings Energy Consumption Survey (CBECS)—see Section 1.5.4 for more background—which teems with continuous variables. Examples include square footage of the sampled building, its kilowatt-hours of electricity consumed, and numerous associated expenditures. A limited amount of consideration is given to dichotomous or nominal variables, which can technically be analyzed in PROC SURVEYMEANS using the CLASS statement or by creating one or more indicator variables. Since virtually any estimate for a categorical variable derived from PROC SURVEYMEANS is replicable with PROC SURVEYFREQ, however, much of the discussion surrounding that variable type will be deferred to Chapter 4.

This chapter is structured by a separate section being devoted to each of the four broad classes of statistics available in PROC SURVEYMEANS: totals, means, ratios, and quantiles. Each estimator is first introduced algebraically, along with a brief mention of how PROC SURVEYMEANS estimates its variance, followed by one or more elementary syntax examples. Each section then concludes with a tabular summarization of the statistical keywords available in the PROC statement.

3.2 Totals

The first statistic we will consider is the estimated total for a survey outcome variable y, which can be expressed in the most general sense as

$$\hat{Y} = \sum_{h=1}^{H} \sum_{i=1}^{n_h} \sum_{j=1}^{m_{hi}} w_{hij} y_{hij} \tag{3.1}$$

where
  H is the number of strata
  n_h is the number of primary sampling units (PSUs) selected from the hth stratum
  m_{hi} is the number of units selected from within the n_hth PSU

In essence, the estimated total is the weighted sum of all observations in the data set. The expression is given in "the most general sense" because it assumes the sample design includes stratification, clustering, and unequal weights. The expression simplifies in the absence of any of those features.
  The variance of the estimated total in Equation 3.1 is

$$\text{var}(\hat{Y}) = \sum_{h=1}^{H} \frac{n_h}{(n_h - 1)} \sum_{i=1}^{n_h} \left(\sum_{j=1}^{m_{hi}} w_{hij} y_{hij} - \frac{\sum_{i=1}^{n_h} \sum_{j=1}^{m_{hi}} w_{hij} y_{hij}}{n_h} \right)^2 \left(1 - \frac{n_h}{N_h} \right) \tag{3.2}$$

Although this equation appears rather complex at first glance, it is also given in the "most general sense" and simplifies in the absence of certain complex survey features. For instance, the rightmost term in parentheses represents the stratum-specific finite population correction factor with respect to the first-stage sample size. If the design involves multiple stages of clustering and the ultimate cluster assumption is implicitly invoked (see discussion in Section 1.4.4), this term vanishes and one would not specify PSU totals using one of the options available in the PROC statement presented in Table 1.2. In a similar vein, when there is no stratification, all subscripts and summations pertaining to the hth stratum are no longer necessary.
  One of the key measurements taken on all buildings sampled as part of CBECS is square footage, which is stored in the variable SQFT8 on the publically available data set. Recall that the CBECS sample design involves three features of complex survey data: stratification, clustering, and unequal weights. These can be accounted for in any SURVEY procedure by pointing SAS to the variables STRATUM8, PAIR8, and ADJWT8, respectively. Program 3.1 illustrates the syntax necessary to estimate the total square footage of all buildings in the population of eligible buildings. (The criteria for determining whether a building is eligible for the survey are rather complex and will not be defined here. More details can be found on the survey's website: http://www.eia.gov/consumption/commercial/).

TABLE 3.1

Summary of PROC SURVEYMEANS Statement Statistical Keywords
Pertaining to Totals

Keyword	Output
SUM	Estimated total of each variable listed in the VAR statement or all levels of each categorical variable listed in the CLASS statement
STD	Standard error of an estimated total, output by default when the SUM keyword is used
VARSUM	Variance of the estimated total(s)
CVSUM	Coefficient of variation (STD divided by SUM) of the estimated total(s)
CLSUM	Confidence interval of the estimated total(s) based on the significance level specified in the ALPHA= option (default is ALPHA=.05)

You can specify the keyword SUM in the PROC SURVEYMEANS statement to estimate the total for any variable listed in the VAR statement and its corresponding standard error, the square root of the quantity in Equation 3.2. This quantity is labeled "Std Dev" in the output, which is somewhat of a misnomer (see discussion of measures of uncertainty in Section 1.2), although there are plans to relabel it in a future release. A comprehensive listing of statistical keywords available in the PROC statement as they relate to totals is given in Table 3.1.

Program 3.1: Estimating the Total of a Continuous Variable

```
proc surveymeans data=CBECS_2003 sum;
  strata STRATUM8;
  cluster PAIR8;
  var SQFT8;
weight ADJWT8;
run;
```

SURVEYMEANS Procedure

Data summary	
Number of strata	44
Number of clusters	88
Number of observations	5,215
Sum of weights	4,858,749.82

Statistics		
Variable	Sum	Std Dev
SQFT8	71,657,900,522	2,227,163,020

The PROC SURVEYMEANS output begins with a "Data Summary" section tabulating a few basic, descriptive statistics about the complex survey features identified on the input data set. It informs us that the 5215 observations in the data set CBECS_2003 were found to be spread amongst 88 PSUs in 44 strata and that the weights sum to approximately 4,858,750. In fact, this value can be interpreted as an estimate of the total number of eligible buildings $\hat{N}$ in the population as of 2003. This is actually an estimate of interest since it is not known elsewhere. Measures of variability associated with this particular statistic are not output by default. If desired, one work-around would be to apply syntax of the form illustrated in Program 3.1 after first creating a user-defined numeric variable equaling 1 for all observations in the data set. Another work-around would be to omit the WEIGHT statement and specify ADJWT8 in the VAR statement. PROC SURVEYFREQ could also be used (see output from Program 4.1).

Quantities pertinent to the sum of the variable SQFT8 are given in the "Statistics" section of the output. Specifically, we find that the estimated total square footage is almost 72 billion with an estimated standard error of 2.2 billion.

Summations are also possible for categorical variables. As an example, the variable STUSED8 indicates whether the building uses district steam. A value of 1 indicates yes, whereas 2 means no. Program 3.2 illustrates how we can estimate the total number of buildings for both conditions. Since the STUSED8 variable is numeric but we want it treated as a categorical variable, we also specify it in the CLASS statement (character variables listed in the VAR statement are treated as categorical by default). For each CLASS statement variable, PROC SURVEYMEANS constructs a sequence of K binary indicator variables, one for each of the $k = 1, 2, \ldots, K$ unique values ("levels" in SAS terminology), defined as $y_{hijk} = 1$ if the observation falls within that category, and $y_{hijk} = 0$ otherwise. From there, Equations 3.1 and 3.2 are employed.

Program 3.2: Estimating the Totals of a Categorical Variable

```
proc format;
  value YESNO
    1='Yes'
    2='No';
run;

proc surveymeans data=CBECS_2003 sum;
  strata STRATUM8;
  cluster PAIR8;
  class STUSED8;
  var STUSED8;
weight ADJWT8;
format STUSED8 YESNO.;
run;
```

Class Level Information		
CLASS Variable	Levels	Values
STUSED8	2	Yes, No

Statistics			
Variable	Level	Sum	Std Dev
STUSED8	Yes	47,106	7,602.878471
	No	4,811,644	239,130

We can observe from the output that an estimated total of 47,106 buildings use district steam and 4,811,644 do not. Because all buildings are characterized by one of these two conditions, the sum of the two matches the sum of weights for the entire data set, 4,858,750.

Table 3.1 summarizes useful statistical keywords pertaining to estimated totals that can be requested in the PROC SURVEYMEANS statement. Most are self-explanatory, although a few words are warranted about the complex survey degrees of freedom. Under the default variance estimation procedure, the SURVEY procedures utilize a widely adopted rule of thumb that the degrees of freedom equal the number of distinct PSUs minus the number of strata. This is often dramatically smaller than the number of observations minus one, the degrees of freedom assumed for an SRSWR sample design. For instance, in Programs 3.1 and 3.2, the effective degrees of freedom are $88 - 44 = 44$, not $5215 - 1 = 5214$. The degrees of freedom do not influence point estimates or their estimated variances, but they can impact confidence intervals because confidence interval widths are contingent upon a reference t distribution with specified degrees of freedom. Valliant and Rust (2010) provide some of the underlying theory behind this approximation and also discuss situations in which it may be inappropriate. (Note that the alternative variance estimation procedures explored in Chapter 9 abide by a different set of rules for determining degrees of freedom.)

Before continuing, it is worth mentioning that there is no built-in mechanism for conducting significance tests on estimated totals. In other words, there is no way to input a null hypothesis total and have PROC SURVEYMEANS compute a t statistic and p value. That said, all of the essential components are output for you to do so without much hassle. The same is true for other statistics estimated by PROC SURVEYMEANS, with the exception of the null hypothesis that the true population mean or ratio is 0 via the statistical keyword T. Additional methods and considerations for conducting significance tests on sample means will be covered in Chapter 8.

3.3 Means

The second statistic considered in this chapter is the sample mean for an outcome variable y, which can be expressed as

$$\hat{\bar{y}} = \frac{\sum_{h=1}^{H}\sum_{i=1}^{n_h}\sum_{j=1}^{m_{hi}} w_{hij} y_{hij}}{\sum_{h=1}^{H}\sum_{i=1}^{n_h}\sum_{j=1}^{m_{hi}} w_{hij}} \tag{3.3}$$

Another name for this estimator is the *weighted mean*. It is the weighted sum of values for the given variable divided by the sum of the weights. The weighted mean complicates the variance estimation task because it is actually a nonlinear function (ratio) of two estimated totals, $\hat{\bar{y}} = \hat{Y}/\hat{N}$, and $\mathrm{var}\,(\hat{Y}/\hat{N}) \neq \mathrm{var}(\hat{Y})/\mathrm{var}(\hat{N})$. As noted by Heeringa et al. (2010), a closed-form variance estimator does not exist. This requires one to make an approximation. A popular method is Taylor series linearization (TSL), which is the default approach used by PROC SURVEYMEANS. The TSL variance estimate for the weighted mean in Equation 3.3 can be written as

$$\mathrm{var}\!\left(\frac{\hat{Y}}{\hat{N}}\right) \approx \frac{1}{\hat{N}^2}\left[\, \mathrm{var}(\hat{Y}) + \left(\frac{\hat{Y}}{\hat{N}}\right)^2 \mathrm{var}(\hat{N}) - 2\!\left(\frac{\hat{Y}}{\hat{N}}\right)\mathrm{cov}(\hat{Y},\hat{N})\right] \tag{3.4}$$

Note that the variance terms in Equation 3.4 are calculated in the same manner as Equation 3.2, factoring in the applicable features of the complex sample design, and that $\mathrm{cov}(\hat{Y},\hat{N})$ represents the covariance of $\hat{Y}$ and $\hat{N}$. Ignoring the FPC, the covariance is

$$\mathrm{cov}(\hat{Y},\hat{N}) = \sum_{h=1}^{H}\frac{n_h}{(n_h-1)}\sum_{i=1}^{n_h}\left(\sum_{j=1}^{m_{hi}} w_{hij} y_{hij} - \frac{\sum_{i=1}^{n_h}\sum_{j=1}^{m_{hi}} w_{hij} y_{hij}}{n_h}\right)$$

$$\times \left(\sum_{j=1}^{m_{hi}} w_{hij} - \frac{\sum_{i=1}^{n_h}\sum_{j=1}^{m_{hi}} w_{hij}}{n_h}\right) \tag{3.5}$$

Further note that in the absence of weights, when we effectively assume a uniform weight of 1 for all units in the sample, $\hat{N} = n$, which is a constant, and we know that the variance of a constant and any covariance term involving a constant is 0. So, all that remains of Equation 3.4 is $\mathrm{var}(\hat{Y})/n^2 = n^2\,\mathrm{var}(\hat{\bar{y}})/n^2 = \mathrm{var}(\hat{\bar{y}})$, and we can reason that the variance formula reverts back to the traditional sample mean variance estimate.

If you consult the PROC SURVEYMEANS documentation, you will find a sample mean variance formula that looks markedly different from the one given in Equation 3.4. Despite their different appearances, the two are computationally equivalent. SAS, like most other competing software, uses the TSL algorithm described in Woodruff (1971). We will expound the algorithm in Chapter 9 prior to contrasting it against an alternative class of variance estimators collectively referred to as *replication techniques*.

Program 3.3 shows how to estimate the average square footage for all buildings in the CBECS population. The syntax mirrors that of Program 3.1; only the keyword MEAN is now specified in the PROC statement. From the output, we find that the estimated mean is 14,748 with an estimated standard error of 625.

Program 3.3: Estimating a Sample Mean

```
proc surveymeans data=CBECS_2003 mean;
  strata STRATUM8;
  cluster PAIR8;
  var SQFT8;
weight ADJWT8;
run;
```

Statistics		
Variable	Mean	Std Error of Mean
SQFT8	14,748	625.460801

Means of categorical variables listed in the CLASS statement are output in the form of *proportions*. In a comparable manner to what occurs when totals are requested, PROC SURVEYMEANS constructs a sequence of binary indicator variables defined as 1 if the observation falls within the kth level ($k = 1,..., K$) and 0 otherwise. From there, Equations 3.3 and 3.4 are applied.

One way of thinking about the estimated proportion for the kth category is that it is the sum of the weights for all observations falling within that category divided by the overall sum of weights. Textbooks frequently express variances of proportions differently from the sample means of continuous variables. For instance, the estimated variance of a sample proportion $\hat{p}$ in an SRSWR design is often written var$(\hat{p}) = \hat{p}\,(1-\hat{p})/(n-1)$. This expression is the result of an algebraic simplification that can be made when summing squared deviations of a variable consisting of either a 0 or a 1 from its average value (i.e., the estimated proportion). Calculating the variance this way is less computationally intensive, but the "standard" equation for a continuous variable still applies with an appropriately constructed 0/1 indicator variable.

Program 3.4 demonstrates these concepts by estimating the proportion of buildings in the CBECS population that use natural gas, a characteristic designated by the variable NGUSED8 in the CBECS_2003 data set. The variable is specified in both the VAR and CLASS statements. From the output, it appears roughly 52.2% of the buildings use natural gas and 47.8% do not. Note that the standard errors of these two point estimates are equivalent, as will always be the case for a dichotomous categorical variable.

Program 3.4: Estimating the Proportions of All Categories of a Variable

```
proc format;
  value YESNO
    1='Yes'
    2='No';
run;

proc surveymeans data=CBECS_2003 mean;
  strata STRATUM8;
  cluster PAIR8;
  class NGUSED8;
  var NGUSED8;
weight ADJWT8;
format NGUSED8 YESNO.;
run;
```

Class Level Information		
CLASS Variable	**Levels**	**Values**
NGUSED8	2	Yes No

Statistics			
Variable	**Level**	**Mean**	**Std Error of Mean**
NGUSED8	Yes	0.522272	0.028083
	No	0.477728	0.028083

Table 3.2 summarizes the relevant statistical keywords available for means.

TABLE 3.2

Summary of PROC SURVEYMEANS Statement Statistical Keywords Pertaining to Means

Keyword	Output
MEAN	Estimated mean of each variable listed in the VAR statement or estimated proportion of all levels of each categorical variable listed in the CLASS statement
STDERR	Standard error of the estimated mean(s), output by default when the MEAN keyword is used
VAR	Variance of the estimated mean(s)
CV	Coefficient of variation (STDERR divided by MEAN) of the estimated mean(s)
CLM	Confidence interval of the estimated mean(s) based on the significance level specified in the ALPHA= option (default is ALPHA=.05)

3.4 Ratios

The third statistic discussed in this chapter is the ratio of two totals, $\hat{Y}$ and $\hat{X}$, which can be expressed as

$$\hat{R} = \frac{\hat{Y}}{\hat{X}} = \frac{\sum_{h=1}^{H} \sum_{i=1}^{n_h} \sum_{j=1}^{m_{hi}} w_{hij} y_{hij}}{\sum_{h=1}^{H} \sum_{i=1}^{n_h} \sum_{j=1}^{m_{hi}} w_{hij} x_{hij}} \qquad (3.6)$$

Recall we remarked how a weighted mean is a ratio in which the numerator consists of a variate equaling the weight times the given outcome variable and the denominator is simply the weight itself. It is instructive to see how these two variates can be created and passed to PROC SURVEYMEANS to replicate the weighted mean estimation and inference process. Observe how the output from Program 3.5 matches what we see in the output from Program 3.3. The two variates are created on the data set RATIO_EXAMPLE and named NUM for numerator and DEN for denominator. The RATIO statistical keyword in the PROC SURVEYMEANS statement requests the point estimate and standard error for the ratio defined in the RATIO statement. The required syntax calls for separating the numerator variable from the denominator variable with a slash. These two variables should also be listed in the VAR statement; without a VAR statement, SURVEYMEANS outputs statistics for all numeric variables in the input data set not already designated for a complex survey design feature. Note that the WEIGHT statement is unnecessary since the NUM and DEN variables already account for the weighting.

Program 3.5: Demonstrating How a Weighted Mean Is Estimated as a Ratio

```
data ratio_example;
  set CBECS_2003;
num=ADJWT8*SQFT8;
den=ADJWT8;
run;

proc surveymeans data=ratio_example ratio;
  strata STRATUM8;
  cluster PAIR8;
  var num den;
  ratio num / den;
run;
```

Ratio Analysis			
Numerator	**Denominator**	**Ratio**	**Std Err**
num	den	14,748	625.460801

The equivalence is attributable to the fact that the variance of any kind of ratio parallels what we have already seen for the weighted mean. Specifically, the TSL variance estimate for the estimator in Equation 3.6 is

$$
\text{var}\left(\frac{\hat{Y}}{\hat{X}}\right) \approx \frac{1}{\hat{X}^2}\left[\text{var}(\hat{Y}) + \left(\frac{\hat{Y}}{\hat{X}}\right)^2 \text{var}(\hat{X}) - 2\left(\frac{\hat{Y}}{\hat{X}}\right)\text{cov}(\hat{X},\hat{Y})\right] \quad (3.7)
$$

An example ratio of interest in the CBECS is *electricity intensity*, which is defined as the average amount of electricity expended per square foot of building space. To get this figure, the total amount of kilowatt-hours of electricity consumed (ELCSN8) is divided by the total square footage (SQFT8). Program 3.6 demonstrates how to estimate this ratio. From the output, we find this statistic is approximately 14.86 with standard error 0.42.

Program 3.6: Estimating a Ratio

```
proc surveymeans data=CBECS_2003 ratio;
   strata STRATUM8;
   cluster PAIR8;
   var ELCNS8 SQFT8;
   ratio ELCNS8 / SQFT8;
weight ADJWT8;
run;
```

Ratio Analysis			
Numerator	Denominator	Ratio	Std Err
ELCNS8	SQFT8	14.864030	0.420386

Ratio estimation is discussed at length in some of the classic sampling texts (cf., Chapter 4 of Lohr, 2009; Chapter 6 of Cochran, 1977). One example application with the potential for significant precision gains is when the total for the denominator is known with certainty, perhaps from the sampling frame or some external source. We can denote this value X (without a hat). The idea is to estimate the ratio $\hat{R} = \hat{Y}/\hat{X}$ from the sample and multiply by X to get an estimate of Y. That is, instead of the estimator shown in Equation 3.1, one would use

$$
\hat{Y}_{ratio} = X\hat{R} = X\left(\frac{\hat{Y}}{\hat{X}}\right) \quad (3.8)
$$

with an estimated variance of

$$\text{var}(\hat{Y}_{ratio}) = X^2 \, \text{var}(\hat{R}) \qquad (3.9)$$

where $\text{var}(\hat{R})$ would be calculated as shown in Equation 3.7. There is no way to have this particular ratio estimator output directly, but an indirect way is to create a new numerator variable equaling the original numerator times X, and then specify this variable in the RATIO statement in tandem with the original denominator. The ratio estimate reported by PROC SURVEYMEANS would be $\hat{Y}_{ratio}$ and the standard error $\text{se}(\hat{Y}_{ratio}) = \sqrt{\text{var}(\hat{Y}_{ratio})}$. Again, the rationale is that this estimator could prove more precise than the estimator given by Equation 3.1, but it requires we know X.

To motivate an example of this alternative total estimator, let us revisit the ratio of total electricity consumed to square footage of the building. From the output of Program 3.1, we found the total square footage estimate to be 71,657,900,522. Suppose that we knew this was the true population total. Program 3.7 obtains the estimator in Equation 3.8 by creating a variable ELCNS8_ (with an underscore) equaling ELCNS8 times the fixed square footage total. This new variable is then specified as the numerator in the RATIO statement with SQFT8 specified as the denominator. From the Ratio Analysis table generated from the first PROC SURVEYMEANS run, we find $\hat{Y}_{ratio} = 1.0651 \times 10^{12}$ with $\text{se}(\hat{Y}_{ratio}) = 3.0124 \times 10^{10}$, whereas the traditional total estimator produced from the second PROC SURVEYMEANS run is $\hat{Y} = 1.0432 \times 10^{12}$ with $\text{se}(\hat{Y}) = 4.1377 \times 10^{10}$. So the two estimators result in a very similar point estimate, but the standard error is $1 - 3.0124/4.1377 \approx 28\%$ smaller for the ratio estimator.

Program 3.7: Estimating a Total Using a Ratio Estimator

```
* ratio estimator of a total;
data ex_ratio_total;
  set CBECS_2003;
ELCNS8_=ELCNS8*71657900522;
run;

proc surveymeans data=ex_ratio_total ratio;
  strata STRATUM8;
  cluster PAIR8;
  var ELCNS8_ SQFT8;
  ratio ELCNS8_ / SQFT8;
weight ADJWT8;
run;
```

```
* traditional estimator of a total;
proc surveymeans data=CBECS_2003 sum;
  strata STRATUM8;
  cluster PAIR8;
  var ELCNS8;
weight ADJWT8;
run;
```

Ratio Analysis			
Numerator	Denominator	Ratio	Std Err
ELCNS8_	SQFT8	1.0651251E12	30,124,001,614

Statistics		
Variable	Sum	Std Dev
ELCNS8	1.0431757E12	41,377,010,767

In fact, one can formulate an analogous ratio estimator for a mean. The only difference is that instead of multiplying the estimated ratio by X it is multiplied by $\bar{x}$, the known population mean for the auxiliary variable. The point estimate for this ratio estimator is $\hat{\bar{y}}_{ratio} = \bar{x}\hat{R} = \bar{x}\left(\hat{Y}/\hat{X}\right)$ with variance $\text{var}(\hat{\bar{y}}_{ratio}) = \bar{x}^2 \text{var}(\hat{R})$. The setup and execution parallel what was shown in Program 3.7; the only necessary modification is to multiply the numerator variable used in the RATIO statement through by $\bar{x}$ instead of X in the preliminary DATA step. For brevity, we will not illustrate an example here.

Table 3.3 summarizes the statistical keywords available in the PROC SURVEYMEANS statement related to estimated ratios. Note that there is no coefficient of variation available like there is with the estimated total and mean, and that confidence intervals are requested using the same syntax as for means.

TABLE 3.3

Summary of PROC SURVEYMEANS Statement Statistical Keywords Pertaining to Ratios

Keyword	Output
RATIO	Estimated ratio(s) specified in the RATIO statement
STDERR	Standard error of the estimated ratio(s)
VAR	Variance of the estimated ratio(s)
CLM	Confidence interval of the estimated ratio(s) based on the significance level specified in the ALPHA= option (default is ALPHA=.05)

3.5 Quantiles

The fourth and final class of statistics we will discuss in this chapter is *quantiles*, which became available with the release of SAS Version 9.2. Although the term "quantiles" may not be immediately familiar to every reader, the concept of them likely is. These include measures such as medians and percentiles.

The formal mathematical definition of population quantiles requires us to first define the *cumulative distribution function* (CDF) for a variable y. For a population of size N, the CDF is defined as

$$F(y) = \frac{\sum_{i=1}^{N} I(y_i \leq y)}{N} \tag{3.10}$$

where $I(y_i \leq y)$ is a 0/1 indicator variable for whether y_i is less than or equal to some specified value y. The γth quantile is defined as the smallest y_i such that the CDF is greater than or equal to γ. We will denote this value q_γ. For instance, $q_{0.50}$ represents the population median or, synonymously, the 50th percentile. This value is interpreted as the midpoint of all y_is in the population, the value for which 50% of the y_is are smaller in magnitude and the complementary 50% larger.

The population CDF in Equation 3.10 can be estimated from a complex survey data set using the weights as follows:

$$\hat{F}(y) = \frac{\sum_{h=1}^{H} \sum_{i=1}^{n_h} \sum_{j=1}^{m_{hi}} w_{hij} I(y_{hij} \leq y)}{\sum_{h=1}^{H} \sum_{i=1}^{n_h} \sum_{j=1}^{m_{hi}} w_{hij}} \tag{3.11}$$

The estimated CDF is typically visualized by plotting the ordered y_{hij}s in the sample data set on the y-axis as a step function of γ on the x-axis, running from 0 to 1. While there is no way to request this particular plot within PROC SURVEYMEANS, one can be generated using the CDFPLOT statement in PROC UNIVARIATE. This statement will not work in conjunction with a WEIGHT statement, but the FREQ statement can be used instead. Note that the FREQ statement truncates the variable specified to the nearest integer. For large weights—in the hundreds or thousands, say—the truncation will be inconsequential, but for smaller weights it would be wise to simply multiply all weights by 100. A common weight inflation will have no impact on the appearance of a CDF.

PROC SURVEYMEANS utilizes the input data set and the estimated CDF to produce $\hat{q}_\gamma$, a sample-based estimate of the γth population quantile q_γ. It does so by first constructing the *sample order statistics* of y or the sorted sequence of the D distinct values of the variable, which we can denote in general terms by $y_{(1)} < y_{(2)} < \cdots < y_{(D)}$. If all values are distinct, $D = n$, but $D < n$ if there are ties, meaning two or more cases share the same value. Next, it finds the value of j such that $\hat{F}(y_{(j)}) \leq \gamma < \hat{F}(y_{(j+1)})$. Then, the γth population quantile is estimated by

$$\hat{q}_\gamma = y_{(j)} + \frac{\gamma - \hat{F}(y_{(j)})}{\hat{F}(y_{(j+1)}) - \hat{F}(y_{(j)})}(y_{(j+1)} - y_{(j)}) \tag{3.12}$$

The second term is an interpolation correction factor that comes into play whenever $\gamma > \hat{F}(y_{(j)})$. If $\gamma = \hat{F}(y_{(j)})$, the term is zero and $\hat{q}_\gamma$ reduces to $y_{(j)}$. The only additional exceptions worth pointing out are that whenever $\gamma < \hat{F}(y_{(1)})$ or $\gamma > \hat{F}(y_{(D)})$ the quantiles reported default to the minimum or maximum values, respectively, in the input data set.

PROC SURVEYMEANS estimates the standard error of an estimated quantile using a three-step method proposed in Woodruff (1952), as discussed in Dorfman and Valliant (1993). We will first define each in mathematical terms and then unify the three steps in the process with the help of an annotated visualization.

The method begins by finding the estimated variance of $\hat{F}(\hat{q}_\gamma)$ with respect to the complex sample design as

$$\text{var}(\hat{F}(\hat{q}_\gamma)) = \sum_{h=1}^{H} \frac{n_h}{(n_h - 1)} \sum_{i=1}^{n_h} \left(e_{hi} - \frac{\sum_{i=1}^{n_h} e_{hi}}{n_h} \right)^2 \left(1 - \frac{n_h}{N_h} \right) \tag{3.13}$$

where

$$e_{hi} = \frac{\sum_{j=1}^{m_{hi}} w_{hij} \left[I(y_{hij} \leq \hat{q}_\gamma) - \hat{F}(\hat{q}_\gamma) \right]}{\sum_{j=1}^{m_{hi}} w_{hij}}$$

The estimated variance in Equation 3.13 is used to form a confidence interval around $\hat{F}(\hat{q}_\gamma)$ using $t_{df,\alpha/2}$, the $100(1 - \alpha/2)$th percentile of a t distribution with df complex survey degrees of freedom. If we denote these end points $\hat{F}^L(\hat{q}_\gamma)$ and $\hat{F}^U(\hat{q}_\gamma)$, respectively, the second step is to invert the estimated CDF at these two points using Equation 3.12. If we label the two resulting values

$\hat{q}_Y^L$ and $\hat{q}_Y^U$, the third step is to solve the implied confidence interval equation about $\hat{q}_Y$ for the standard error as follows:

$$\text{se}(\hat{q}_Y) = \frac{\hat{q}_Y^U - \hat{q}_Y^L}{2 \times t_{df,\alpha/2}} \tag{3.14}$$

There is no option available to request the variance of the quantile to be output, but it can be obtained simply by squaring the quantity in Equation 3.14. As was done for the three other statistics covered in this chapter, there will be a table given toward the end of this section summarizing all statistical keywords in the PROC SURVEYMEANS statement as they pertain to quantiles.

A visualization of the Woodruff procedure is extremely helpful, if not imperative, for comprehending the sequence of steps involved. This is the aim of Figure 3.1, an annotated plot of the estimated CDF for an example continuous variable in a complex survey data set. Again, the first step is to form a confidence interval around $\hat{F}(\hat{q}_Y)$, and the second step is to translate those two interval end points back to the variable scale to get the end points $\hat{q}_Y^L$ and $\hat{q}_Y^U$. The distance between these two end points is the basis for solving for $\text{se}(q_Y)$, which is used by PROC SURVEYMEANS to form confidence limits on the estimated quantile as $\hat{q}_Y \pm t_{df,\alpha/2}\text{se}(\hat{q}_Y)$. (You can output the asymmetric

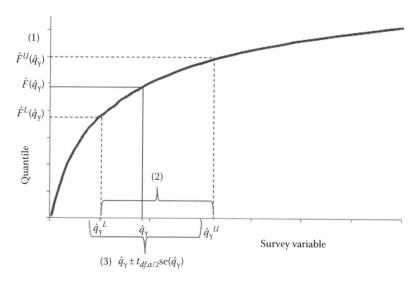

FIGURE 3.1
Illustration of the steps involved in the Woodruff method for quantifying uncertainty in an estimated quantile.

confidence limits labeled (2) in Figure 3.1 by specifying the NONSYMCL option in the PROC SURVEYMEANS statement.)

Suppose we were interested in estimating the *quartiles* of annual electricity expenditures (ELEXP8) in U.S. dollars of all buildings in the 2003 CBECS population. Quartiles are defined as γ = 0.25, 0.50, and 0.75 quantiles. Program 3.8 demonstrates the syntax to have PROC SURVEYMEANS output these estimates as well as their standard errors and 95% confidence limits. In addition to the QUARTILES keyword in the PROC statement, MEAN is also specified to permit a comparison of the estimated median with the mean. We observe the mean ($17,930) is significantly larger than the median ($4,301), an indication that the electricity expenditures distribution is right skewed.

Program 3.8: Estimating Quartiles of the Distribution of a Continuous Variable

```
proc surveymeans data=CBECS_2003 quartiles mean;
  strata STRATUM8;
  cluster PAIR8;
  var ELEXP8;
weight ADJWT8;
run;
```

Statistics		
Variable	Mean	Std Error of Mean
ELEXP8	17,930	995.072145

Quantiles					
Variable	Percentile		Estimate	Std Error	95% Confidence Limits
ELEXP8	25%	Q1	1,619.675091	100.746908	1,416.6330	1,822.7171
	50%	Median	4,300.740709	298.693274	3,698.7640	4,902.7174
	75%	Q3	12,664	897.759484	10,855.1304	14,473.7611

In Program 3.8, placing the keyword QUARTILES in the PROC statement is equivalent to specifying QUANTILE=(.25 .50 .75), PERCENTILE=(25 50 75), or Q1 MEDIAN Q3. Hence, there are several ways to request estimated quantiles. These are summarized in Table 3.4 alongside other pertinent statistical keywords. With the exception of the minimum and maximum value, standard errors and confidence limits are output by default any time quantiles are requested. At the time of this writing, there are no additional keywords available pertaining to measures of uncertainty.

TABLE 3.4

Summary of PROC SURVEYMEANS Statement Statistical
Keywords Pertaining to Quantiles

Keyword	Output
MIN	Minimum value
MAX	Maximum value
RANGE	MAX–MIN
Q1	Lower quartile (25th percentile)
MEDIAN	Median (50th percentile)
Q3	Upper quartile (75th percentile)
DECILES	The 10th, 20th,…, 90th percentiles
PERCENTILES = (*values*)	User-defined percentiles specified as whole numbers between 0 and 100, separated by a space or comma
QUANTILES = (*values*)	User-defined quantiles specified as decimals between 0 and 100, separated by a space or comma

3.6 Summary

This chapter was composed of four sections, one for each of the four classes of statistics that can be estimated using PROC SURVEYMEANS—namely, totals, means, ratios, and quantiles. The syntax examples were intentionally rudimentary to illustrate the fundamental concepts with minimal distraction. Although it would have been superfluous to output and comment on all of the available statistical keywords, a tabular summarization of those pertaining to the given statistic is provided at the end of each section. If you ever want to simply canvass all reportable statistics associated with the underlying analysis, you can specify the ALL keyword in the PROC SURVEYMEANS statement.

Note how all examples involved estimation at the population level. *Domain analysis* is the term describing estimation tasks targeting a subset of the population (e.g., a particular region of the country or building type). Examples of this type are covered exclusively and extensively in Chapter 8. Indeed, we will revisit a number of population-level analyses in this and the forthcoming chapters, replicating for one or more domains of interest.

4

Analyzing Categorical Variables Using PROC SURVEYFREQ

4.1 Introduction

This chapter showcases features of PROC SURVEYFREQ to facilitate categorical variables analyses. Whereas PROC SURVEYMEANS demonstrated in the previous chapter is the complex survey companion procedure to PROC MEANS, PROC SURVEYFREQ is the companion procedure to PROC FREQ. The chapter begins with a simple example pointing out how PROC SURVEYMEANS and PROC SURVEYFREQ produce equivalent results for certain univariate, descriptive statistics. The only difference is one of scaling: PROC SURVEYMEANS reports in terms of proportions, while PROC SURVEYFREQ reports in terms of percentages. We then segue into bivariate analyses and complex survey data adaptations to certain popular tests of association such as chi-square and likelihood ratio tests. Following that is a section devoted to odds ratios and a class of "risk" statistics that are useful for data that can be summarized in a 2 × 2 table. The concluding section illustrates multivariate capabilities of PROC SURVEYFREQ, albeit briefly, because most researchers prefer utilizing modeling procedures such as PROC SURVEYREG, PROC SURVEYLOGISTIC, and PROC SURVEYPHREG for multivariate analyses. Those three modeling procedures are covered in Chapters 5 through 7.

Examples in this chapter are drawn from the responses of 12,279 women aged 15–44 interviewed as part of the 2006–2010 National Survey of Family Growth (NSFG). More background on the NSFG can be found in Section 1.5.3. The public-use data website contains links to preformatted SAS code that was used to read in the raw data. The questions on the survey instrument differ somewhat depending on the respondent's gender. As such, there are separate programming files for female and male respondents. For the purpose of the present analyses, the female responses have been read in and saved to a SAS data set named NSFG_0610_F, which also contains variables reflecting the stratification (SEST), clustering (SECU), and unequal weights (WGTQ1Q16).

4.2 Univariate Analyses

4.2.1 Descriptive Statistics

We start this section with a brief example demonstrating the equivalence of PROC SURVEYFREQ and PROC SURVEYMEANS when estimating the totals and proportions of all *K* distinct values of a categorical variable. For simplicity, this is illustrated in Program 4.1 using the dichotomous indicator variable EVRMARRY for which a code of 1 indicates the respondent has been married at least once and 0 indicates the respondent has never married. For all intents and purposes, specifying a variable in the CLASS and VAR statements of PROC SURVEYMEANS is equivalent to specifying the variable in the TABLES statement of PROC SURVEYFREQ. All else equal, however, PROC SURVEYMEANS tends to run faster than PROC SURVEYFREQ, particularly with a large input data set.

Both SURVEY procedures' output begins with a summary of the complex survey features. As with the CBECS examples in the previous chapter, we can interpret the sum of weights as an estimate of the number of eligible population units. But here, instead of commercial buildings, the 61.8 million figure corresponds to the number of females aged 15–44. This estimate and its standard error also appear in the "Total" row of the PROC SURVEYFREQ output. Observe how the statistics labeled "Sum" and "Std Dev" in the SURVEYMEANS output match the statistics labeled "Weighted Frequency" and "Std Dev of Wgt Freq" in the SURVEYFREQ output. The same is true for "Mean" and "Std Err" and "Percent" and "Std Err of Percent," only the latter two are specified as percentages, while the former two are specified as proportions.

Program 4.1: Demonstrating the Equivalence of PROC SURVEYMEANS and PROC SURVEYFREQ for Descriptive Statistics of Categorical Variables

```
proc format;
  value EVRMARRY
    0='NEVER MARRIED'
    1='MARRIED';
run;

proc surveymeans data=NSFG_0610_F nobs sum mean;
  strata SEST;
  cluster SECU;
  class EVRMARRY;
  var EVRMARRY;
format EVRMARRY EVRMARRY.;
```

```
weight WGTQ1Q16;
run;

proc surveyfreq data=NSFG_0610_F;
   strata SEST;
   cluster SECU;
   tables EVRMARRY;
format EVRMARRY EVRMARRY.;
weight WGTQ1Q16;
run;
```

SURVEYMEANS Procedure

Data Summary	
Number of strata	56
Number of clusters	152
Number of observations	12,279
Sum of weights	61,754,741.1

Class Level Information		
CLASS Variable	Levels	Values
EVRMARRY	2	NEVER MARRIED MARRIED

Statistics						
Variable	Level	N	Mean	Std Error of Mean	Sum	Std Dev
EVRMARRY	NEVER MARRIED	6745	0.467186	0.009175	28,885,0952	941,508
	MARRIED	5534	0.532814	0.009175	32,903,789	1,294,136

SURVEYFREQ Procedure

Data Summary	
Number of strata	56
Number of clusters	152
Number of observations	12,279
Sum of weights	61,754,741.1

Table of EVRMARRY					
EVRMARRY	Frequency	Weighted Frequency	Std Dev of Wgt Freq	Percent	Std Err of Percent
NEVER MARRIED	6,745	28,850,952	941,508	46.7186	0.9175
MARRIED	5,534	32,903,789	1,294,136	53.2814	0.9175
Total	12,279	61,754,741	1,936,743	100.000	

From the output, we find that approximately 46.7% of women aged 15–44, or a total of about 28.9 million, have never been married. Notice how the estimated standard errors for these two statistics are the same as reported by PROC SURVEYMEANS and PROC SURVEYFREQ. Although the formulas in the documentation for point estimates and associated measures of uncertainty appear different at first glance, they are algebraically equivalent. And because this is true for virtually all descriptive statistics, to avoid being redundant, no specific formulas will be given here. The reader seeking more information on how these quantities are estimated can refer to Sections 3.2 and 3.3 or to the documentation.

PROC SURVEYFREQ produces a lot of statistics by default, but additional statistical keywords can be specified after the slash in the TABLES statement. Many are counterparts to those available in the PROC SURVEYMEANS statement listed in Tables 3.1 and 3.2. Table 4.1 summarizes those that are directly interchangeable—in fact, PROC SURVEYFREQ will recognize the PROC SURVEYMEANS keywords, but not vice versa.

The volume of statistics output by PROC SURVEYFREQ can be overwhelming at times. For instance, any of the keywords specified from Table 4.1 are output alongside all default statistics. To reduce clutter in the listing, you can call upon a battery of suppression options prefixed with NO (e.g., NOWT, NOFREQ). Table 4.2 summarizes some of the more useful ones. As with the statistical keywords, you specify them after the slash in the TABLES statement.

To motivate use of some of the options displayed in Tables 4.1 and 4.2, suppose we were interested in estimating the distribution of the number of pregnancies a woman has experienced. This is maintained in

TABLE 4.1

Summary of Comparable Statistical Keywords for Categorical Variables: PROC SURVEYFREQ TABLES Statement versus the PROC SURVEYMEANS Statement

PROC SURVEYFREQ TABLES Statement	PROC SURVEYMEANS Statement	Output
VAR	VAR	Variance of an estimated proportion/percentage
VARWT	VARSUM	Variance of an estimated total
CV	CV	Coefficient of variation of an estimated proportion/percentage
CVWT	CVSUM	Coefficient of variation of an estimated total
CL	CLM	Confidence interval of an estimated proportion/percentage based on the significance level specified in the ALPHA= option (default is ALPHA=.05)
CLWT	CLSUM	Confidence interval of an estimated total based on the significance level specified in the ALPHA= option (default is ALPHA=.05)

TABLE 4.2

Summary of Output Suppression Options Available after the Slash in the TABLES Statement of PROC SURVEYFREQ

Option	Effect
NOFREQ	Suppresses unweighted counts
NOWT	Suppresses weighted counts
NOCELLPERCENT	Suppresses overall cell percentages
NOPERCENT	Suppresses all percentages
NOTOTAL	Suppresses row and column totals (when applicable)
NOSTD	Suppresses standard errors

the variable PREGNUM, which ranges from 0 to 19. Note that this tally reflects all pregnancies that occurred, not strictly those culminating in a live birth (although the data can be filtered accordingly if desired). Let us suppose that we were less interested in counts than we were percentages. Program 4.2 demonstrates how to restrict output to only percentages and 95% confidence limits.

Program 4.2: Univariate Analysis of a Categorical Variable

```
proc surveyfreq data=NSFG_0610_F;
   strata SEST;
   cluster SECU;
   tables PREGNUM / nowt nofreq cl;
weight WGTQ1Q16;
run;
```

Table of PREGNUM				
PREGNUM	Percent	Std Err of Percent	95% Confidence Limits for Percent	
0	37.8801	1.0944	35.7076	40.0525
1	14.3533	0.4942	13.3724	15.3342
2	17.8230	0.6021	16.6277	19.0182
3	13.9525	0.5886	12.7841	15.1208
4	8.1313	0.3525	7.4316	8.8311
5	3.7789	0.2510	3.2806	4.2771
6	1.9862	0.2142	1.5610	2.4114
7	0.9768	0.1407	0.6975	1.2561
8	0.5992	0.1100	0.3809	0.8174
9	0.2445	0.0783	0.0891	0.4000
10	0.0478	0.0165	0.0150	0.0807
11	0.1046	0.0598	0.0000	0.2234
12	0.0192	0.0092	0.0009	0.0375

(Continued)

Table of PREGNUM				
PREGNUM	Percent	Std Err of Percent	95% Confidence Limits for Percent	
13	0.0618	0.0409	0.0000	0.1429
15	0.0034	0.0034	0.0000	0.0102
16	0.0075	0.0074	0.0000	0.0222
18	0.0298	0.0298	0.0000	0.0889
19	0.0002	0.0002	0.0000	0.0007
Total	100.000			

The analysis indicates that the vast majority of women have experienced four or fewer pregnancies. In fact, the prevalence of any distinct pregnancy total greater than four is less than 5%. Toward the tail end of the distribution, percentages become extremely small and the standard errors become almost as large as the estimate itself (i.e., the CV approaches 1). Note how the lower bound of several of the confidence intervals gets truncated at 0 since negative values would be nonsensical. The problem of making proper inferences on extreme proportions has received considerable attention in the statistical literature (Vollset, 1993; Brown et al., 2001). The crux of the issue is that the sampling distribution of rare proportions can be far from normal. In lieu of the standard linear (or "Wald") confidence intervals, a number of alternatives have been proposed (Wilson, 1927; Clopper and Pearson, 1934) and adapted to complex surveys (Korn and Graubard, 1998, 1999). Several of these alternatives were introduced in PROC SURVEYFREQ with the release of SAS Version 9.3. They are discussed in the next section.

4.2.2 Alternative Methods of Constructing Confidence Intervals for Extreme Proportions

One appealing method for constructing confidence intervals around extreme proportions is to make use of a log-odds, or *logit*, transformation, because the sampling distribution of this transformed value tends to be closer to normal than the sampling distribution of the proportion itself. (We will see this transformation again in Chapter 6 when we explore logistic regression models for categorical outcome variables.) If we denote the estimated proportion $\hat{p}$, the logit transformation is $\hat{l} = \ln(\hat{p}/(1-\hat{p}))$, where $\ln(x)$ represents the natural logarithm function, the logarithm of x with base $e \approx 2.7183$ (e stands for Euler's number). The fundamental property to be aware of is that e raised the power of the natural logarithm of any quantity returns the quantity inside the natural logarithm function—that is, $e^{\ln(x)} = x$. A common way of expressing e^x is $\exp(x)$, which is the notation we will use throughout this book.

The logit-transformed confidence interval is calculated in two steps. The first step is to compute the end points on the logit scale as

$$CI(\hat{l}) = \ln\left(\frac{\hat{p}}{1-\hat{p}}\right) \pm \frac{t_{df,\alpha/2} \times se(\hat{p})}{\left(\hat{p}(1-\hat{p})\right)} \tag{4.1}$$

where

$t_{df,\alpha/2}$ is the $100(1 - \alpha/2)$th percentile of a t distribution with df complex survey degrees of freedom and significance level α

$se(\hat{p})$ is the standard error of the proportion accounting for the complex survey design (i.e., using Equation 3.4)

Denoting the two end points of the interval in Equation 4.1 $\hat{l}_L$ and $\hat{l}_u$, the second step is to transform them back to proportions and create the confidence interval

$$CI(\hat{p}) = \left\{\frac{\exp(\hat{l}_L)}{1+\exp(\hat{l}_L)}, \frac{\exp(\hat{l}_u)}{1+\exp(\hat{l}_u)}\right\} \tag{4.2}$$

which will always fall within the interval (0,1). Note, however, that the interval is generally asymmetric about the point estimate, which is very different from the traditional confidence intervals to which most analysts are accustomed. In general, the point estimate will be situated nearer the more extreme end point of the interval.

Returning to the analysis in Program 4.2, we can have PROC SURVEYFREQ calculate logit-transformed 95% confidence intervals by specifying CL(TYPE=LOGIT) after the slash in the TABLES statement. The PSMALL<−p> option available within the parentheses is also handy as it will call upon the alternative method specified in the TYPE= option whenever the estimated proportion is less than p or greater than $1 - p$. Note that specifying PSMALL without providing a threshold invokes the default threshold of 0.25, and that omitting the PSMALL option altogether forces the alternative method be applied to all confidence intervals. Program 4.3 requests traditional confidence intervals unless the proportion is less than 0.1 or greater than 0.9, in which case logit-transformed confidence intervals are calculated. Asterisks and a corresponding footnote in the output flag instances where the alternative method was employed.

Program 4.3: Requesting Logit-Transformed Confidence Limits for Extreme Proportions

```
proc surveyfreq data=NSFG_0610_F;
   strata SEST;
   cluster SECU;
   tables PREGNUM / nowt nofreq cl(psmall=.1 type=logit);
weight WGTQ1Q16;
run;
```

Table of PREGNUM				
PREGNUM	Percent	Std Err of Percent	95% Confidence Limits for Percent	
0	37.8801	1.0944	35.7076	40.0525
1	14.3533	0.4942	13.3724	15.3342
2	17.8230	0.6021	16.6277	19.0182
3	13.9525	0.5886	12.7841	15.1208
4	8.1313	0.3525	7.4585	8.8591*
5	3.7789	0.2510	3.3110	4.3099*
6	1.9862	0.2142	1.6027	2.4591*
7	0.9768	0.1407	0.7337	1.2995*
8	0.5992	0.1100	0.4161	0.8621*
9	0.2445	0.0783	0.1294	0.4615*
10	0.0478	0.0165	0.0241	0.0950*
11	0.1046	0.0598	0.0336	0.3251*
12	0.0192	0.0092	0.0074	0.0497*
13	0.0618	0.0409	0.0166	0.2297*
15	0.0034	0.0034	0.0005	0.0247*
16	0.0075	0.0074	0.0010	0.0536*
18	0.0298	0.0298	0.0041	0.2162*
19	0.0002	0.0002	0.0000	0.0017*
Total	100.000			

* Logit confidence limits are computed for percents outside the PSMALL range, 10%–90%.

The logit transformation is just one of several methods available in PROC SURVEYFREQ. Two other adjustment techniques available are TYPE=WILSON, a complex survey adjustment to the method of Wilson (1927) discussed in Korn and Graubard (1999), and TYPE=CP, an adaptation of the method originally proposed by Clopper and Pearson (1934) to complex survey data as described in Korn and Graubard (1998). Because there is no convincing evidence in the literature that any one particular method is superior (Rust and Hsu, 2007), we will not devote separate examples to these techniques. Consult the PROC SURVEYFREQ documentation for more details.

Finally, be aware that these alternatives are not restricted to univariate analyses like the one illustrated in Program 4.3. Indeed, you can request any of the three in multivariate analyses, such as those to be demonstrated in Programs 4.5 and 4.8, where the objective is to analyze the cross-classification of two or more categorical variables.

4.2.3 Goodness-of-Fit Tests

In addition to making inferences on the proportions or totals for any of the K distinct values of a categorical variable, we may wish to conduct a joint hypothesis test on a vector of null proportions. To motivate an example of this, let us simplify the analysis of pregnancy numbers begun in the previous section by concerning ourselves with a collapsed version consisting of $K = 5$ categories: 0, 1, 2, 3, or 4 or more pregnancies. Examining the output from Program 4.3, perhaps the goal is to test whether this set of observed proportions $\hat{\mathbf{p}} = \{\hat{p}_1 = 0.3788, \hat{p}_2 = 0.1435, \hat{p}_3 = 0.1782, \hat{p}_4 = 0.1395, \hat{p}_5 = 0.1600\}$ provides sufficient evidence that the true set of proportions in the population is not equal to $\mathbf{p}_0 = \{p_{01} = 0.4, p_{02} = 0.15, p_{03} = 0.15, p_{04} = 0.15, p_{05} = 0.15\}$. That is, we were interested in testing $H_0: \mathbf{p} = \mathbf{p}_0$ vs. $H_1: \mathbf{p} \neq \mathbf{p}_0$.

In general, given a simple random sample of size n, one can construct the following Pearson chi-square *goodness-of-fit* test statistic as

$$\chi^2_{SRS} = n \sum_{k=1}^{K} \frac{\left(\hat{p}_k - p_{0k}\right)^2}{p_{0k}} \tag{4.3}$$

and assess significance by comparing it to $\chi^2_{df,\alpha}$, the $100(1 - \alpha)$th percentile of a chi-square distribution with $df = K - 1$ degrees of freedom and significance level α. If $\chi^2_{SRS} > \chi^2_{df,\alpha}$, the null hypothesis is rejected. An asymptotically equivalent test statistic is the *likelihood ratio* test statistic

$$G^2_{SRS} = 2n \sum_{k=1}^{K} \hat{p}_k \ln\left(\frac{\hat{p}_k}{p_{0k}}\right) \tag{4.4}$$

As with the chi-square statistic, the likelihood ratio test statistic can be referenced against a chi-square distribution with $df = K - 1$.

When data are collected via a complex survey design, these test statistics no longer abide by the same chi-square distribution under the null hypothesis. Rao and Scott (1981, 1984) proposed methods for adjusting them based on a generalized design effect of sorts (for a discussion of the design effect, see Section 1.4.4). The qualifier "generalized" is used because there are actually K distinct design effects, one for each proportion. Although the mechanics for computing this factor, which we can denote *gdeff*, are rather complex, the remedy is straightforward to apply: simply divide the standard chi-square goodness-of-fit test statistic by this factor as follows:

$$\chi^2_{R-S} = \frac{\chi^2_{SRS}}{gdeff} \tag{4.5}$$

or

$$G^2_{R-S} = \frac{G^2_{SRS}}{gdeff} \tag{4.6}$$

From there, the adjusted test statistics are properly rescaled such that they can be referenced against a chi-square distribution with $df = K - 1$.

PROC SURVEYFREQ will compute the standard and adjusted statistics if we specify the CHISQ and LRCHISQ options after the slash in the TABLES statement as well as provide the null proportions using syntax TESTP=(*values*), where *values* are specified either as proportions or percentages. (The TESTP= syntax is technically optional; without a formal declaration, SAS will assume the test is for equal proportions, or that $p_{0k} = 1/K$ for all K categories.) Program 4.4 demonstrates syntax to carry out the tests on the collapsed pregnancy number example with null vector $\mathbf{p}_0 = \{p_{01} = 0.4, p_{02} = 0.15, p_{03} = 0.15, p_{04} = 0.15, p_{05} = 0.15\}$.

Program 4.4: Requesting Univariate Goodness-of-Fit Tests Adjusted for the Complex Survey Design

```
proc format;
  value PRG4PLUS
    0='NONE'
    1='1 PREGNANCY'
    2='2 PREGNANCIES'
    3='3 PREGNANCIES'
    other='4 OR MORE PREGNANCIES' ;
run;

proc surveyfreq data=NSFG_0610_F;
  strata SEST;
  cluster SECU;
  tables PREGNUM / chisq lrchisq testp=(.4 .15 .15 .15 .15);
format PREGNUM PRG4PLUS.;
weight WGTQ1Q16;
run;
```

Table of PREGNUM						
PREGNUM	Frequency	Weighted Frequency	Std Dev of Wgt Freq	Percent	Test Percent	Std Err of Percent
NONE	4,741	23,392,728	1,014,910	37.8801	40.00	1.0944
1 PREGNANCY	1,928	8,863,834	408,854	14.3533	15.00	0.4942
2 PREGNANCIES	2,095	11,006,517	492,924	17.8230	15.00	0.6021
3 PREGNANCIES	1,616	8,616,306	469,753	13.9525	15.00	0.5886
4 OR MORE PREGNANCIES	1,899	9,875,356	437,134	15.9913	15.00	0.5399
Total	12,279	61,754,741	1,936,743	100.000		

Rao–Scott Chi-Square Test	
Pearson chi-square	99.4805
Design correction	3.4389
Rao–Scott chi-square	28.9277
DF	4
Pr > ChiSq	<.0001
F Value	7.2319
Num DF	4
Den DF	384
Pr > F	<.0001
Sample size = 12,279	

Rao–Scott Likelihood Ratio Test	
Likelihood ratio chi-square	96.0805
Design correction	3.4389
Rao–Scott chi-square	27.9390
DF	4
Pr > ChiSq	<.0001
F Value	6.9848
Num DF	4
Den DF	384
Pr > F	<.0001
Sample size = 12,279	

From the Rao–Scott Chi-Square Test table of the output, we observe that the unadjusted chi-square statistic is $\chi^2_{SRS} = 99.4805$, but with *gdeff* = 3.4389, the Rao–Scott adjusted test statistic is $\chi^2_{R-S} = 99.4805/3.4389 = 28.9277$. Although the adjusted test statistic is notably smaller than what would be reported if we ignored the complex survey design, it is still highly significant, suggesting the null hypothesis should be rejected. In practice, especially if the sample design involves clustering, we can expect *gdeff* to be larger than 1, and so neglecting to make the adjustment could lead to erroneous, anticonservative statistical conclusions. A similar line of reasoning applies for the likelihood ratio version of the goodness-of-fit test.

The only component of the output thus far not discussed is the "F Value" row. This is an *F*-distribution transformation of the Rao–Scott chi-square statistics Thomas and Rao (1987) and Lohr (2009) assert can exhibit improved stability in certain circumstances. The test statistic is found by dividing the Rao–Scott test statistic by $K - 1$ and referencing that value against an *F* distribution with $K - 1$ numerator degrees of freedom and denominator degrees of freedom equaling $K - 1$ times the complex survey degrees of freedom. So, under the default Taylor series linearization approach, where the complex survey degrees of freedom are the number of distinct PSUs minus the number of distinct strata, this explains why PROC SURVEYFREQ is refer-encing an F distribution with $K - 1 = 4$ numerator degrees of freedom and $(152 - 56) * (5 - 1) = 384$ denominator degrees of freedom.

The default adjustments that occur when the CHISQ and LRCHISQ options are specified are *first-order* corrections. Thomas and Rao (1987) also discuss *second-order* corrections, which go one step further by matching not only the mean of the chi-square distribution, but also the variance. We will not go into the computational details here, but the gist is that the first-order adjusted test statistic is divided through once more by a second adjustment factor. PROC SURVEYFREQ will construct these second-order corrections if you specify CHISQ(SECONDORDER) or LRCHISQ(SECONDORDER). Lohr (2009) suggests these are warranted if the K design effects vary appreciably. Although there are no guidelines for how much variability is too much vari-ability, you can inspect the range of design effects in the output by specify-ing the DEFF option after the slash in TABLES statement.

4.3 Bivariate Analyses

4.3.1 Introduction

Examples up through this point in the chapter have been exclusively uni-variate in nature. In this section, we consider bivariate analyses in which two variables separated by an asterisk are specified in the TABLES state-ment. Before launching into a statistical analysis, it is instructive to first see how the PROC SURVEYFREQ output is oriented. The default appearance is not the default grid layout of PROC FREQ; rather, it more closely resembles PROC FREQ output when the CROSSLIST option is specified after the slash in the TABLES statement.

Program 4.5 is a simple bivariate analysis of current religious affiliation (RELIGION) and an indicator variable of whether the female respondent has ever been married (EVRMARRY). The raw codes for these two variables are assigned formats in order for more meaningful labels to appear in the output. Observe how the output consists of a tabular summarization of the $R = 4$ distinct categories of the row variable and the $C = 2$ distinct categories of the column variable. The two dimensions combine for $4 \times 2 = 8$ distinct *cells*, but PROC SURVEYFREQ also outputs the marginal statistics for the row and column factors. For instance, the third row of output, the first row labeled "Total" under the EVRMARRY column, provides marginal informa-tion about the first row, respondents indicating no religious affiliation. The last two rows provide marginal information on the two columns. Aside from these rows, the table summarizes cell-specific frequencies and weighted totals as well as table-wide (overall) cell percentages.

Program 4.5: Bivariate Analysis of Two Categorical Variables

```
proc format;
  value EVRMARRY
    0='NEVER MARRIED'
    1='MARRIED';

  value RELIGION
    1='NO RELIGION'
    2='CATHOLIC'
    3='PROTESTANT'
    4='OTHER RELIGIONS';
run;

proc surveyfreq data=NSFG_0610_F;
  strata SEST;
  cluster SECU;
  tables RELIGION*EVRMARRY;
format RELIGION RELIGION.
       EVRMARRY EVRMARRY.;
weight WGTQ1Q16;
run;
```

Table of RELIGION by EVRMARRY						
RELIGION	**EVRMARRY**	**Frequency**	**Weighted Frequency**	**Std Dev of Wgt Freq**	**Percent**	**Std Err of Percent**
NO RELIGION	NEVER MARRIED	1,461	6,085,509	319,257	9.8543	0.5051
	MARRIED	890	4,997,620	326,499	8.0927	0.5079
	Total	2,351	11,083,129	544,368	17.9470	0.8409
CATHOLIC	NEVER MARRIED	1,633	7,171,300	444,581	11.6126	0.6515
	MARRIED	1,502	8,228,192	522,667	13.3240	0.7801
	Total	3,135	15,399,492	879,057	24.9365	1.2740
PROTESTANT	NEVER MARRIED	3,166	13,231,597	547,467	21.4260	0.8001
	MARRIED	2,590	16,261,335	884,708	26.3321	1.0946
	Total	5,756	29,492,932	1,215,576	47.7582	1.4143
OTHER RELIGIONS	NEVER MARRIED	485	2,362,546	517,054	3.8257	0.8008
	MARRIED	552	3,416,643	643,641	5.5326	0.9781
	Total	1,037	5,779,189	1,128,340	9.3583	1.7242
Total	NEVER MARRIED	6,745	28,850,952	941,508	46.7186	0.9175
	MARRIED	5,534	32,903,789	1,294,136	53.2814	0.9175
	Total	12,279	61,754,741	1,936,743	100.000	

Similar to what was argued previously, it would be redundant to provide formulas for the standard errors output for totals and proportions in bivariate analyses. PROC SURVEYFREQ uses the same principles laid out in Sections 3.2 and 3.3 with indicator variables defining a particular cell, row, or column. For example, the overall percentage and standard error of Catholics who have never married could be ascertained by creating an indicator variable equaling 1 if the respondent was both Catholic and never married and 0 otherwise, and then specifying this variable in the VAR statement of PROC SURVEYMEANS.

Again, the purpose of this brief syntax example was to acquaint the reader with the default output orientation for bivariate analyses. In the next section, we will formalize the two-way table notation and explore some of the additional statistical tests available within PROC SURVEYFREQ.

4.3.2 Tests of Association

The goodness-of-fit tests presented in Section 4.2.3 extend to two-way tables, although in this context they are more commonly referred to as "tests of association." Instead of comparing the observed proportions to a user-defined null vector of proportions, these tests seek to determine whether the row factor and column factor are *independent*, which is to say the two factors are not associated with one another.

Suppose a row factor consisting of $r = 1,..., R$ distinct categories is crossed with a column factor consisting of $c = 1,..., C$ distinct categories. Table 4.3 offers a visualization of how data might be tabulated under an SRS sample design. The $R \times C$ cell counts are subscripted with the row and column category indicators (e.g., n_{12} is the count of cases defined by the first category of the row factor and second category of the column factor), and a dot symbolizes summing over the dimension placeholder (e.g., $n_{.2}$ represents the second column factor category summed over all rows).

Any of the $R \times C$ cells can be converted to proportions by dividing through by n, as can any of the $R + C$ marginal counts. The *Pearson chi-square test of association* begins by calculating each cell's *expected* value as the product of its row and column proportions. For example, the expected proportion for the cell in row r and column c is $p_{rc} = (n_{r.}/n)(n_{.c}/n)$. If we denote the *observed* cell proportion as $\hat{p}_{rc} = n_{rc}/n$, the Pearson chi-square test of association is calculated as

TABLE 4.3

Two-Way Table Layout under Simple Random Sampling

		Column Factor				
		1	2	...	C	Row Totals
Row Factor	1	n_{11}	n_{12}	...	n_{1C}	$n_{1.}$
	2	n_{21}	n_{22}	...	n_{2C}	$n_{2.}$
	...	...	...	...	...	...
	R	n_{R1}	n_{R2}	...	n_{RC}	$n_{R.}$
Column Totals		$n_{.1}$	$n_{.2}$	...	$n_{.c}$	n

$$\chi^2_{SRS} = n \sum_{r=1}^{R} \sum_{c=1}^{C} \frac{\left(\hat{p}_{rc} - p_{rc} \right)^2}{p_{rc}} \tag{4.7}$$

It has the same general structure as the goodness-of-fit test, only the summation terms and associated subscripts differ slightly. The observed test statistic is referenced against $\chi^2_{df,\alpha}$, a chi-square random variate with $df = (R - 1) \times (C - 1)$ degrees of freedom and significance level α. If $\chi^2_{SRS} > \chi^2_{df,\alpha}$ there is evidence that the two factors are not independent. Similar reasoning applies for the *likelihood ratio test of association*, which is calculated as

$$G^2_{SRS} = 2n \sum_{r=1}^{R} \sum_{c=1}^{C} \hat{p}_{rc} \ln \left(\frac{\hat{p}_{rc}}{p_{rc}} \right) \tag{4.8}$$

Table 4.4 shows the tabulation comparable to Table 4.3 when data are collected via a complex survey. Instead of sample counts, cells and margins are populated with weighted counts (i.e., estimated totals), so we symbolize them by a capital letter and top with a hat. The weighted counts are used to formulate the observed and expected proportions. For example, the expected proportion for the row r and column c is found by $p_{rc} = (\hat{N}_{r.}/N)(\hat{N}_{.c}/\hat{N})$, whereas the observed proportion is $\hat{p}_{rc} = \hat{N}_{rc}/\hat{N}$.

Even after substituting the weighted versions of the expected and observed proportions, the test statistics as defined in Equations 4.7 and 4.8 still require rescaling. SAS uses a similar algorithm as discussed in the previous section regarding goodness-of-fit tests attributable to Rao and Scott (1981, 1984) and Thomas and Rao (1987). As seen in Equations 4.5 and 4.6, the adjusted test statistics are formed by dividing the original test statistics by a generalized design effect.

TABLE 4.4

Two-Way Table Layout under a Complex Survey Design

		Column Factor				
		1	2	...	C	Row Totals
Row Factor	1	$\hat{N}_{11}$	$\hat{N}_{12}$	...	$\hat{N}_{1C}$	$\hat{N}_{1.}$
	2	$\hat{N}_{21}$	$\hat{N}_{22}$	...	$\hat{N}_{2C}$	$\hat{N}_{2.}$
	...	...	...	...	...	...
	R	$\hat{N}_{R1}$	$\hat{N}_{R2}$	...	$\hat{N}_{RC}$	$\hat{N}_{R.}$
Column Totals		$\hat{N}_{.1}$	$\hat{N}_{.2}$	...	$\hat{N}_{.C}$	$\hat{N}$

To see how PROC SURVEYFREQ carries out these calculations, let us return to the NSFG analysis investigating whether the current religious affiliation of a woman has any bearing on whether she has ever been married. Program 4.6 demonstrates syntax to perform the chi-square and likelihood ratio tests of association with a first-order design correction. It is structured similar to what we saw in Program 4.5. The two tests of association are requested with the CHISQ and LRCHISQ options after the slash in the TABLES statement. Although not demonstrated here, as before, second-order corrections can be requested by specifying CHISQ(SECONDORDER) or LRCHISQ(SECONDORDER)—see the documentation for more details.

Program 4.6: Requesting Bivariate Tests of Association Adjusted for the Complex Survey Design

```
proc format;
  value EVRMARRY
    0='NEVER MARRIED'
    1='MARRIED';

  value RELIGION
    1='NO RELIGION'
    2='CATHOLIC'
    3='PROTESTANT'
    4='OTHER RELIGIONS' ;
run;

proc surveyfreq data=NSFG_0610_F;
  strata SEST;
  cluster SECU;
  tables RELIGION*EVRMARRY / chisq lrchisq;
format RELIGION RELIGION.
       EVRMARRY EVRMARRY.;
weight WGTQ1Q16;
run;
```

Rao–Scott Chi-Square Test	
Pearson chi-square	83.2411
Design correction	2.3749
Rao–Scott chi-square	35.0500
DF	3
Pr > ChiSq	<.0001
F Value	11.6833
Num DF	3
Den DF	288
Pr > F	<.0001
Sample size = 12,279	

Rao–Scott Likelihood Ratio Test	
Likelihood ratio chi-square	83.2142
Design correction	2.3749
Rao–Scott chi-square	35.0387
DF	3
Pr > ChiSq	<.0001
F Value	11.6796
Num DF	3
Den DF	288
Pr > F	<.0001
Sample size = 12,279	

The Rao–Scott chi-square statistic reported (35.05) is the Pearson chi-square statistic (83.2411) divided by the generalized design effect (2.3749). The likelihood ratio chi-square statistic is also adjusted accordingly. Both are referenced against a chi-square distribution with $df = (4 - 1) \times (2 - 1) = 3$. A rescaled test statistic based on an underlying F distribution is also given. Even after adjusting for the complex survey design, all tests are in agreement concluding that there is evidence of an association between these two factors.

We close this section making brief mention that there is also a *Wald chi-square test of association* available in PROC SURVEYFREQ, which can be output using the keyword WCHISQ. Heeringa et al. (2010, pp. 167–168) cite a few references suggesting that the tests of association presented here tend to perform better, although we will see other uses of Wald chi-square statistics in later chapters.

4.3.3 Risk Statistics and Odds Ratios

For categorical data that can be displayed as a 2 × 2 table, PROC SURVEYFREQ offers a class of "risk" statistics (see Section 2.2 of Agresti, 2013) that have an appealingly simple and intuitive interpretation. The necessary orientation for these types of analyses is portrayed in Table 4.5. Once again, weighted totals are the building blocks for these statistics.

TABLE 4.5

2 × 2 Table Layout Permitting the Calculation of Risk Statistics and Odds Ratios

	Column 1	Column 2	Row Totals
Row 1	$\hat{N}_{11}$	$\hat{N}_{12}$	$\hat{N}_{1.}$
Row 2	$\hat{N}_{21}$	$\hat{N}_{22}$	$\hat{N}_{2.}$
Column Totals	$\hat{N}_{.1}$	$\hat{N}_{.2}$	$\hat{N}$

The estimated probability of a case falling in the first column ($c = 1$) given it falls within the first row ($r = 1$) can be written as $\Pr(c=1 \mid r=1) = \hat{p}_{c=1|r=1} = \hat{N}_{11}/\hat{N}_1$. Borrowing terminology from the field of epidemiology, this is referred to as the estimated Column 1 *risk* for the first row. Alternatively, this can be interpreted as the estimated proportion of cases characterized by the first column amongst cases in the first row; hence, this is an estimated proportion for the *domain* defined by the first row (see Chapter 8). We can express the estimated Column 1 risk similarly for the second row as $\Pr(c=1 \mid r=2) = \hat{p}_{c=1|r=2} = \hat{N}_{21}/\hat{N}_2$. If we differentiate some risk factor using the row dimension (e.g., smoker/nonsmoker, treatment/placebo) and some target outcome using the first column, one estimator of interest would be the estimated risk difference defined as

$$\hat{r}^1_{diff} = \hat{p}_{c=1|r=1} - \hat{p}_{c=1|r=2} = \hat{N}_{11}/\hat{N}_1 - \hat{N}_{21}/\hat{N}_2 \qquad (4.9)$$

PROC SURVEYFREQ will compute row-specific estimated risks and the estimator in Equation 4.9 when the statistical keyword RISK is specified after the slash in the TABLES statement. It will also estimate a standard error and form a confidence interval around the difference. Significance of the two estimated risks' difference can be assessed by determining whether or not this interval contains 0. As we will see in Program 8.9, this is equivalent to testing whether the Column 1 proportions of the two domains defined by the two rows are significantly different from one another.

Another useful statistic is the estimated *relative risk*, which is defined as the ratio of the two estimated risks, or

$$\hat{r}^1_{rel} = \frac{\hat{p}_{c=1|r=1}}{\hat{p}_{c=1|r=2}} \qquad (4.10)$$

A ratio of 1 implies that the two estimated risks are equivalent, which also means that the estimated risk difference is 0. A ratio of, say, 1.3, means that the first row estimated risk is 30% greater than the second row estimated risk. This statistic can be output by specifying RELRISK after the slash in the TABLES statement. PROC SURVEYFREQ always computes this statistic as the first row over the second row, whereas the estimated risks and estimated risk difference are calculated for both columns when the RISK keyword is specified.

Program 4.7 illustrates how these statistics can offer insight into the relationship between ever being married (EVRMARRY) and ever using the birth control pill (PILLR) for NSFG-eligible females. Although the terminology sounds a bit strange in this context, we can compare the estimated "risk" of ever using the birth control pill (cases where PILLR=1) based on whether or not one has been married by specifying EVRMARRY as the row factor and PILLR as the column factor. The ORDER=FORMATTED option in the PROC statement forces the row and column factors to be ordered according to their respective formats. (In general, assigning format labels that start with 1 or 2 is a convenient way to control which appears first and which second.)

Program 4.7: Risk Statistics and the Odds Ratio of a 2 × 2 Table

```
proc format;
  value EVRMARRY
    1='1. MARRIED'
    0='2. NEVER MARRIED';

  value PILLR
    1='1. YES'
    2='2. NO';
run;

proc surveyfreq data=NSFG_0610_F order=formatted;
  strata SEST;
  cluster SECU;
  tables EVRMARRY*PILLR / nofreq nocellpercent row risk
  relrisk;
format EVRMARRY EVRMARRY.
       PILLR PILLR.;
weight WGTQ1Q16;
run;
```

			Table of EVRMARRY by PILLR			
EVRMARRY	**PILLR**	**Weighted Frequency**	**Std Dev of Wgt Freq**	**Row Percent**	**Std Err of Row Percent**
1. MARRIED	1. YES	28,475,997	1,180,305	86.5432	0.8180
	2. NO	4,427,792	304,082	13.4568	0.8180
	Total	32,903,789	1,294,136	100.000	
2. NEVER MARRIED	1. YES	16,545,294	632,194	57.3475	1.3355
	2. NO	12,305,658	582,853	42.6525	1.3355
	Total	28,850,952	941,508	100.000	
Total	1. YES	45,021,292	1,490,396		
	2. NO	16,733,450	720,287		
	Total	61,754,741	1936,743		

	Column 1 Risk Estimates			
	Risk	**Standard Error**	**95% Confidence Limits**	
Row 1	0.8654	0.0082	0.8492	0.8817
Row 2	0.5735	0.0134	0.5470	0.6000
Total	0.7290	0.0079	0.7133	0.7448
Difference	0.2920	0.0158	0.2606	0.3233
Difference is (Row 1 − Row 2)				
Sample size = 12,279				

(Continued)

Column 2 Risk Estimates				
	Risk	Standard Error	95% Confidence Limits	
Row 1	0.1346	0.0082	0.1183	0.1508
Row 2	0.4265	0.0134	0.4000	0.4530
Total	0.2710	0.0079	0.2552	0.2867
Difference	−0.2920	0.0158	−0.3233	−0.2606
Difference is (Row 1 − Row 2)				
Sample size = 12,279				

Odds Ratio and Relative Risks (Row1/Row2)			
	Estimate	95% Confidence Limits	
Odds ratio	4.7832	4.0034	5.7149
Column 1 relative risk	1.5091	1.4353	1.5867
Column 2 relative risk	0.3155	0.2752	0.3617
Sample size = 12,279			

Observe how the estimated risk statistics for the first and second rows in the Column 1 Risk Estimates output component effectively match values reported under the "Row Percent" heading of the tabular summary of the raw data for lines where the formatted value of PILLR is "1. Yes." Approximately 86.5% of females who have ever married have used the birth control pill, whereas that figure is only 57.3% for females who have never married. The 95% confidence limits reported are the same end points of a confidence interval that would appear in the tabular summary if the CL option were specified after the slash in the TABLES statement. Similar reasoning translates to the Column 2 Risk Estimates table and the PILLR line with formatted value "2. No" in the tabular summary.

The line labeled "Total" in the risk estimates portion of the output is an estimate of the overall risk. For example, the overall risk for the first column is estimated as $\hat{N}_{.1}/\hat{N} = (45,021,292)/(61,754,741) = 0.729$, which is to say the marginal percentage of females who have ever used birth control is 72.9%. The last line given is the estimated difference between the first and second row estimated risks (0.865 − 0.573 = 0.292). The 95% confidence interval around this statistic does not contain zero, so we conclude that the difference is significant. Females who have married at least once are significantly more likely than females who have never married to report ever using the birth control pill. The estimated relative risk is $\hat{r}_{rel}^{1} = 0.865/0.573 = 1.51$, which is provided toward the bottom of the output alongside its 95% confidence interval. That is, females who have married at least once are 51% more likely to have taken the birth control pill at some point than females who have yet to marry.

Another measure of association output is the estimated *odds ratio*, which is the ratio of the odds of falling in the first column given being in the first row to the comparable odds given being in the second row. The odds ratio goes by numerous algebraic representations, including either

$$\hat{O} = \frac{(\hat{p}_{c=1|r=1})/(1 - \hat{p}_{c=1|r=1})}{(\hat{p}_{c=1|r=2})/(1 - \hat{p}_{c=1|r=2})} \tag{4.11}$$

or

$$\hat{O} = \frac{\hat{N}_{11}\hat{N}_{22}}{\hat{N}_{12}\hat{N}_{21}} \tag{4.12}$$

Since odds are always nonnegative, so is any ratio of them. A value of 1 indicates equivalence with respect to the odds, suggesting independence of the row and column factors. Values nearer 0 or much greater than 1 indicate a strong association. Note that the odds ratio is inverted if the order of the columns or rows is reversed. For instance, we note the estimated odds ratio in the output of Program 4.7 is (28,475,997 * 12,305,658)/(4,427,792 * 16,545,294) = 4.78. If the two rows factors were reversed, the estimated odds ratio would become 1/(4.78) ≈ 0.21.

The odds ratio is often confused with and/or misinterpreted as the relative risk. Specifically, a common misstatement is to interpret an odds ratio of, say, 4.78 to mean that the probability of falling in the first column is 4.78 times greater for the first row cases relative to the second. Actually, that is the interpretation of the relative risk. As Agresti (2013) notes, the two quantities are related to one another, but approximate equivalence only holds if the risks in both rows are close to 0.

A formal way to use the odds ratio in tests of association is to form a confidence interval around the estimated odds ratio and determine whether it encompasses 1. PROC SURVEYFREQ outputs this interval by default. Note how the interval is asymmetric about the point estimate as it is with the relative risk—in contrast, the confidence intervals formed around the risk and risk difference point estimates are symmetric. Similar in spirit to the logit-transformed extreme proportion confidence intervals discussed in Section 4.2.2, a symmetric interval is first formed with respect to a natural logarithm transformation of the odds ratio and then the results are back-transformed.

4.4 Multiway Tables

PROC SURVEYFREQ can be utilized with three or more dimensions, but the result is a series of bivariate analyses, one for each distinct value of the first dimension(s) identified. For example, specifying TABLES VAR1*VAR2*VAR3 will produce a series of two-way tables of VAR2 crossed with VAR3 independently for each distinct value of VAR1. We can think of VAR1 as the *control factor* or *page factor*. If four variables were provided in the TABLES statement, the cross-classification of the first two variables would play this role.

Suppose we sought to replicate the relative risk analysis in Program 4.7, only now controlling for religiosity. This is accomplished with the syntax in Program 4.8. First, we create a format consisting of dichotomized categories of RELIGION. The second step is to add this variable as the leading factor in the TABLES statement. As we see from the output, estimated relative risk statistics and odds ratios are reported for both 2 × 2 tables defined by the two formatted values of RELIGION.

Whereas the overall estimated relative risk for the first column factor (ever using the birth control pill) was determined in Program 4.7 to be approximately 1.51, the figure is somewhat less for those who do not affiliate with any particular religion (1.35) and slightly higher for those who do (1.56). PROC SURVEYFREQ is equipped with data visualization capabilities to facilitate comparisons of two or more relative risks. Namely, specifying PLOTS=RELRISKPLOT after the slash in the TABLES statement produces a side-by-side plot of relative risks and their associated confidence intervals for each distinct value of the page factor(s) in a multiway table. The only added requirement is that ODS GRAPHICS must be enabled beforehand, as shown in Program 4.8.

Examining the estimated risks from the "Row Percent" column of the tabular summary portion of the output, we can ascribe the relative risk imparity to how the risks for those who have never been married differ substantively based on religiosity. Specifically, those unaffiliated with any religion have a risk of 0.652, whereas those who are religious have a risk of 0.553. The corresponding risks for those who have been married are much closer: 0.879 versus 0.863.

Program 4.8: Multivariate Analysis Example: A Three-Way Table

```
proc format;
  value RELIGION
    1='1. NOT RELIGIOUS'
    2,3,4='2. RELIGIOUS';

  value EVRMARRY
    1='1. MARRIED'
    0='2. NEVER MARRIED';

  value PILLR
    1='1. YES'
    2='2. NO';
run;

ods graphics on;
proc surveyfreq data=NSFG_0610_F order=formatted;
```

```
    strata SEST;
    cluster SECU;
    tables RELIGION*EVRMARRY*PILLR/nofreq nocellpercent row relrisk
                                plots=relriskplot;
format RELIGION RELIGION.
       EVRMARRY EVRMARRY.
       PILLR PILLR.;
weight WGTQ1Q16;
run;
ods graphics off;
```

Table of EVRMARRY by PILLR					
Controlling for RELIGION=1. NOT RELIGIOUS					
EVRMARRY	PILLR	Weighted Frequency	Std Dev of Wgt Freq	Row Percent	Std Err of Row Percent
1. MARRIED	1. YES	4,394,379	298,137	87.9294	1.6778
	2. NO	603,241	93,070	12.0706	1.6778
	Total	4,997,620	326,499	100.000	
2. NEVER MARRIED	1. YES	3,965,343	243,061	65.1604	1.9271
	2. NO	2,120,166	158,204	34.8396	1.9271
	Total	6,085,509	319,257	100.000	
Total	1. YES	8,359,722	464,881		
	2. NO	2,723,407	190,781		
	Total	11,083,129	544,368		

Odds Ratio and Relative Risks (Row1/Row2)			
	Estimate	95% Confidence Limits	
Odds ratio	3.8949	2.7673	5.4819
Column 1 relative risk	1.3494	1.2623	1.4426
Column 2 relative risk	0.3465	0.2600	0.4617
Sample size = 12,279			

Table of EVRMARRY by PILLR					
Controlling for RELIGION=2. RELIGIOUS					
EVRMARRY	PILLR	Weighted Frequency	Std Dev of Wgt Freq	Row Percent	Std Err of Row Percent
1. MARRIED	1. YES	24,081,618	1,109,402	86.2950	0.8687
	2. NO	3,824,551	271,231	13.7050	0.8687
	Total	27,906,169	1,210,848	100.000	
2. NEVER MARRIED	1. YES	12,579,952	550,313	55.2590	1.5192
	2. NO	10,185,492	531,540	44.7410	1.5192
	Total	22,765,443	834,085	100.000	

(Continued)

Table of EVRMARRY by PILLR					
Controlling for RELIGION=2. RELIGIOUS					
EVRMARRY	PILLR	Weighted Frequency	Std Dev of Wgt Freq	Row Percent	Std Err of Row Percent
Total	1. YES	36,661,569	1,370,203		
	2. NO	14,010,043	655,677		
	Total	50,671,612	1,795,977		

Odds Ratio and Relative Risks (Row1/Row2)			
	Estimate	95% Confidence Limits	
Odds ratio	5.0981	4.1980	6.1911
Column 1 relative risk	1.5616	1.4722	1.6565
Column 2 relative risk	0.3063	0.2649	0.3543
Sample size = 12,279			

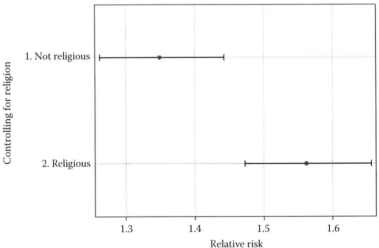

Relative risks with 95% confidence limits
EVRMARRY by PILLR

Risks computed for column 1 (PILLR = 1. yes)

For brevity, the analysis here was restricted to relative risks, but the same ideas translate to other bivariate analyses. For example, resubmitting the PROC SURVEYFREQ step with the CHISQ option specified after the slash in the TABLES statement would result in two Rao–Scott-adjusted chi-square test statistics, one for each 2 × 2 table.

4.5 Summary

This chapter demonstrates the capabilities of PROC SURVEYFREQ for a variety of categorical variable analyses. For basic descriptive statistics, we saw how PROC SURVEYFREQ and PROC SURVEYMEANS produce equivalent results, albeit on a different scale (i.e., percentages as opposed to proportions). Because of this, no specific measures of variability were defined; they are the same as those defined in Sections 3.2 and 3.3. It was stressed that goodness-of-fit tests and tests of association discussed in general statistics texts such as Agresti (2013) can be used with weighted proportions generated from complex survey designs, but only after one or more adjustments factors proposed by Rao and Scott (1981, 1984) and Thomas and Rao (1987) are applied. These adjustments are available within PROC SURVEYFREQ and are reported by default in the output.

Measures of variability were intentionally omitted during the discussion of risk statistics and odds ratios. This is because risk statistics are simply proportions, or means of 0/1 indicator variables, and so the formulas specified in Section 3.3 are directly applicable. We will defer the estimation of variability for a risk difference to Section 8.6, when we tie that particular analysis task to the broader concept of domain mean significance tests (see Program 8.9). Measures of variability for odds ratios will be discussed further in Chapter 6 when we explore the concept of logistic regression, which entails modeling logit-transformed odds ratios.

This chapter concluded with a brief section on multiway tables, those consisting of three or more dimensions. Although these types of analyses can be useful for detecting intricate patterns underlying categorical data, a more common method for multivariate analyses is to fit a regression model. Depending on the nature and scale of the outcome variable, three procedures you can use when complex survey data are at hand are PROC SURVEYREG, PROC SURVEYLOGISTIC, and PROC SURVEYPHREG. These procedures are the topics of the next three respective chapters.

5

Fitting Linear Regression Models Using PROC SURVEYREG

5.1 Introduction

This chapter covers linear regression, a tool for relating one or more explanatory, or predictor, variables to a continuous outcome variable. Regression models enable us to estimate the impact of a particular variable on the expected value of some outcome variable while simultaneously accounting for other covariates. For example, the primary research goal of a clinical trial might be to evaluate the effect of some treatment on a disease. Given random assignment of subjects to either an experimental or control group, the estimated treatment effect can be ascertained by the difference in the outcome measure across the two groups, but this only gives the overall picture. A natural follow-up question is whether the difference, if any, varies according to the subject's gender, race, or age. Fitting a multiple linear regression model allows one to answer that question.

This chapter begins with a section overviewing the fundamentals of *textbook* linear regression, where simple random sampling is implied. We formally define the regression model and its assumptions and introduce the matrix notation needed for estimation and hypothesis testing. We also discuss how SAS constructs indicator variables to represent categorical variable effects. Developing a solid grasp of these foundational concepts is essential because we will repeatedly draw upon them when we discuss other types of regression models in Chapters 6 and 7.

In Section 5.3, we introduce the formulaic idiosyncrasies associated with fitting linear regression models to complex survey data. While there are numerous SAS/STAT procedures available to fit linear regression models in a simple random sampling setting (e.g., PROC REG, PROC GLM), the procedure developed exclusively for complex survey data is PROC SURVEYREG. Syntax examples make use of publicly available data released as part of the 2010 National Ambulatory Medical Care Survey (NAMCS). Described in more detail in Section 1.5.2, the ultimate sampling unit in the NAMCS is a visit to a physician's office. We will suppose that the research goal is to build and interpret a model regressing the average duration of the patient–physician encounter on a set of characteristics about the patient, the physician's practice, and other information collected about the visit, such as the method of payment and whether medication was prescribed.

5.2 Linear Regression in a Simple Random Sampling Setting

Consider the scatterplot in Figure 5.1 depicting the relationship between two continuous variables, x and y. Larger values of x tend to be associated with larger values of y. That is, there is a *positive association* between the two variables as opposed to a *negative association*, if y were to tend to decrease as x increased. An intuitive method for summarizing this kind of relationship is to overlay a line of best fit within the scatterplot, such as the one shown.

While there is more than one method to determine "best fit," a popular criterion is to find the line that minimizes the sum of the squared vertical distances between all data points and the line. This technique is referred to as "least squares regression," and the resulting line is referred to as a "regression line." The approach begins by postulating a model governing the bivariate relationship. For example, the model might look something like

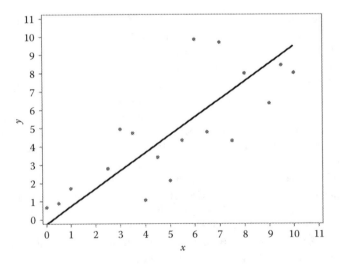

FIGURE 5.1
Scatterplot of two continuous variables overlaid with a least squares regression line.

$$y_i = \beta_0 + \beta_1 x_i + \varepsilon_i \tag{5.1}$$

where

β_0 represents the intercept, or expected value of y when $x=0$

β_1 represents the slope of the regression line, or expected change in y given a one-unit increase in x

Collectively, the beta terms are called model "parameters" or "coefficients." The term ε_i represents a *residual* or deviation of the observed value y_i from the regression line, which represents the expected value of y_i given x_i. Another way of thinking about β_0 and β_1 is that they are the values for which the sum of the squared ε_is is minimized.

A critical assumption underlying least squares regression theory is that the residuals are independently and identically distributed with a constant variance, typically denoted σ_ε^2 and referred to as the "mean squared error" (MSE) of the model. A measure of model precision, the MSE can be interpreted as the average squared deviation between the y_is and the regression line. The constant variance assumption means that if we were to plot the residuals along the y-axis as a function of the predicted values, there would be no decipherable pattern(s). Figure 5.2 illustrates three example violations of this assumption.

The first plot depicts the concept of *heteroscedasticity* or a nonconstant variance amongst the residuals. Larger predicted values are associated with residuals exhibiting more variability than smaller predicted values. A remedy often pursued in this situation is to transform either the predictor variable(s) or outcome variables. The second plot suggests an outright misspecification of the model as the mean function is not adequately explained by the regression line. One might consider reformulating the model by adding a quadratic term or seeking additional predictor variables that better explain y. The third plot illustrates *serial correlation* in the residual terms. This issue crops up frequently in timeseries analysis (Milhøj, 2013). Most textbooks covering linear regression expend a substantial amount of effort discussing residual diagnostics. We will do little of that in this book not because it is deemed unimportant but because the literature is still in its nascent stage and such diagnostics are not widely available in software accounting for complex survey features. Jianzhu Li and Richard Valliant have pioneered many of the recent developments in this area. The reader seeking more information is referred to Li and Valliant (2009, 2011a,b).

The model in Equation 5.1 is specified in terms of unknown population quantities. Given a sample random sample of size n drawn from the population, we can estimate those quantities by

$$\hat{\beta}_1 = \frac{\sum_{i=1}^{n} \left(x_i - \hat{\bar{x}}\right)\left(y_i - \hat{\bar{y}}\right)}{\sum_{i=1}^{n} \left(x_i - \hat{\bar{x}}\right)^2} \tag{5.2}$$

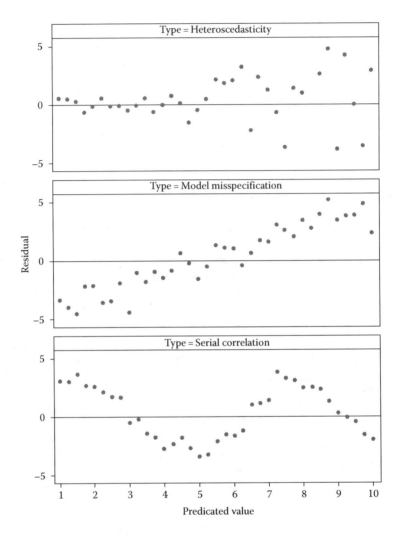

FIGURE 5.2
Three example residual distribution violations in a linear regression model.

and

$$\hat{\beta}_0 = \hat{\bar{y}} - \hat{\beta}_1 \hat{\bar{x}} \qquad (5.3)$$

where
    $\hat{\bar{x}}$ denotes the sample mean of x
    $\hat{\bar{y}}$ denotes the sample mean of y

The MSE can be estimated by

$$\hat{\sigma}_\varepsilon^2 = \frac{\sum_{i=1}^{n} \left(y_i - \hat{y}_i \right)^2}{n-2} \tag{5.4}$$

where $\hat{y}_i$ represents the predicted value of y_i, or $\hat{y}_i = \hat{\beta}_0 + \hat{\beta}_1 x_i$. In words, the estimated MSE is the sum of the squared residuals divided by the degrees of freedom for the model, defined as $n - p$, where p is the number of model parameters. In the simple linear regression setting with an intercept, there are $p=2$ parameters (β_0 and β_1). The square root of this quantity is called the "root-mean-squared error" (RMSE) and can be interpreted as a standard deviation of sorts reflecting the expected distance between a given y_i and $\hat{y}_i$.

Alternatively, matrix algebra can be employed to obtain the estimators in Equations 5.2 through 5.4. The first step is to construct an $n \times 2$ matrix $\mathbf{X}$ consisting of a column of ones for the intercept and a second column with predictor variable values. The matrix $\mathbf{X}$ is commonly referred to as the "design matrix" and in the simple linear regression case would look like

$$\mathbf{X} = \begin{bmatrix} 1 & x_1 \\ 1 & x_2 \\ \cdot & \cdot \\ \cdot & \cdot \\ \cdot & \cdot \\ 1 & x_n \end{bmatrix}$$

The second step is to construct an $n \times 1$ column vector $\mathbf{Y}$ containing the n values of the outcome variable as follows:

$$\mathbf{Y} = \begin{bmatrix} y_1 \\ y_2 \\ \cdot \\ \cdot \\ \cdot \\ y_n \end{bmatrix}$$

Once these have been constructed, estimates of $\hat{\beta}_0$ and $\hat{\beta}_1$ can be obtained by calculating

$$\hat{\beta} = (\mathbf{X}^T \mathbf{X})^{-1} \mathbf{X}^T \mathbf{Y} \tag{5.5}$$

where
T is the superscript that denotes the matrix transpose
−1 denotes the matrix inverse

The result is a 2×1 vector of "beta hats" $\hat{\beta} = \begin{bmatrix} \hat{\beta}_0 \\ \hat{\beta}_1 \end{bmatrix}$. The estimated MSE of the

model $\hat{\sigma}_\varepsilon^2$ is found by summing the squared entries of the $n \times 1$ residual vector $\mathbf{e} = \mathbf{Y} - \mathbf{X}\hat{\beta} = \mathbf{Y} - \hat{\mathbf{Y}}$ and dividing the result by $n - 2$. SAS performs the matrix algebra on our behalf, but the same results can be obtained using, say, PROC IML (Wicklin, 2010). The estimated model can be expressed in terms of these matrices as follows:

$$\mathbf{Y} = \mathbf{X}\hat{\beta} + \mathbf{e} \tag{5.6}$$

The process generalizes to *multiple* linear regression models, those with more than one predictor variable. All we need to do is append additional columns to the design matrix. The matrix algebra approach is especially convenient for multiple linear regression because the closed-form equations for individual model parameters (e.g., as shown in Equations 5.2 and 5.3) become much more complex.

We can motivate an illustration of multiple linear regression by considering how categorical variables are typically handled. Suppose that the goal is to analyze results from a clinical trial of $n=6$ patients, of whom 4 were randomly assigned to a treatment group and 2 were assigned to a control group receiving no treatment. Perhaps the primary objective is to estimate the overall treatment effect. One way to do so is by parameterizing treatment as an indicator variable in a simple linear regression model, where a case is coded as 1 if part of the treatment group and 0 if part of the control group. If the first four cases in the data set correspond to the experimental group, the design matrix would be defined as

$$\mathbf{X} = \begin{bmatrix} 1 & 1 \\ 1 & 1 \\ 1 & 1 \\ 1 & 1 \\ 1 & 0 \\ 1 & 0 \end{bmatrix}$$

Given a corresponding 6×1 vector $\mathbf{Y}$ housing the key (continuous) outcome measure for the six subjects in the study, the first row of $\hat{\beta} = (\mathbf{X}^T\mathbf{X})^{-1}\mathbf{X}^T\mathbf{Y}$ would be the intercept of the linear model, which under this parameterization is actually the mean of the control group, the average outcome for cases having $x_i = 0$. The second row would house the treatment effect or the average change in the outcome variable transitioning from $x_i = 0$ to $x_i = 1$.

The sum of the two terms in $\hat{\beta}$ would represent the average outcome of the treatment group.

Now suppose that the four subjects in the treatment group actually received one of two possible treatments the researchers were investigating. Two individuals received Treatment A and two received Treatment B, and we sought to differentiate the mean outcome amongst all three possible combinations. This converts a binomial predictor variable setting into a trinomial setting, and to account for the third category, we must append an additional indicator variable column to the design matrix. Generally speaking, when a categorical predictor consists of K *classes* or *levels*, we must construct $(K - 1)$ indicator variables to exploit the *reference parameterization* of categorical variable effects. The reason one column must be omitted is that a linear dependency can develop between two columns in X (i.e., one column is a linear combination of one or more other columns), causing the matrix X^TX to be uninvertible, which means a unique solution for $\hat{\beta}$ cannot be found. Technically, other parameterizations are possible, but in this author's opinion the reference parameterization allows for the most straightforward interpretation. It is so named because the omitted level serves as the *reference category*, and the model parameter(s) linked to the nonomitted levels represent the expected change in the outcome relative to the reference category, all other covariates held constant.

Returning to the clinical trial example, it would be convenient to designate the control group as the reference category using the design matrix

$$X = \begin{bmatrix} 1 & 1 & 0 \\ 1 & 1 & 0 \\ 1 & 0 & 1 \\ 1 & 0 & 1 \\ 1 & 0 & 0 \\ 1 & 0 & 0 \end{bmatrix}$$

The first two rows correspond to cases assigned Treatment A, the second two rows to cases assigned to Treatment B, and the last two rows to cases in the control group. Under this parameterization, $\hat{\beta}$ is a 3×1 vector. The term associated with the first row is the mean outcome for the control group as it represents the mean of y where both treatment indicator variables are zero. The term in the second row can be interpreted as the expected change in the outcome attributable to Treatment A *relative to the control*, and the term corresponding to the third row is the like for Treatment B. The difference between these two parameters is an estimate of the mean change in the outcome between Treatments A and B.

SAS automatically creates a set of indicator variables for any variable listed in the CLASS statement of PROC SURVEYREG. It assigns the reference category as the last in the sort-ordered list of distinct values. There are a few options to override this default via the ORDER= option in the PROC SURVEYREG statement, but a general work-around is to define a format with labels assigned such that the desired reference category's label appears last in the sort-ordered list. Rest assured that changing the parameterization by modifying the reference category has no effect on overall impact of that categorical variable; it merely alters the interpretation of its parameters.

For any linear regression model, constructing an analysis of variance (ANOVA) table is a concise way to summarize certain key measures of variability and assess the overall explanatory power of the predictor variables. Table 5.1 outlines the general structure of an ANOVA table with the underlying calculations provided. The notion behind an ANOVA table is to partition the total sum of squares, $\text{SST} = \sum_{i=1}^{n} (y_i - \hat{\bar{y}})^2$, or the sum of the n squared deviations between the y_is and the overall sample mean $\hat{\bar{y}}$, into two components: (1) the sum of squares for the model, $\text{SSM} = \sum_{i=1}^{n} (\hat{y}_i - \hat{\bar{y}})^2$, or the sum of squared deviations of the predicted values from the overall sample mean, and (2) the sum of squares for error, $\text{SSE} = \sum_{i=1}^{n} (y_i - \hat{y}_i)^2$, or the sum of squared residuals. The key property is that $\text{SST} = \text{SSM} + \text{SSE}$.

The statistic

$$R^2 = \frac{\text{SSM}}{\text{SST}} \tag{5.7}$$

is called the "coefficient of determination" and offers a very popular way to relate these quantities due to its intuitive interpretation: It represents the proportion of the total variance observed in the outcome variable that

TABLE 5.1

Components of an Analysis of Variance Table

Source	Sum of Squares	Degrees of Freedom	Mean of Squares	F
Model	$\text{SSM} = \sum_{i=1}^{n}(\hat{y}_i - \hat{\bar{y}})^2$	$p - 1$	$\text{MSM} = \text{SSM}/(p-1)$	MSM/MSE
Error	$\text{SSE} = \sum_{i=1}^{n}(y_i - \hat{y}_i)^2$	$n - p$	$\text{MSE} = \text{SSE}/(n-p)$	
Total	$\text{SST} = \sum_{i=1}^{n}(y_i - \hat{\bar{y}})^2$	$n - 1$		

is explained by the regression model. For example, a model with $R^2 = 0.75$ implies that the model explains 75% of the total variability in y. There is no universally accepted minimum R^2 value to strive for, but based on their combined decades of experience, Heeringa et al. (2010, p. 194) comment:

> Analysts who are new to regression modeling of social science, education, or epidemiological data should not fret if the achieved R^2 values are lower than those seen in their textbook training. Physicists may be disappointed with $R^2 < 0.98 - 0.99$ and chemists with $R^2 < 0.90$, but social scientists and others who work with human populations will find that their best regression model will often explain only 20%–40% of the variation in the dependent variable.

The SSM has $p - 1$ degrees of freedom and the SSE has $n - p$ degrees of freedom. Dividing a sum of squares by its respective degrees of freedom returns the mean sum of squares. The value under the column labeled "F" is ratio of MSM to MSE, which is the test statistic for H_0: $\beta_1 = \beta_2 = \cdots = \beta_{(p-1)} = 0$ versus H_1: $\beta_j \neq 0$ for one or more values of $j = 1, 2, \ldots, (p - 1)$. Note how the intercept is excluded. Essentially, this is a test to gauge whether including the given set of predictor variables is any better at explaining the overall variability in y than an intercept-only model. The observed value can be referenced against an F distribution with p numerator degrees of freedom and $n - p$ denominator degrees of freedom. When the observed value exceeds the critical value determined by the desired confidence level α, it suggests the null hypothesis should be rejected or that at least one of the model coefficients is significantly different from zero.

Another important piece to any model is the estimated $p \times p$ model parameter *covariance matrix*. For the simple linear regression case, this is defined as $\text{cov}(\hat{\boldsymbol{\beta}}) = \hat{\sigma}_\varepsilon^2 (\mathbf{X}^T\mathbf{X})^{-1}$, which produces a matrix that looks like the following:

$$\text{cov}(\hat{\boldsymbol{\beta}}) = \begin{bmatrix} \text{var}(\hat{\beta}_0) & \text{cov}(\hat{\beta}_0, \hat{\beta}_1) \\ \text{cov}(\hat{\beta}_0, \hat{\beta}_1) & \text{var}(\hat{\beta}_1) \end{bmatrix} \tag{5.8}$$

Generally speaking, entries along the diagonal of $\text{cov}(\hat{\boldsymbol{\beta}})$ are the estimated variances for the model parameters, while off-diagonal entries are covariances between two particular model parameters. The $p(p - 1)/2$ distinct covariances are symmetric about the diagonal. The covariance matrix is used to test hypotheses about individual parameters, such as H_0: $\beta_j = 0$ versus H_1: $\beta_j \neq 0$ for some j. For example, if β_j is the coefficient for a treatment indicator variable, conducting this test is tantamount to assessing whether there is a statistically significant treatment effect. On the other hand, if β_j is the coefficient for a continuous variable, this test is tantamount to assessing whether the given variable can be used to meaningfully differentiate the expected value of y. One way to carry out this test is to find $t = \hat{\beta}_j / \sqrt{\text{var}(\hat{\beta}_j)}$

and reference against the $100(1 - \alpha/2)$th percentile of a student t distribution with $n - p$ degrees of freedom. Most software outputs the results of these p parameter-specific tests by default.

Suppose that the model at hand is $y_i = \beta_0 + \beta_1 x_i + \varepsilon_i$ and we have used the sample data to estimate $\hat{\beta} = \begin{bmatrix} \hat{\beta}_0 \\ \hat{\beta}_1 \end{bmatrix}$. Another method of conducting the test

$H_0: \beta_1 = 0$ versus $H_1: \beta_1 \neq 0$ begins by defining the *contrast vector* $\mathbf{C} = \begin{bmatrix} 0 \\ 1 \end{bmatrix}$, then

finding $F = (\mathbf{C}^\mathsf{T} \hat{\beta})^\mathsf{T} (\mathbf{C}^\mathsf{T} \text{cov}(\hat{\beta}) \mathbf{C})^{-1} (\mathbf{C}^\mathsf{T} \hat{\beta})$ and referencing against an F distribution with 1 numerator degree of freedom and $n - 2$ denominator degrees of freedom. Although the formula seems rather involved, it can be interpreted as the contrast squared divided by the variance of the contrast. The value returned is a scalar that equals the t statistic squared. In the special case of a simple linear regression model, this statistic is also equivalent to the F statistic provided in an ANOVA table.

As we will see in a forthcoming example, the technique generalizes to multicolumn *contrast matrices* as follows:

$$F = (\mathbf{C}^\mathsf{T} \hat{\beta})^\mathsf{T} (\mathbf{C}^\mathsf{T} \text{cov}(\hat{\beta}) \mathbf{C})^{-1} (\mathbf{C}^\mathsf{T} \hat{\beta}) / \text{rank}(\mathbf{C}) \qquad (5.9)$$

where $\text{rank}(\mathbf{C})$ denotes the number of linearly independent columns of the contrast matrix $\mathbf{C}$. Contrast matrices are handy for testing whether models can be reduced by eliminating one or more terms simultaneously. For example, suppose that the "full" model is $y_i = \beta_0 + \beta_1 x_{1i} + \beta_2 x_{2i} + \varepsilon_i$, but we are considering simplifying it to the "reduced" model $y_i = \beta_0 + \beta_1 x_{1i} + \varepsilon_i$. That is, we wish to test $H_0: \beta_2 = 0$ versus $H_1: \beta_2 \neq 0$. Each additional variable incorporated into a linear regression model will only serve to reduce the SSE, so if we denote the residual sum of squares for the full model SSE_{full} and the same quantity for the reduced model SSE_{red}, we know that $\text{SSE}_{\text{full}} < \text{SSE}_{\text{red}}$. To assess the merit of x_2 in the model, we can evaluate whether the residual sum of squares is substantively reduced by its inclusion. This line of reasoning forms the basis of the following general F test for simplifying a model:

$$F = \frac{(\text{SSE}_{\text{red}} - \text{SSE}_{\text{full}})/q}{\hat{\sigma}^2_{\varepsilon(\text{full})}} \qquad (5.10)$$

where
   q is the number of parameters dropped
   $\hat{\sigma}^2_{\varepsilon(\text{full})}$ is the MSE of the full model

In the present example, $q = 1$, but the core structure is applicable to situations where we are considering reducing the model by eliminating two or more parameters.

We can certainly glean all quantities called for by Equation 5.10 from the two ANOVA tables produced after fitting both the full and reduced models, but there is an alternative, more efficient method. Specifically, we can fit only the full model and compute

$$F = (\mathbf{C}^T \hat{\beta})^T (\mathbf{C}^T \text{cov}(\hat{\beta}) \mathbf{C})^{-1} (\mathbf{C}^T \hat{\beta}) / \text{rank}(\mathbf{C})$$

where

$$\hat{\beta} = \begin{bmatrix} \hat{\beta}_0 \\ \hat{\beta}_1 \\ \hat{\beta}_2 \end{bmatrix}$$

$\text{cov}(\hat{\beta})$ is the 3×3 model parameter covariance matrix

$$\mathbf{C} = \begin{bmatrix} 0 \\ 0 \\ 1 \end{bmatrix}$$

$\text{rank}(\mathbf{C}) = 1$

One clear advantage of the matrix algebra technique is that we need only fit the full model, not both the full and reduced.

In fact, any number of hypotheses can be accommodated using this approach, not just those testing whether one or more parameters are significantly different from zero. For instance, recall the clinical trial parameterization example where β_1 represented the effect of Treatment A relative to the control and β_2 represented the comparable effect of Treatment B (β_0 represented the mean effect for the control). Testing whether these two treatment effects are significantly different is equivalent to testing H_0: $\beta_1 - \beta_2 = 0$ versus H_1: $\beta_1 - \beta_2 \neq 0$, so we could define the contrast matrix $\mathbf{C} = \begin{bmatrix} 0 \\ 1 \\ -1 \end{bmatrix}$ and proceed in the same manner.

Another important extension worth discussing is when the goal is to conduct two or more hypotheses jointly. Continuing with the clinical trial example, a test closely related to the one described earlier is H_0: $\beta_1 = 0$, $\beta_2 = 0$ versus H_1: $\beta_1 \neq 0$, $\beta_2 \neq 0$. This test's null hypothesis states that both treatment effects are indistinguishable from zero, whereas the previous null hypothesis only stipulated that the treatment effects are equivalent (but possibly nonzero). To test this new hypothesis, we can augment the contrast matrix with an additional column by constructing $\mathbf{C} = \begin{bmatrix} 0 & 0 \\ 1 & 0 \\ 0 & 1 \end{bmatrix}$. The structure of the F test is the same and will again produce a scalar. An important modification, however, is that the numerator degrees of freedom for the reference F distribution are now 2 instead of 1. Again, this is because the numerator degrees of freedom are defined by the *rank* of **C** or the number of linearly independent columns. The denominator degrees of freedom are still $n - p$.

5.3 Linear Regression with Complex Survey Data

Shah et al. (1977) discuss how complex survey data can violate many of the underlying assumptions in traditional least squares regression. In contrast to simple random sampling from an infinite population, survey data are typically collected to make generalizations about finite populations. As such, model parameters have a subtly different interpretation, which is often reflected in the survey research literature by denoting them with a Roman letter such as a B_0 or B_1 instead of a Greek letter. Using the weights w_i for each of the $i = 1, \ldots, n$ units in the sample, the finite population parameters of a simple linear regression model are estimated by

$$\hat{B}_1 = \frac{\sum_{i=1}^{n} w_i \left(x_i - \hat{\bar{x}}_w \right) \left(y_i - \hat{\bar{y}}_w \right)}{\sum_{i=1}^{n} w_i \left(x_i - \hat{\bar{x}}_w \right)^2} \tag{5.11}$$

and

$$\hat{B}_0 = \hat{\bar{y}}_w - \hat{B}_1 \hat{\bar{x}}_w \tag{5.12}$$

where the subscript "w" signifies weighted estimates (i.e., $\hat{\bar{y}}_w$ is the weighted mean of the y_is and $\hat{\bar{x}}_w$ is the weighted mean of the x_is).

The two frameworks for thinking about the model parameters from Equations 5.2 and 5.3 as well as those in Equations 5.11 and 5.12 can be unified through the notion of a *superpopulation* (Heeringa et al., 2010, p. 184), which posits that although the N units in the survey population represent a finite set, the given regression model corresponds to some overarching superpopulation model. Put another way, one might view the finite population as itself a sample from an infinite population, acknowledging there may be variability in the jth parameter between one realized finite population and another. Heeringa et al. (2010) point out, however, that any such variability is negligible for large populations (i.e., large N), such that $\hat{B}_j$ effectively approximates the corresponding superpopulation parameter.

The estimates $\hat{B}_0$ and $\hat{B}_1$ are *weighted least squares estimates* that approximate finite population quantities. For example, $\hat{B}_1$ estimates $B_1 = \dfrac{\sum_{i=1}^{N}(x_i - \bar{x})(y_i - \bar{y})}{\sum_{i=1}^{N}(x_i - \bar{x})^2}$, the slope coefficient that would be computed if we had data for the entire population. In a similar vein, recall that the function we seek to minimize in ordinary least squares regression is the residual sum of squares $\text{SSE} = \sum_{i=1}^{n}\left(y_i - \left(\hat{\beta}_0 + \hat{\beta}_1 x_i\right)\right)^2$. In weighted least squares estimation, that function is $\text{SSE}_w = \sum_{i=1}^{n} w_i\left(y_i - \left(\hat{B}_0 + \hat{B}_1 x_i\right)\right)^2$, which can be conceptualized as an estimate of the residual sum of squares for the model fit using the entire finite population or an estimate of $\text{SSE}_{pop} = \sum_{i=1}^{N}\left(y_i - \left(B_0 + B_1 x_i\right)\right)^2$.

Table 5.2 summarizes how weights are incorporated into the standard quantities appearing in an ANOVA table. As before, y_i represents the observed data and $\hat{y}_i$ the predicted value for the ith sampling unit using the weighted least squares estimates. The quantity $\bar{\hat{y}}_w$ represents the overall (weighted) sample mean. Since weights are typically positive, the sums of squares are often much larger in magnitude than those from a comparably constructed table ignoring the weights.

One can compute a weighted version of the R^2 statistic given in Equation 5.7 using the weighted sums of squares as follows:

$$R_w^2 = \frac{\text{SSM}_w}{\text{SST}_w} \tag{5.13}$$

TABLE 5.2

Components of an Analysis of Variance Table Reported by PROC SURVEYREG when a WEIGHT Statement Is Used

Source	Sum of Squares	Degrees of Freedom	Mean of Squares	F
Model	$\text{SSM}_w = \sum_{i=1}^{n} w_i(\hat{y}_i - \bar{\hat{y}})^2$	$p - 1$	$\text{MSM} = \text{SSM}/(p - 1)$	MSM/MSE
Error	$\text{SSE}_w = \sum_{i=1}^{n} w_i(y_i - \hat{y}_i)^2$	$n - p$	$\text{MSE} = \text{SSE}/(n - p)$	
Total	$\text{SST}_w = \sum_{i=1}^{n} w_i(y_i - \bar{\hat{y}})^2$	$n - 1$		

PROC SURVEYREG will report this quantity in the presence of a WEIGHT statement; in the absence of a WEIGHT statement, the quantity reported is the same unweighted version specified in Equation 5.7.

The MSE appearing in the ANOVA table is not the actual MSE used by PROC SURVEYREG when estimating one of the various statistics that calls for it. Instead, the quantity PROC SURVEYREG uses is

$$\hat{\sigma}_\varepsilon^2 = \left(\frac{n}{\hat{N}} \right) \left(\frac{\mathrm{SSE_w}}{n-p} \right) \tag{5.14}$$

which is a rescaled version accounting for the *weighted* squared deviations in $\mathrm{SSE_w}$ by dividing through by the sum of the weights, $\sum_{i=1}^{n} w_i = \hat{N}$. This distinction is important to realize because, with complex sample designs, the F statistic in the ANOVA table reported by PROC SURVEYREG does not carry the same interpretation discussed previously and should be ignored. That is, it no longer represents the test statistic for the null hypothesis that all parameters are jointly equal to zero except for the intercept. The result of that particular test, however, is output by PROC SURVEYREG in the "Model" line of the Tests of Model Effects table, as we will see in Program 5.1.

With respect to the matrix algebra used to estimate model coefficients, weights are incorporated by introducing an $n \times n$ diagonal matrix **W**, or a matrix with a 0 as all off-diagonal entries but the ith sampling unit's weight in the ith row and ith column. Specifically, the vector of weighted least squares estimates is found by calculating

$$\hat{\mathbf{B}} = (\mathbf{X}^\mathrm{T}\mathbf{W}\mathbf{X})^{-1}\mathbf{X}^\mathrm{T}\mathbf{W}\mathbf{Y} \tag{5.15}$$

where
  X is the $n \times p$ design matrix discussed in the previous section
  Y is the $n \times 1$ outcome variable vector

This structure parallels the weighted least squares estimator used to account for violations in the homogeneity of residuals assumption under simple random sampling (Weisberg, 2013). In that context, however, weights are assigned inversely proportional to the estimated residual-specific variance.

This is an apt point to acknowledge the considerable ongoing debate amongst survey researchers as to whether weights are truly needed when modeling survey data. Binder and Roberts (2009) eloquently summarize the merits of both sides of the argument. The side opposing the use of weights consists of researchers adopting the *model-based* perspective to survey inferences, an alternative to the *design-based* perspective prevailing throughout this book. Their core argument centers on the fact that, if the model is correctly specified, the weighted and unweighted regressions both estimate the same finite population parameters. The implication is that if one is confident the given model correctly describes the true state of affairs in the population, then using the weights is unnecessary and potentially inefficient since arbitrarily variable weights can inflate variances (Kish, 1992). Pfeffermann (1996) argues, however, that using the weights provides robustness against a misspecified model, such as one missing certain influential predictor variables, and Skinner et al. (1989) assert weights can protect against biases introduced by *nonignorable* sample designs, those in which the sample inclusion indicator is strongly related to the outcome variable.

Specifying the weight variable in the WEIGHT statement of PROC SURVEYREG is all that is required to have SAS calculate weighted least squares estimates. These will match what is output from PROC REG when a comparable WEIGHT statement appears, but in that procedure $\text{cov}(\hat{\mathbf{B}})$ will generally not be correct if there are any additional complex survey features involved. Variances estimated using PROC SURVEYREG properly account for these features using a multivariate application of Taylor series linearization. We report only the main result of the derivation here, but the reader seeking more detail is referred to Fuller (1975), Shah et al. (1977), Binder (1983), and Kott (1991). Following notation similar in spirit to that used by Lohr (2009), the Taylor series linearization estimator of the model parameter covariance matrix is

$$\text{cov}(\hat{\mathbf{B}}) = (\mathbf{X}^T\mathbf{W}\mathbf{X})^{-1}\,\text{var}\left(\sum_{i=1}^{n} w_i\mathbf{u_i}\right)(\mathbf{X}^T\mathbf{W}\mathbf{X})^{-1} \qquad (5.16)$$

where

$\mathbf{u_i} = \mathbf{x_i}(y_i - \mathbf{x_i}\hat{\mathbf{B}})$ with $\mathbf{x_i}$ denoting the $1 \times p$ covariate vector for the *i*th sampling unit

$\text{var}\left(\sum_{i=1}^{n} w_i\mathbf{u_i}\right)$ is calculated with respect to the complex survey design

Note that $\text{var}\left(\sum_{i=1}^{n} w_i\mathbf{u_i}\right)$ is essentially shorthand notation to represent a $p \times p$ matrix containing estimated variances of weighted totals along the diagonal—the weighted total of a variate u_i, defined as the product of the given design matrix column and the residual for the *i*th sampling unit (i.e., substituting this variate for the *y* term in Equation 3.2)—and the respective covariances of these weighted totals in the off-diagonal entries (i.e., making comparable substitutions to Equation 3.5). Because both $(\mathbf{X}^T\mathbf{W}\mathbf{X})^{-1}$ and $\text{var}\left(\sum_{i=1}^{n} w_i\mathbf{u_i}\right)$ are $p \times p$ matrices, $\text{cov}(\hat{\mathbf{B}})$ is also a $p \times p$ matrix.

Note that all three SURVEY procedures with modeling capabilities (PROC SURVEYREG, PROC SURVEYLOGISTIC, and PROC SURVEYPHREG) apply the adjustment factor $(n-1)/(n-p)$ when estimating $\text{cov}(\hat{B})$ based on a recommendation by Hidiroglou et al. (1980) to reduce bias that can occur in small samples. If you do not want this adjustment factor to be used, specify VADJUST=NONE after the slash in the MODEL statement.

A critical difference to be aware of is that the degrees of freedom for error is no longer the sample size minus the number of parameters in the model. In fact, it is no longer even a function of the number of parameters in the model. It is the complex survey degrees of freedom, the number of PSUs minus the number of strata.

Finally, another noteworthy difference when modeling complex survey data is that the general F test assessing whether one or more model parameters are significantly different from zero does not adhere to the same distribution as it does in simple random sampling. Korn and Graubard (1990) suggest the modified test statistic:

$$F_{ADJ} = F \times \frac{df - q + 1}{df} \tag{5.17}$$

where
   F is the unadjusted test statistic (e.g., computed in the same manner as shown Equation 5.9)
   df is the complex survey degrees of freedom
   q is the number of parameters dropped

F_{ADJ} is then referenced against an F distribution with q and $df - q + 1$ degrees of freedom for the numerator and denominator, respectively.

Let us now edify these concepts using the NAMCS 2010 publicly available microdata file, which has been read into the temporary SAS data set named NAMCS_2010. Recall from Section 1.5.2 that the NAMCS 2010 is characterized by the following three complex survey design features:

1. Stratification—strata are identifiable by distinct codes of the variable CSTRATM

2. Clustering—each PSU (nested within a stratum) is identifiable by distinct codes of the variable CPSUM

3. Unequal weights—maintained by the variable PATWT

Suppose the objective is to model time spent with the physician (TIMEMD) as a function of whether any medication was prescribed or renewed (MED), the patient's gender (SEX), age (AGE), and race (RACER), the total number of chronic conditions afflicting the patient (TOTCHRON), primary reason for the visit (MAJOR), and whether the doctor is the lone physician in the

TABLE 5.3

Names and Coding Structure of Variables Used in the Example Linear
Regression Model

Variable Name	Description	Coding Structure
TIMEMD	Time spent with physician in minutes	Continuous
MED	Indicator of medication prescribed or renewed	0 = No 1 = Yes
SEX	Patient gender	1 = Female 2 = Male
AGE	Patient age in years	Continuous
RACER	Patient race	1 = White 2 = Black 3 = Other
TOTCHRON	Patient's total number of chronic conditions	Count
MAJOR	Primary reason for the visit	1 = New problem (<3 months onset) 2 = Chronic problem, routine 3 = Chronic problem, flare up 4 = Pre-/post-surgery 5 = Preventive care (e.g., routine prenatal, well-baby, screening, insurance, and general exams)
SOLO	Indicator of solo physician practice	1 = Yes 2 = No

practice (SOLO). Table 5.3 summarizes the scale and coding structure of these variables.

Program 5.1 fits this model using PROC SURVEYREG. The portion of the output germane to the current discussion is provided immediately thereafter.

Program 5.1: Fitting a Linear Regression Model Accounting for Complex Survey Design Features

```
proc surveyreg data=NAMCS_2010;
  strata CSTRATM;
  cluster CPSUM;
  class MED SEX RACER SOLO MAJOR;
  model TIMEMD = MED SEX RACER SOLO MAJOR
                AGE TOTCHRON / solution;
weight PATWT;
run;
```

Class Level Information		
CLASS Variable	Levels	Values
MED	2	0 1
SEX	2	1 2
RACER	3	1 2 3
SOLO	2	1 2
MAJOR	5	1 2 3 4 5

Tests of Model Effects			
Effect	Hum DF	F Value	Pr>F
Model	11	5.20	<.0001
Intercept	1	824.94	<.0001
HIED	1	0.04	0.8486
SEX	1	0.00	0.9669
RACER	2	1 11	0.3294
SOLO	1	2.34	0.1267
MAJOR	4	7.84	<.0001
AGE	1	3.28	0.0706
TOTCHROH	1	13.53	0.0003

Note: The denominator degree of freedom for the F tests is 552.

Estimated Regression Coefficients				
Parameter	Estimate	Standard Error	t Value	Pr > |t|
Intercept	19.1251704	1.11637316	17.13	<.0001
MED 0	0.0876849	0.45899764	0.19	0.8486
MED 1	0.0000000	0.00000000		
SEX 1	−0.0114210	0.27513318	−0.04	0.9669
SEX 2	0.0000000	0.00000000		
RACER 1	0.2436353	1.08408259	0.22	0.8223
RACER 2	−0.6642309	1.18409324	−0.56	0.5751
RACER 3	0.0000000	0.00000000		
SOLO 1	1.2315973	0.80516484	1.53	0.1267
SOLO 2	0.0000000	0.00000000		
MAJOR 1	−0.7737222	0.59559493	−1.30	0.1945
MAJOR 2	−0.8589240	0.77258999	−1.11	0.2667
MAJOR 3	0.6710944	0.81024046	0.83	0.4079
MAJOR 4	−3.6219092	0.87574082	−4.14	<.0001
MAJOR 5	0.0000000	0.00000000		
AGE	0.0216502	0.01195026	1.81	0.0706
TOTCHRON	0.3639201	0.09893484	3.68	0.0003

Note: Matrix X′WX is singular and a generalized inverse was used to solve the normal equations.

The note in the listing regarding singularity of the $\mathbf{X^TWX}$ matrix informs us that the indicator variable for the last level in the sort-ordered list of distinct levels was set to zero (i.e., dropped). The sort-ordered listing of values for each variable listed in the CLASS statement is summarized in the Class Level Information table. As is commented within the syntax, note that the SOLUTION option must be specified after the slash in the MODEL statement when a CLASS statement is used to have the table of model parameter estimates output. Estimates of each model parameter and its associated estimated standard error are reported in the Estimated Regression Coefficients table. Alongside these figures are test statistics and p values corresponding to the parameter-specific two-sided t test with null hypothesis that the given parameter is equal to 0. Hence, a small p value is indicative of a significant effect for the given variable or level of a categorical variable. Recall the property that parameter estimates for a particular level of a categorical variable measure the expected change in the mean of the outcome variable relative to the reference category, all other predictor variables held constant. So, for instance, we can gather from the output that the model suggests a visit instigated by pre-/post-surgery (MAJOR=4) results in the average time spent with the physician being about 3.62 min less than a visit addressing some kind of preventative care (MAJOR=5), all else equal. The t statistic reported alongside the parameter for MAJOR=4 is −4.14 ($p<0.0001$), which strongly suggests the difference is significant.

There are also significance tests appearing in the Tests of Model Effects table. These are the result of a comprehensive series of F tests: one for the overall model and one for each "effect," a single parameter for a continuous or dichotomous predictor variable or the set of $K-1$ parameters associated with a categorical variable consisting of K distinct levels. The former test is carried out following Equation 5.9, assigning $\mathbf{C}$ to be a $p \times 1$ vector of zeros for all rows except that corresponding to the given parameter, which is assigned a value of 1. For the latter, $\mathbf{C}$ is a $p \times (K-1)$ vector consisting of a separate column for each level of the categorical variable constructed in the same manner.

For effects comprised of a standalone parameter (e.g., AGE, SEX), the F test reported equals the squared value of the t test, but the interpretation is somewhat different for categorical variables with $K>2$ because the F test simultaneously assesses whether all $K-1$ parameters are significantly different from zero. To see how this could prove useful, observe how only one of the parameters capturing of the effect of reason for the visit race is significant. Perhaps, this raised suspicion as to whether the effect, as a whole, was a significant factor in the model. The highly significant F statistic of 7.84 ($p<0.0001$) should eradicate any such doubt.

Although we have already touched on several of these over the course of the chapter, Table 5.4 is a summary of useful MODEL statement options that can be specified after the slash to modify or supplement the default statistics reported by PROC SURVEYREG.

TABLE 5.4

Summary of Useful MODEL Statement Options Available after the Slash in PROC SURVEYREG

Option	Effect
ANOVA	Outputs an ANOVA table
NOINT	Excludes the intercept from the model by eliminating the columns of ones in the design matrix
COVB	Outputs the model parameter covariance matrix
SOLUTION	Outputs the model parameter estimates, which are suppressed by default when a CLASS statement is present
VADJUST= NONE	Omits the model parameter covariance matrix adjustment factor applied by default as discussed in Hidiroglou et al. (1980)
CLPARM	Reports confidence limits for model parameters based on the significance level specified in the ALPHA= option in the PROC SURVEYREG statement (default is ALPHA=.05)
DEFF	Outputs estimated design effects for model parameter estimates

5.4 Testing for a Reduced Model

Scrutinizing the output generated from Program 5.1, we can observe that none of the effects attributable to medication being prescribed or renewed (MED), patient gender (SEX), or patient race (RACER) are even marginally significant. Suppose we wanted to determine whether all terms related to those three effects can be removed from the model. This can be accomplished in more than one way, but we illustrate in Program 5.2 how to do so via a general F test using the CONTRAST statement in PROC SURVEYREG. The CONTRAST statement requires a label to be specified. The next step is to populate the contrast matrix **C** with *contrast coefficients*. We name the effect and then provide the coefficient(s) immediately afterwards. For a continuous effect, one value is sufficient, but K values are needed for categorical variable effects with K distinct levels. Be advised that the function of the commas is to distinguish columns of **C**, not to separate effect/coefficient clauses; multiple effect/coefficient clauses within the same column of **C** can be separated by spaces. The last coefficient for each CLASS variable corresponds to the indicator variable column that was dropped from the design matrix. Because there is a requirement that the coefficients for any categorical effect sum to 0, we must assign a coefficient of −1.

There is no need to define contrast coefficients for all $p = 12$ rows of $\hat{\mathbf{B}}$ (intercept + 11 nonzero parameters) in this instance because any effect not listed is assigned a contrast coefficient(s) of 0. The contrast matrix is not output by default, but you can specify the option E after the slash in the CONTRAST statement to have it printed to the listing. This offers an opportunity to verify **C** was defined as intended, which is a good idea when the model or contrast is complex, such as when interaction terms are involved.

The test results are displayed in the output under the heading Analysis of Contrasts. At the time of this writing, PROC SURVEYREG does not adjust the F statistic per the recommendation of Korn and Graubard (1990). In terms of Equation 5.17, F is reported, not F_{ADJ}. This is inconsequential in the present example since the adjustment factor would be $(df - q + 1)/df = (552 - 4 + 1)/552 \approx 1$, but the discrepancy between the two could be greater for other models and/or sample designs. Recall that the null hypothesis of the test states that all model parameters associated with MED, SEX, and RACER are jointly equal to zero. Since the p value associated with this F statistic is relatively large, we fail to reject the null hypothesis. It seems we could safely eliminate from the model all terms related to these three predictor variables.

Program 5.2: Testing for a Reduced Linear Regression Model in PROC SURVEYREG

```
proc surveyreg data=NAMCS_2010;
  strata CSTRATM;
  cluster CPSUM;
  class MED SEX RACER SOLO MAJOR;
  model TIMEMD = MED SEX RACER SOLO MAJOR
                 AGE TOTCHRON / solution;
weight PATWT;
contrast 'Test of Reduced Model'
         MED 1 -1, SEX 1 -1, RACER 1 0 -1, RACER 0 1 -1;
run;
```

Analysis of Contrasts			
Contrast	Num DF	F Value	Pr>F
Test of reduced model	4	0.57	0.6832

5.5 Computing Unit-Level Statistics

Apart from global analyses on model parameters, we often desire unit-level statistics that are a function of the final model on which we have settled. For instance, we may wish to formulate expected, or predicted, values for each sampling unit as well as one or more associated measures of uncertainty. The OUTPUT statement is capable of generating a host of these measures and will append them to the original data set and output to the data set specified in the OUT= option.

We begin with a few formal definitions. Let x_i denote the row of the design matrix X associated with the ith sampling unit, which we can think of as the "covariate profile" for that unit. The expected value of the outcome variable y_i given x_i can be written as

$$E(y_i | x_i) = x_i \hat{B} \tag{5.18}$$

and the variance of this quantity can be written as

$$\text{var}\left(E(y_i | x_i)\right) = x_i \, \text{cov}(\hat{B}) x_i^T \tag{5.19}$$

Therefore, a confidence interval for the expected value of the outcome variable can be constructed by

$$x_i \hat{B} \pm t_{df, \alpha/2} \sqrt{x_i \, \text{cov}(\hat{B}) x_i^T} \tag{5.20}$$

where $t_{df, \alpha/2}$ represents the $100(1 - \alpha/2)$th percentile of a t distribution with df complex survey degrees of freedom and significance level α. The OUTPUT statement keyword P= will populate a column in the data set named in the OUT= option with the quantity in Equation 5.18 for all observations in the input data set. The keyword STD= can be used to output square root of the quantity defined by Equation 5.19, and the keywords L= and U= can be used for the lower and upper bounds, respectively, of the confidence interval given by Equation 5.20.

There is a difference between the confidence interval in Equation 5.20 and a *prediction interval*. The former can be interpreted as a confidence interval for the mean of units sharing the same covariate profile x_i, but for any single observation we would anticipate a larger deviation. Relative to the expected value, the variance of a single observation is found by adding on the estimated MSE, and so the prediction interval is

$$x_i \hat{B} \pm t_{df, \alpha/2} \sqrt{x_i \, \text{cov}(\hat{B}) x_i^T + \hat{\sigma}_\varepsilon^2} \tag{5.21}$$

There are no options in the OUTPUT statement to explicitly request bounds for a prediction interval, but the DATA step in Program 5.3 demonstrates a work-around given other quantities output by PROC SURVEYREG. The program fits the reduced model following results of the contrast in Program 5.2 and uses the OUTPUT statement to generate a new data set called PREDS containing all variables from the input data set NAMCS_2010 plus four additional ones: (1) Y_HAT, the expected value of y_i given covariate vector x_i; (2) SE_Y_HAT, the standard error of the expected value; (3) Y_HAT_LOWER, the lower bound of a confidence interval about the expected value; and (4) Y_HAT_UPPER, the upper bound of a confidence interval about the expected value. These latter two were constructed with the default $\alpha = 0.05$, a value that can be modified using the ALPHA= option after the slash in the OUTPUT statement.

The subsequent DATA step adds to the data set PREDS the two bounds of a prediction interval for each sampling unit. These are stored in the variables Y_PRED_LOWER and Y_PRED_UPPER, respectively. Not shown, the MSE was retrieved from listing ($\hat{\sigma}_\varepsilon^2 = 13.8772$). The critical t value was assigned using the TINV function. Although we will not investigate these bounds any further here, note that the prediction intervals are much wider than the associated confidence intervals.

Program 5.3: Outputting and Creating Supplemental Unit-Level Statistics from a Linear Regression Model

```
proc surveyreg data=NAMCS_2010;
  strata CSTRATM;
  cluster CPSUM;
  class SOLO MAJOR;
  model TIMEMD = SOLO MAJOR
                AGE TOTCHRON / ANOVA;
weight PATWT;
output out=preds p=y_hat std=se_y_hat l=y_hat_lower
u=y_hat_upper;
run;

* create prediction intervals for each sampling unit;
data preds;
  set preds;
* RMSE obtained from output of PROC SURVEYREG run above;
MSE=13.8772;
t_crit=TINV(0.975,552);
y_pred_lower=y_hat - t_crit*SQRT(se_y_hat**2 + MSE);
y_pred_upper=y_hat + t_crit*SQRT(se_y_hat**2 + MSE);
run;
```

Another popular strategy after fine-tuning the parameters in a regression model is to calculate and discuss the expected values for one or more targeted covariate profiles. For instance, we might be interested in comparing the mean time spent with the doctor for a 77-year-old patient who has one chronic condition visiting a sole practitioner for preventative care with that of a 45-year-old patient, also with one chronic condition, on a routine visit to a group practitioner related to the chronic condition. It may be possible to locate these two covariate profiles in the PREDS data set and extract the corresponding values of Y_HAT. If the profile cannot be located, however, we next discuss two ways to obtain these estimates.

The first is to use the ESTIMATE statement. Generally speaking, the ESTIMATE statement is designed to calculate $\mathbf{L^T \hat{B}}$ and associated measures of uncertainty, such as $\text{se}(\mathbf{L^T \hat{B}}) = \sqrt{\mathbf{L^T} \text{cov}(\hat{\mathbf{B}})\mathbf{L}}$, where $\mathbf{L}$ is a $p \times 1$ vector and

$\hat{\mathbf{B}}$ is the $p \times 1$ vector of parameter estimates. One difference relative to the CONTRAST statement, however, is that the ESTIMATE statement returns a t statistic as opposed to an F statistic, but the underlying hypotheses are the same, H_0: $\mathbf{L}^T\mathbf{B}=0$ versus H_1: $\mathbf{L}^T\mathbf{B} \neq 0$.

Program 5.4 demonstrates syntax to estimate the mean time spent with the doctor for the two covariate profiles defined earlier. We begin by specifying a label and then assigning a coefficient for all effects represented in $\hat{\mathbf{B}}$, including the intercept, which simply gets assigned a value of 1. We specify a 1 for the associated level of a class variable and 0 for all other levels (note how this differs from the CONTRAST statement in which we assigned –1 for the reference category).

Program 5.4: Finding the Expected Values of Two Example Covariate Profiles in a Linear Regression Model

```
proc surveyreg data=NAMCS_2010;
  strata CSTRATM;
  cluster CPSUM;
  class SOLO MAJOR;
  model TIMEMD = SOLO MAJOR
                AGE TOTCHRON;
weight PATWT;
estimate 'Case 1' INTERCEPT 1 AGE 77 TOTCHRON 1
                  SOLO 1 0 MAJOR 0 0 0 0 1;
estimate 'Case 2' INTERCEPT 1 AGE 45 TOTCHRON 1
                  SOLO 0 1 MAJOR 0 1 0 0 0;
run;
```

Estimate					
Label	**Estimate**	**Standard Error**	**DF**	**t Value**	**Pr > $|t|$**
Case 1	22.5506	0.9213	552	24.48	<.0001

Estimate					
Label	**Estimate**	**Standard Error**	**DF**	**t Value**	**Pr > $|t|$**
Case 2	19.7455	0.4598	552	42.95	<.0001

Executing one or more ESTIMATE statements is feasible with a manageable number of targeted covariate profiles. When faced with a lengthier list, an efficient alternative is to supplement the input data set with placeholder records having nonmissing values for independent variables listed the MODEL statement but missing values for all variables, including the outcome variable. These observations are ignored by PROC SURVEYREG during the estimation process, but any statistical keywords requested in the OUTPUT statement are populated in the data set named in the OUT= option.

Program 5.5 demonstrates syntax for this method to replicate the estimate and standard error output from Program 5.4. The placeholder observations are read into a data set named ADDONS, which is concatenated to
the NAMCS_2010 and saved as a temporary data set named EXAMPLE_
EXPECTATION. Upon submitting this as the input data set to PROC
SURVEYREG, the following message is issued in the log:

```
NOTE: In data set EXAMPLE_EXPECTATION, total 31,231
observations read, 2 observations with missing values
or non-positive weights are omitted.
```

As the PROC PRINT output shows, however, the desired statistics have
been calculated and stored in the output data set. Note that this is a general
method that can be utilized with other modeling procedures, and not only
those designed to accommodate complex survey data.

Program 5.5: Creating Supplemental Unit-Level Statistics from a Linear Regression Model for User-Specified Covariate Profiles

```
data addons;
  input AGE TOTCHRON SOLO MAJOR;
datalines;
77 1 1 5
45 1 2 2
;
run;

data example_expectation;
  set NAMCS_2010
      addons;
run;

proc surveyreg data=example_expectation;
  strata CSTRATM;
  cluster CPSUM;
  class SOLO MAJOR;
  model TIMEMD=SOLO MAJOR
               AGE TOTCHRON;
weight PATWT;
output out=PREDS p=Y_HAT std=SE_Y_HAT;
run;

proc print data=PREDS;
  where TIMEMD=.;
  var TIMEMD SOLO MAJOR AGE TOTCHRON Y_HAT SE_Y_HAT;
run;
```

Obs	TIMEMD	SOLO	MAJOR	AGE	TOTCHRON	Y_HAT	SE_Y_HAT
31230	.	1	5	77	1	22.5506	0.92133
31231	.	2	2	45	1	19.7455	0.45978

A natural follow-up question is: "Do the two patients' expected times spent with the doctor differ significantly from one another?" (We are actually jumping ahead a bit to the concept of domain mean comparisons, which are covered in more detail in Section 8.6.) If we denote the first patient's expected value $\bar{y}_1$ and the second patient's $\bar{y}_2$, we can test $H_0: \bar{y}_1 = \bar{y}_2$ versus $H_1: \bar{y}_1 \neq \bar{y}_2$ by comparing $t = (\hat{\bar{y}}_1 - \hat{\bar{y}}_2)/se(\hat{\bar{y}}_1 - \hat{\bar{y}}_2)$, where $se(\hat{\bar{y}}_1 - \hat{\bar{y}}_2) = \sqrt{var(\hat{\bar{y}}_1) + var(\hat{\bar{y}}_2) - 2cov(\hat{\bar{y}}_1, \hat{\bar{y}}_2)}$, to a reference t distribution with $df = 552$ degrees of freedom. The ESTIMATE statement can be exploited to carry out this test. The notion is to first substitute the model parameter estimates into the difference $\hat{\bar{y}}_1 - \hat{\bar{y}}_2$. The intercept and any other parameter for which the two cases' underlying covariate is identical difference out, so we are left with a positive difference $77 - 45 = 32$ for AGE, a value of $+1$ for the indicator variable corresponding to SOLO=1, and a -1 for the indicator variable corresponding to MAJOR=2. Actually, the **L** matrix in this instance can be found by subtracting coefficients of the second ESTIMATE statement in Program 5.4 from coefficients of the first. Terms that reduce to a zero (e.g., INTERCEPT, TOTCHRON) are technically unnecessary to include because any effect not listed is assigned a coefficient(s) of 0 in **L**. Program 5.6 demonstrates syntax to assign the nonzero coefficients.

Program 5.6: Testing whether the Difference between the Expected Values of Two Example Covariate Profiles in a Linear Regression Model Is Statistically Significant

```
proc surveyreg data=NAMCS_2010;
  strata CSTRATM;
  cluster CPSUM;
  class SOLO MAJOR;
  model TIMEMD = SOLO MAJOR AGE TOTCHRON;
weight PATWT;
estimate 'Case 1 - Case 2' AGE 32 SOLO 1 -1 MAJOR 0 -1 0 0 1;
run;
```

Estimate					
Label	Estimate	Standard Error	DF	t Value	Pr > |t|
Case 1 – Case 2	2.8052	1.1326	552	2.48	0.0136

There are a few points to be made from examining the output. First, we can observe that the estimate reported matches the difference between the two estimates output by Programs 5.5. Second, we can reason that $cov(\hat{\bar{y}}_1, \hat{\bar{y}}_2)$ must have been negative since the estimated standard error reported from

Program 5.6, $1.1326 = \sqrt{\text{var}(\hat{\bar{y}}_1) + \text{var}(\hat{\bar{y}}_2) - 2\,\text{cov}(\hat{\bar{y}}_1, \hat{\bar{y}}_2)}$, is larger than the standard error we would have obtained assuming the two estimates were independent, $1.0297 = \sqrt{\text{var}(\hat{\bar{y}}_1) + \text{var}(\hat{\bar{y}}_2)} = \sqrt{(0.9213)^2 + (0.4598)^2}$. The column labeled "t Value" column is the test statistic $t = (\hat{\bar{y}}_1 - \hat{\bar{y}}_2)/\text{se}(\hat{\bar{y}}_1 - \hat{\bar{y}}_2)$. The value of 2.48 appears significant at the $\alpha = 0.05$ level given the complex survey degrees of freedom, and so we have evidence that the two expected times spent with the physician are not equal.

5.6 Summary

In this chapter, we delved into the topic of linear regression in the presence of complex survey data. We began by reviewing the familiar, textbook theory of linear regression, featuring prominently the matrix notation and matrix algebra inherent in estimating model parameters and conducting hypothesis tests. Although the modeling techniques explored in the following two chapters may appear markedly different at first glance, we will draw upon many of these core concepts as their application is functionally quite similar within the typology of generalized linear models.

6

Fitting Logistic Regression Models Using PROC SURVEYLOGISTIC

6.1 Introduction

All too commonly, analysts naively apply the modeling techniques of Chapter 5 to dichotomous outcome variables. This chapter begins by commenting on the perils of that practice and then introduces logistic regression, the technique analysts should use. There are numerous textbooks devoted exclusively to the topic of logistic regression. Two of this author's favorites are Allison (2012) and Hosmer et al. (2013). Instead of modeling the outcome directly as a 0/1 indicator variable of some event, status, or characteristic, the notion behind logistic regression is to model a transformation of it—namely, the natural logarithm of the odds. This quantity is also known as the "log-odds" or the "logit." While this results in changes to the estimation process and interpretation of model parameters, the basic principles and objectives behind fitting a logistic regression model are comparable to those laid out in the previous chapter. Moreover, as was the premise of Section 5.3, there are a number of idiosyncrasies when fitted using data collected as part of a complex survey design.

Examples in this chapter feature PROC SURVEYLOGISTIC. It operates very similarly to PROC LOGISTIC, its companion procedure you can use in a simple random sampling setting. As in Chapter 5, examples in this chapter utilize publicly released data collected from the 2010 National Ambulatory Medical Care Survey (NAMCS). Rather than modeling the time spent with the physician during the visit, however, we will assume that the research goal is to predict whether medication was prescribed. In Section 6.7, we discuss extensions of the logistic regression model to handle outcome variables with three or more possible values.

6.2 Logistic Regression in a Simple Random Sampling Setting

Consider the scatterplot in Figure 6.1 portraying the relationship between a continuous predictor variable x and dichotomous outcome variable y. Suppose $y=1$ indicates some event occurred and a $y=0$ indicates it did not occur. Larger values of x tend to be associated with a higher probability of the event occurring and vice versa. In other words, the expectation that y equals 1 tends to increase as x increases. The scatterplot is overlaid with a least squares regression line using the formulas presented in Section 5.2. We can actually conceive of the underlying model posited as the following *linear probability model*:

$$\Pr(y_i = 1 \mid x_i) = p_i = \beta_0 + \beta_1 x_i + \varepsilon_i \qquad (6.1)$$

While this approach may seem sensible at first, it violates several critical assumptions underpinning linear regression modeling, particularly those pertaining to the distribution of residuals. The most obvious violation is that, for any given x_i, ε_i can be only one of two possible values. Therefore, the errors are distributed binomially, not normally. Another violation is that the variance of the ε_is is greater for predicted probabilities around 0.5 than for probabilities closer to 0 or 1. This implies heteroscedasticity or a nonconstant residual variance.

Aside from contravening the theoretical assumptions, a potential interpretive dilemma that can surface is when a predicted probability falls outside

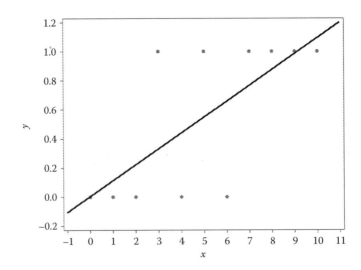

FIGURE 6.1
Scatterplot showing a least squares regression line fitted to a dichotomous outcome variable.

the range of 0 and 1. Indeed, for the hypothetical relationship depicted in Figure 6.1, this happens whenever $x < 0$ or $x > 9$. As we will see, this is not an issue in logistic regression models because the logit transformation has a distribution that can range from $-\infty$ to $+\infty$.

Sticking with the single-predictor scenario for the time being, in lieu of directly modeling p_i, the probability of some event occurring, logistic regression involves modeling

$$\ln\left(\frac{p_i}{1-p_i}\right) = \beta_0 + \beta_1 x_i \tag{6.2}$$

where $\ln(p_i/(1 - p_i))$ represents the natural logarithm of the odds of the event, more succinctly referred to as the "logit" or "log-odds." As Allison (2012) notes, converting p_i to the odds $p_i/(1 - p_i)$ removes the upper bound associated with p_i and taking the natural logarithm of the odds removes any kind of lower bound because the asymptote of the natural logarithm function is 0. Using the fundamental property that $\exp(\ln(x)) = x$, we can convert a logit to a probability at any time by exponentiating both sides of Equation 6.2 and solving for p_i, which returns the following:

$$p_i = \frac{\exp(\beta_0 + \beta_1 x_i)}{1 + \exp(\beta_0 + \beta_1 x_i)} \tag{6.3}$$

While the computations will be detailed shortly, suppose for the moment that we have estimated the parameters (i.e., beta terms) of the model specified in Equation 6.2 for the data plotted in Figure 6.1 and then used Equation 6.3 to convert each estimated logit into a predicted probability. Figure 6.2 overlays these predicted probabilities with the raw data. The line connecting the predicted probabilities reveals the inherent curvature on the probability scale. (If the y-axis were specified on the logit scale, no such curvature would be exhibited, at least under this particular model, one with a lone continuous predictor variable.) We can observe how predicted probabilities approach 0 for smaller values of x and approach 1 for larger values of x, but never actually breach those boundaries like predictions generated from the linear probability model do. The quantity in Equation 6.3 always falls between 0 and 1.

A facet of the logit transformation uninitiated analysts often have difficulty grasping is that the model parameters are interpreted differently than in linear regression. Instead of interpreting the term β_1 as the expected increase in the outcome (or probability) given a one-unit increase in x, it represents the expected change in the *logit* given a one-unit change in x. Few researchers this author interacts with regularly are adept at interpreting changes in the logit, but we can exponentiate the beta term to convert it to an odds ratio, which is easier to deal with.

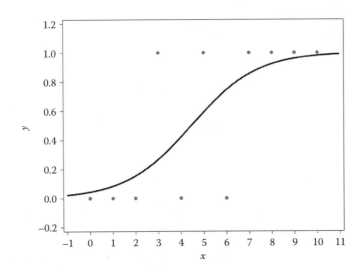

FIGURE 6.2
Scatterplot overlaid with the trend line of predicted probabilities derived from a logistic regression model fitted to a dichotomous outcome variable.

To see why the slope coefficient can be interpreted this way, consider a case with $x = 6$ relative to a case with $x = 5$. We can write the logit of the former as $l_1 = \beta_0 + 6\beta_1$ and the logit of the latter as $l_2 = \beta_0 + 5\beta_1$, and so the difference between the two is $l_1 - l_2 = (\beta_0 + 6\beta_1) - (\beta_0 + 5\beta_1) = \beta_1$. But because of the property that $\ln(x) - \ln(y) = \ln(x/y)$, β_1 can be expressed as

$$l_1 - l_2 = \ln\left(\frac{p_1}{1 - p_1}\right) - \ln\left(\frac{p_2}{1 - p_2}\right) = \ln\left(\frac{p_1/(1 - p_1)}{p_2/(1 - p_2)}\right)$$

Exponentiating this quantity returns the odds ratio of the first case relative to the second—again, two cases that are equivalent except for a one-unit difference in x.

To provide a numerical example, suppose that the estimated slope parameter for the logistic regression model fit using the data from Figure 6.2 is $\hat{\beta}_1 = 0.5$. This implies an estimated odds ratio of $\hat{O} = \exp(0.5) = 1.6847$. An odds ratio derived from a logistic regression model is often colloquially interpreted with a statement such as "a one-unit increase in x is associated with a 68.47% higher likelihood of the event occurring." In reality, such an interpretation is more befitting of the relative risk statistic defined in Section 4.3.3.

When the odds of the event are constant for all values of x, the odds ratio of any two cases one unit of x apart is 1, in which case we the log-odds ratio, as reflected by β_1, is 0 (since $\ln(1) = 0$). Therefore, conducting a hypothesis test to determine whether a logistic regression parameter is significantly different from 0 is just as relevant in the logistic regression setting as it is in traditional linear regression. The only subtle difference is one of interpretation: failing to conclude significance implies the variable associated with the parameter does not sufficiently differentiate the expected change in the log-odds of the outcome.

More specifics regarding the various methods to conduct hypothesis tests on model parameters will be discussed shortly, but first a few words are in order pertaining to how categorical explanatory variables are to be interpreted. We will begin by restricting our attention to a predictor variable with $K = 2$ categories, and then briefly discuss how the concepts generalize to variables with $K > 2$ categories.

The effect of a categorical predictor variable is best explained with the help of a 2×2 contingency table. To motivate an example, suppose that an epidemiologist collected data for a sample of $n = 10{,}000$ adults to study the impact of smoking cigarettes on the risk of developing lung cancer, and the findings have been summarized in Table 6.1. An intuitive way to parameterize the effect of smoking is to create an indicator variable x_i equaling 1 if the ith adult smokes and 0 otherwise. The parameter linked to this indicator variable would then represent the log-odds ratio of a smoker versus a nonsmoker. Exponentiating the parameter converts that quantity into an odds ratio. Therefore, prior to even fitting the logistic regression model to the data summarized in Table 6.1, we can reason that the slope term will equal the natural logarithm of the odds ratio, or $\hat{\beta}_1 = \ln[(200 \times 7840)/(160 \times 1800)] = \ln(5.4444) = 1.6946$.

The extension to handle a variable with $K > 2$ categories is straightforward. To briefly consider an example, suppose we further classified smokers based on the number of cigarettes smoked per day: less than one pack and more than one pack. This renders the overall effect of smoking to be a categorical variable with $K = 3$ distinct levels. Similarly to how we handled categorical effects in linear regression, when faced with K distinct levels, we create $K - 1$ indicator variables. The omitted level is called the baseline, or "reference," category. It serves as the denominator in the implied log-odds ratios associated with parameters summarizing the respective impacts of the $K - 1$ nonomitted levels.

Table 6.2 illustrates how the contingency table might look under the more fleshed out smoking classification. In terms of the $K - 1 = 2$ parameters

TABLE 6.1

Example Counts in a 2×2 Table Relating a Dichotomous Predictor Variable to a Dichotomous Outcome Variable

	Lung Cancer	No Lung Cancer	Row Totals
Smoker	200	1800	2,000
Nonsmoker	160	7840	8,000
Column totals	360	9640	10,000

TABLE 6.2

Example Counts in a Table Relating a Trichotomous Predictor
Variable to a Dichotomous Outcome Variable

	Lung Cancer	No Lung Cancer	Row Totals
Smoker: <1 pack/day	100	1400	1,500
Smoker: >1 pack/day	100	400	500
Nonsmoker	160	7840	8,000
Column totals	360	9640	10,000

required to represent the effect, the first would be calculated as $\ln[(100 \times 7840)/(160 \times 1400)] = \ln(3.5) = 1.2528$, and the second would be calculated as $\ln[(100 \times 7840)/(160 \times 400)] = \ln(12.25) = 2.5056$. Note that these two parameters are simply the log-odds ratios of two 2×2 tables, one for each of the two smoking frequency designations versus the nonsmoker designation.

Now that we have covered the rudiments of the logistic regression model specification and how to interpret parameters, we can delve deeper into the process of estimating model parameters. As opposed to the least squares method employed to estimate parameters of the linear regression model, logistic regression parameters are estimated using the technique of *maximum likelihood* (ML). The notion is to derive a *likelihood function*, which involves expressing the probability of the observed data as a function of unknown parameters. One then solves this expression for the parameter values that maximize the likelihood function.

Let **X** be the $n \times p$ design matrix consisting of one row for each of the n sampling units, a column of ones to represent the intercept, and a set of $p-1$ additional columns housing either continuous variables or 0/1 indicator variables representing categorical effects. Let x_i signify the row of the design matrix associated with the ith sampling unit and let β signify the $p \times 1$ vector of model parameters, such that $p_i = \exp(x_i\beta)/(1 + \exp(x_i\beta))$. Assuming simple random sampling with replacement, the likelihood function is the product of all possible outcomes and their respective probabilities for each of the n units sampled. Since there are only two outcomes coded $y_i = 1$ for an event and $y_i = 0$ for a nonevent, this function can be expressed as

$$L = \prod_{i=1}^{n} p_i^{y_i} (1-p_i)^{1-y_i} \tag{6.4}$$

Taking the natural logarithm of this likelihood function converts the power terms into coefficients and products into summations as follows:

$$\ln L = \sum_{i=1}^{n} y_i \ln(p_i) + (1-y_i)\ln(1-p_i) \tag{6.5}$$

The quantity in Equation 6.5 is called the "log-likelihood function," and since the natural logarithm is a continuously increasing function, maximizing with respect to that is equivalent to maximizing with respect to the original likelihood function in Equation 6.4.

The idea at this point is to express Equation 6.5 in terms of the underlying parameters in $\boldsymbol{\beta}$ and set all partial derivatives equal to 0. After some algebra and a bit of rearranging, this leaves a system of equations, one for each of the $j = 1,\ldots, p$ columns of the design matrix $\mathbf{X}$

$$\sum_{i=1}^{n} x_{ji} y_i - \sum_{i=1}^{n} x_{ji} p_i = 0 \qquad (6.6)$$

where, again, $p_i = \exp(\mathbf{x_i}\boldsymbol{\beta})/(1 + \exp(\mathbf{x_i}\boldsymbol{\beta}))$. A noteworthy corollary is that the sum of predicted probabilities equals the number of events observed. Because the p equations are nonlinear, there are no closed-form solutions (except in a few special situations). The work-around is to iterate through plausible values of $\boldsymbol{\beta}$ until it converges (i.e., stabilizes) to a value $\hat{\boldsymbol{\beta}}$, termed the ML estimate.

The default algorithm utilized by PROC LOGISTIC and PROC SURVEYLOGISTIC is the "Fisher scoring technique," also known as "iteratively reweighted least squares." Another popular technique used in practice is the *Newton–Raphson algorithm*, which can be explicitly requested with the option TECHNIQUE=NEWTON specified after the slash in the MODEL statement. We will not walk through the details of these algorithms here. The reader seeking more detail is referred to the documentation Allison (2012) or Hosmer et al. (2013). In most instances, either algorithm converges to the same unique solution of $\hat{\boldsymbol{\beta}}$ within a few iterations. There are occasions, however, when the algorithms fail to converge. The two most commonly encountered are *complete separation* and *quasi-complete separation*. These are discussed extensively in Section 3.4 of Allison (2012) and in Allison (2008).

Once we have employed one of the ML algorithms to arrive at an estimate of the $p \times 1$ vector $\hat{\boldsymbol{\beta}}$, the next step might be to make inferences on one or more components or test whether certain terms are significantly different from 0. This requires an estimate of $\text{cov}(\hat{\boldsymbol{\beta}})$, the symmetric $p \times p$ model covariance matrix with variances of parameter estimates along the diagonal and covariances on the off-diagonal. An estimate of this matrix can only be formulated once all $i = 1,\ldots, n$ predicted probabilities $\hat{p}_i = \exp(x_i\hat{\boldsymbol{\beta}}) / (1 + \exp(x_i\hat{\boldsymbol{\beta}}))$ have been determined because it is found by calculating

$$\text{cov}(\hat{\boldsymbol{\beta}}) = (\mathbf{X}^T\mathbf{V}\mathbf{X})^{-1} \qquad (6.7)$$

where
   $\mathbf{X}$ is the $n \times p$ design matrix
   $\mathbf{V}$ is a diagonal $n \times n$ matrix with entries $\hat{p}_i(1 - \hat{p}_i)$

To test the hypothesis $H_0: \beta_j = 0$ versus $H_1: \beta_j \neq 0$ for any one of the p parameters, we form a *Wald chi-square statistic*

$$W = \left(\frac{\hat{\beta}_j}{\sqrt{\text{var}(\hat{\beta}_j)}} \right)^2 \tag{6.8}$$

where $\sqrt{\text{var}(\hat{\beta}_j)}$ denotes the standard error of the jth parameter, the square root of the jth diagonal term of $\text{cov}(\hat{\beta})$, and reference against a chi-square distribution with 1 degree of freedom. As we will see from the output generated by Program 6.1, these tests are automatically conducted by PROC SURVEYLOGISTIC and are reported in the Analysis of Maximum Likelihood Estimates table.

Similar to the linear regression case, an alternative method for conducting these tests is to form a $p \times 1$ *contrast vector* $\mathbf{C}$ comprised of all zeros except the row corresponding to the parameter being tested, which is assigned as 1, and then find

$$W = (\mathbf{C}^T\hat{\beta})^T (\mathbf{C}^T\text{cov}(\hat{\beta})\mathbf{C})^{-1}(\mathbf{C}^T\hat{\beta}) \tag{6.9}$$

Note the resemblance between Equations 5.9 and 6.9. The only difference is that the quantity in Equation 5.9 is an F statistic, whereas Equation 6.9 is a Wald chi-square statistic. Recall that an F statistic is a chi-square statistic divided by its degrees of freedom. For the present case with 1 degree of freedom, there is effectively no difference, but one version of the statistic can always be converted to the other by either dividing by or multiplying by q the degrees of freedom as determined from the number of linearly independent columns of $\mathbf{C}$ (i.e., the rank of $\mathbf{C}$). As with the linear regression case, $\mathbf{C}$ can be augmented with more columns to accommodate multiparameter contrasts. The result is always a scalar, but the reference distribution changes to account for the increased degrees of freedom.

As an example, suppose that the full model posited is $\ln(p_i/(1 - p_i)) = \beta_0 + \beta_1 x_{1i} + \beta_2 x_{2i} + \beta_3 x_{3i}$ and we wish to test $H_0: \beta_2 = 0, \beta_3 = 0$ versus $H_1: \beta_j \neq 0$ for one of $j = 2, 3$. Since $\hat{\beta}$ is a 4×1 vector and $\text{cov}(\hat{\beta})$ is a 4×4 matrix, this test can be conducted by forming $\mathbf{C} = \begin{bmatrix} 0 & 0 \\ 0 & 0 \\ 1 & 0 \\ 0 & 1 \end{bmatrix}$ and plugging all three of these matrices

into Equation 6.9. The result would be compared against $\chi^2_{2,\alpha}$, the $100(1 - \alpha)$ th percentile of a chi-square distribution with 2 degrees of freedom. If the observed test statistic exceeds $\chi^2_{2,\alpha}$, we have evidence that at least one of β_2, β_3 is significantly different from zero.

One example setting in which this test would be especially pertinent is if the two parameters represented the log-odds ratios of two nonreference levels of a categorical predictor variable with $K = 3$ distinct levels. Independently testing the two parameters would offer only a partial picture for assessing the variable's impact in the model. SAS reports the results of these effect-specific tests by default for any categorical variable with $K > 2$ listed in the CLASS statement. They appear in the Type 3 Analysis of Effects table. Equivalent results can be obtained using the CONTRAST statement, which we will demonstrate later in the chapter.

Notice how there is no error term in the logistic regression model like the ε_i term in the linear regression model. There is no ANOVA table either. Rather than basing test statistics for whether the model can be reduced on a function of the sum of squares for error (e.g., as in Equation 5.10), in logistic regression we can construct a *likelihood ratio test* using a comparable quantity referred to as the model *deviance*. The deviance of a model is defined as minus two times the log-likelihood

$$D = -2\ln(L) \tag{6.10}$$

and the likelihood ratio test is defined as

$$G = D_{red} - D_{full} \tag{6.11}$$

where
D_{red} denotes the deviance of the reduced model
D_{full} denotes the deviance of the full model

The quantity in Equation 6.11 is always positive because adding more terms to the model can only serve to reduce the deviance, just like adding more terms to a linear regression model can only serve to reduce the sum of squares for error.

Under the null hypothesis that the q parameters to be omitted from the full model are truly zero, the likelihood ratio test statistic is distributed as a random chi-square variate with q degrees of freedom. A small observed test statistic coupled with a large p value suggests that the null hypothesis should be rejected or that the model can be reduced by eliminating the terms in question. The two tests defined by Equations 6.9 and 6.11 may not match precisely, but they are asymptotically equivalent.

In the table labeled "Testing Global Null Hypothesis: BETA=0," SAS summarizes the result of the likelihood ratio test for the overall model with underlying hypotheses H_0: $\beta_1 = \beta_2 = \cdots = \beta_{(p-1)} = 0$ versus H_1: $\beta_j \neq 0$ for one or more values of $j = 1, 2, \ldots, (p - 1)$. This is the analog to the overall F test reported in an ANOVA table for a linear regression model assessing whether an intercept-only model is sufficient. Also reported in the table is the Wald chi-square version of the test, one based on a $p \times (p-1)$ matrix **C** consisting of all zeros except a series of 1s running diagonally from the first entry in the second row to the pth row and $(p-1)$th column. Deviances of the full and intercept-only models are reported in the "-2LogL" row of the column labeled "Criterion" in the Model Fit Statistics table.

Another popular criterion to evaluate during the model-building process is how well the model fits the data. A model that fits well is one yielding predicted values in close proximity to the observed outcomes—in other words, where some aggregate distance metric relating the $\hat{p}_i$s to the y_is is small. A host of measures have been proposed in the literature, but arguably the most popular is the *Hosmer–Lemeshow test statistic* (Hosmer and Lemeshow, 1980). Results from this test are output by PROC LOGISTIC, whenever the LACKFIT option is specified after the slash in the MODEL statement—that option is not available in PROC SURVEYLOGISTIC, but we will discuss a work-around in Section 6.4. The test is intuitive. The first step is to sort the predicted probabilities for all n sampling units into g groups. A value of g often assigned is 10, in which case the groups are termed "deciles of risk." From there, one creates $C = g \times 2$ cells of events and nonevents, and within each an expected frequency E_c is found by summing the predicted probabilities (for events) or 1 minus the predicted probabilities (for nonevents). If we denote the observed frequency O_c, the test statistic is

$$\chi^2_{H-L} = \sum_{c=1}^{C} \frac{(O_c - E_c)^2}{E_c} \tag{6.12}$$

If the model provides a suitable fit, χ^2_{H-L} will follow a random chi-square distribution with $g-2$ degrees of freedom. Hence, if the observed test statistic exceeds the $100(1-\alpha)$th percentile of that distribution, there is evidence of poor model fit.

6.3 Logistic Regression with Complex Survey Data

This section discusses the theoretical and conceptual differences when fitting logistic regression models to data collected via a complex survey design. The reader seeking even more detail is referred to Roberts et al. (1987), Morel (1989), Skinner et al. (1989), Section 6.4 of Hosmer et al. (2013), and Chapter 8 of Heeringa et al. (2010). As in Chapter 5, we will distinguish a finite population model parameter with a Roman letter (e.g., B_0 or B_1) as opposed to a Greek letter (e.g., β_0 or β_1).

The first fundamental difference to point out is that in the presence of sample weights w_i for all $i = 1, \ldots, n$ observations in the sample, the likelihood function specified in Equation 6.4 is modified to the following:

$$PL = \prod_{i=1}^{n} \left[p_i^{y_i} (1 - p_i)^{1-y_i} \right]^{w_i} \tag{6.13}$$

The quantity in this equation is denoted *PL* instead of *L* because it represents the *pseudo-likelihood* (Binder, 1981, 1983), an approximation of the full-population likelihood.

$$L = \prod_{i=1}^{N} p_i^{y_i} (1 - p_i)^{1-y_i} \tag{6.14}$$

Similarly in spirit to the ML process in the simple random sampling setting, the objective is to maximize the natural logarithm of the pseudo-likelihood function

$$\ln PL = \sum_{i=1}^{n} w_i y_i \ln(p_i) + w_i(1 - y_i)\ln(1 - p_i) \tag{6.15}$$

with respect to each of the model parameters. We are still left with a system p equations, one for each of the $j = 1, \ldots, p$ columns of the design matrix $\mathbf{X}$, but instead of the unweighted system defined in Equation 6.6, we have

$$\sum_{i=1}^{n} w_i x_{ji} y_i - \sum_{i=1}^{n} w_i x_{ji} p_i = 0 \tag{6.16}$$

where $p_i = \exp(\mathbf{x}_i \mathbf{B})/(1 + \exp(\mathbf{x}_i \mathbf{B}))$. A corollary is that the sum of the *weighted* predicted probabilities matches the weighted sum of events. The p equations are still nonlinear, and so we must still use an iterative method to find plausible values of $\mathbf{B}$ until it converges to a value $\hat{\mathbf{B}}$ referred to as the "pseudo-maximum likelihood estimate." This vector can then be used to obtain predicted probabilities for any sampling unit by $\hat{p}_i = \exp(\mathbf{x}_i \hat{\mathbf{B}})/(1 + \exp(\mathbf{x}_i \hat{\mathbf{B}}))$.

Stratification and clustering, when present, introduce additional complexities with respect to estimating $\text{cov}(\hat{\mathbf{B}})$. One way to tackle the problem is to use a replication approach (see Chapter 9), but the default technique employed in most software, including SAS, is the multivariate application of Taylor series linearization to implicit functions proposed by Binder (1983). While the formula for implementing the Binder approach can take on various forms in textbooks and software documentation, following the exposition of Section 6.4 of Hosmer et al. (2013), the expression this author finds easiest to follow is

$$\mathrm{cov}(\hat{\mathbf{B}}) = (\mathbf{X}^{\mathsf{T}}\mathbf{W}\mathbf{V}\mathbf{X})^{-1}\,\mathrm{var}\left(\sum_{i=1}^{n}w_i\mathbf{u_i}\right)(\mathbf{X}^{\mathsf{T}}\mathbf{W}\mathbf{V}\mathbf{X})^{-1} \tag{6.17}$$

where
  $\mathbf{X}$ is the $n \times p$ design matrix
  $\mathbf{W}$ is an $n \times n$ diagonal matrix of weights
  $\mathbf{V}$ is an $n \times n$ diagonal matrix of $\hat{p}_i(1-\hat{p}_i)$ terms
  $\mathbf{u_i} = \mathbf{x_i}(y_i - \hat{p}_i)$, where $\mathbf{x_i}$ denotes the $1 \times p$ covariate vector for the ith sampling unit
  $\mathrm{var}\left(\sum_{i=1}^{n}w_i\mathbf{u_i}\right)$ is calculated with respect to the complex survey design

Notice how the structure of Equation 6.17 is very similar to that of Equation 5.16 for the linear regression case. These variance estimators are sometimes referred to in the statistical literature as "sandwich estimators."

Note that all three SURVEY procedures with modeling capabilities (PROC SURVEYREG, PROC SURVEYLOGISTIC, and PROC SURVEYPHREG) include a multiplicative adjustment factor of $(n-1)/(n-p)$ in the calculation of $\mathrm{cov}(\hat{\mathbf{B}})$ based on a recommendation by Hidiroglou et al. (1980) to reduce bias that can occur in small samples. If you do not want this adjustment factor to be used, specify VADJUST=NONE after the slash in the MODEL statement.

To test whether a model can be reduced by eliminating one or more parameters, you should only use the Wald chi-square tests of the form given in Equation 6.8 or 6.9. Although all necessary ingredients for a likelihood ratio test are still reported by PROC SURVEYLOGISTIC, Heeringa et al. (2010, p. 241) assert that the key distributional assumptions behind that test are invalidated in the presence of complex survey design features. However, the same principles motivating the modified F statistic proposed by Korn and Graubard (1990) (see Equation 5.17) can be used to derive a test statistic adhering closer to the stated distribution under the null hypothesis. The first step is to transform the Wald chi-square statistic to an F statistic by dividing through by q, the rank of $\mathbf{C}$, or the number of linearly independent columns. If we denote the Wald chi-square statistic W, given df complex survey degrees of freedom, the modified test statistic is

$$F_{ADJ} = \frac{W}{q} \times \frac{(df - q + 1)}{df} \tag{6.18}$$

which can be referenced against an F distribution with q and $(df - q + 1)$ degrees of freedom for the numerator and denominator, respectively.

In the presence of complex survey data, the standard Hosmer–Lemeshow test statistic as reported in Equation 6.12 is no longer guaranteed to abide by the stated chi-square distribution. Based on the findings of Archer (2001), Archer and Lemeshow (2006) discuss an alternative version that looks quite different at first glance but is rooted in an analogous objective. To obtain the test statistic, we must first sort the complex survey data set in ascending order of the predicted probabilities based on the pseudo-maximum likelihood estimated parameters. Using a technique described in Graubard et al. (1997), the next step is to define the g groups such that the sums of the weights for units falling in each group are approximately equal. If we define $\hat{r}_{gi} = y_{gi} - \hat{p}_{gi}$ as the estimated residual for the ith sampling unit in group g and w_{gi} the given unit's weight, the idea is to form a $g \times 1$ vector $\hat{\mathbf{R}}$ comprised of g weighted mean residuals $\Sigma w_{gi}\hat{r}_{gi}/\Sigma w_{gi}$, estimate its $g \times g$ variance–covariance matrix $\text{cov}(\hat{\mathbf{R}})$, and then form a Wald chi-square test statistic $W_{A-L} = \hat{\mathbf{R}}^T \text{cov}(\hat{\mathbf{R}})^{-1}\hat{\mathbf{R}}$. This is then rescaled to an F statistic

$$F_{A-L} = \frac{W_{A-L}}{g} \times \frac{(df - g + 2)}{df} \tag{6.19}$$

which can be referenced against an F distribution with $g - 1$ numerator degrees of freedom and $df - g + 2$ denominator degrees of freedom, where df denotes the complex survey degrees of freedom. Although this test is not yet available in PROC SURVEYLOGISTIC, a work-around is demonstrated in Program 6.3.

Now that we have sufficiently delineated the differences in logistic regression modeling with respect to the underlying theory and interpretation of parameters, let us explore some of the PROC SURVEYLOGISTIC syntax to fit these models and the associated output it generates. Continuing with the same data set utilized for examples appearing in Chapter 5, we will fit a model using the NAMCS 2010 public-use data file, which contains the following three features of complex survey data:

1. Stratification—strata are identifiable by distinct codes of the variable CSTRATM
2. Clustering—each PSU (nested within a stratum) is identifiable by distinct codes of the variable CPSUM
3. Unequal weights—maintained by the variable PATWT

Instead of treating the key outcome variable as the time spent with the physician (TIMEMD) during the visit, we will model MED, a 0/1 indicator variable for whether medication was prescribed. We will consider TIMEMD as a potential predictor variable along with patient gender (SEX), patient age (AGE), and patient race (RACER), the total number of chronic

TABLE 6.3

Names and Coding Structure of Variables Used in the Example Logistic Regression Model

Variable Name	Description	Coding Structure
MED	Indicator of medication prescribed or renewed	0 = No 1 = Yes
TIMEMD	Time spent with physician in minutes	Continuous
SEX	Patient gender	1 = Female 2 = Male
AGE	Patient age in years	Continuous
RACER	Patient race	1 = White 2 = Black 3 = Other
TOTCHRON	Patient's total number of chronic conditions	Count
MAJOR	Primary reason for the visit	1 = New problem (<3 months onset) 2 = Chronic problem, routine 3 = Chronic problem, flare up 4 = Pre-/post surgery 5 = Preventive care (e.g., routine prenatal, well-baby, screening, insurance, and general exams)

conditions afflicting the patient (TOTCHRON), and the primary reason for the visit (MAJOR). Despite Table 5.3 summarizing much of the same information, for quick reference Table 6.3 summarizes the scale and coding structure of these variables.

Program 6.1 illustrates PROC SURVEYLOGISTIC syntax to fit the model. The PARAM=REF option after the slash in the CLASS statement overrides the default *effect parameterization* of the categorical variables to instead request the *reference parameterization*. Lewis (2007) compares and contrasts these two parameterization schemes. As discussed previously, to implement the reference parameterization, we must eliminate (by setting to 0) one of the K parameters associated with the 0/1 indicator variables representing each distinct level of a categorical variable. Consequently, the parameters representing the nonomitted levels are interpreted as the log-odds ratio relative to the omitted level, or the *reference category*. By default, the last value in the sort-ordered list of distinct values—summarized in the Class Level Information table—is assigned as the reference category. There is more than one way to override this default, but the most direct method is to declare the desired reference category in parentheses immediately after naming the variable in the CLASS statement. For example, specifying MAJOR (REF='4') would change the reference category of MAJOR from preventative care to pre-/post surgery. As was remarked in Chapter 5, changing the reference category does not impact the overall significance of the categorical variable in the model; it merely alters the interpretation of its associated parameters.

The syntax EVENT='1' in parentheses immediately after declaring the dependent variable in the MODEL statement is technically optional, but good practice to include, since it ensures the procedure will model the log-odds of the intended outcome. The default approach utilized by PROC SURVEYLOGISTIC is to model the log-odds of the first value in the sort-ordered list of values. A very common coding structure, as is presently the case, is to indicate the event with a 1 and the nonevent with a 0, so accepting the default would actually lead one to model the log-odds of a *non*event. The parameters obtained would be of the same absolute value, but with the opposite sign (e.g., a log-odds ratio would be reported as −0.5 instead of 0.5).

Program 6.1: Fitting a Logistic Regression Model Accounting for Complex Survey Design Features

```
proc surveylogistic data=NAMCS_2010;
  strata CSTRATM;
  cluster CPSUM;
  class SEX RACER MAJOR / param=ref;
  model MED (event='1') = SEX RACER MAJOR
                          AGE TOTCHRON TIMEMD;
weight PATWT;
run;
```

SURVEYLOGISTIC Procedure

Model Information	
Data set	WORK.NAMCS_2010
Response variable	MED
Number of response levels	2
Stratum variable	CSTRATM
Number of strata	74
Cluster variable	CPSUM
Number of clusters	626
Weight variable	PATWT
Model	Binary logit
Optimization technique	Fisher's scoring
Variance adjustment	Degrees of freedom (DF)

Variance Estimation	
Method	Taylor series
Variance adjustment	Degrees of freedom (DF)

Number of observations read	31,229
Number of observations used	31,229
Sum of weights read	1.0088E9
Sum of weights used	1.0088E9

Response Profile			
Ordered Value	MED	Total Frequency	Total Weight
1	0	8,248	251,274,996
2	1	22,981	757,527,009
Probability modeled is MED = 1.			

Class Level Information					
Class	Value	Design Variables			
SEX	1	1			
	2	0			
RACER	1	1	0		
	2	0	1		
	3	0	0		
MAJOR	1	1	0	0	0
	2	0	1	0	0
	3	0	0	1	0
	4	0	0	0	1
	5	0	0	0	0

Model Convergence Status
Convergence criterion (GCONV = 1E-8) satisfied

Model Fit Statistics		
Criterion	Intercept Only	Intercept and Covariates
AIC	1.13253E9	1.07682E9
SC	1.13253E9	1.07692E9
−2LogL	1.13253E9	1.07692E9

Testing Global Null Hypothesis: BETA=0			
Test	Chi-Square	DF	Pr>ChiSq
Likelihood Ratio	55,608,888.1	10	<.0001
Score	57,369,698.7	10	<.0001
Wald	279.2490	10	<.0001

Type 3 Analysis of Effects			
Effect	DF	Wald Chi-Square	Pr>ChiSq
SEX	1	0.1770	0.6740
RACER	2	13.8646	0.0010
MAJOR	4	137.8999	<.0001
AGE	1	3.2606	0.0710
TOTCHRON	1	79.2365	<.0001
TIMEMD	1	0.1454	0.7029

Analysis of Maximum Likelihood Estimates						
Parameter		DF	Estimate	Standard Error	Wald Chi-Square	Pr>ChiSq
Intercept		1	0.1114	0.2070	0.2898	0.5904
SEX	1	1	0.0183	0.0435	0.1770	0.6740
RACER	1	1	0.4837	0.1860	6.7587	0.0093
RACER	2	1	0.6960	0.2009	12.0078	0.0005
MAJOR	1	1	0.3501	0.0861	16.5503	<.0001
MAJOR	2	1	0.5611	0.1190	22.2340	<.0001
MAJOR	3	1	0.4515	0.1043	18.7457	<.0001
MAJOR	4	1	−0.7982	0.1186	45.3271	<.0001
AGE		1	0.00306	0.00169	3.2606	0.0710
TOTCHRON		1	0.1749	0.0197	79.2365	<.0001
TIMEMD		1	−0.00092	0.00241	0.1454	0.7029

Odds Ratio Estimates			
Effect	Point Estimate	95% Wald Confidence Limits	
SEX 1 vs 2	1.018	0.935	1.109
RACER 1 vs 3	1.622	1.126	2.336
RACER 2 vs 3	2.006	1.353	2.973
MAJOR 1 vs 5	1.419	1.199	1.680
MAJOR 2 vs 5	1.753	1.388	2.213
MAJOR 3 vs 5	1.571	1.280	1.927
MAJOR 4 vs 5	0.450	0.357	0.568
AGE	1.003	1.000	1.006
TOTCHROH	1.191	1.146	1.238
TIMEMD	0.999	0.994	1.004

PROC SURVEYLOGISTIC generates a lot of output by default, which can be a little daunting initially, but each statistical summarization has a purpose. (Of course, you can use the ODS SELECT or ODS EXCLUDE statements for more control over this—see the documentation for more details.) The Response Profile table includes an unweighted and weighted summary of the two outcomes of MED. We can interpret the weighted counts as we do any finite population total estimated from a survey (see Section 3.2). For instance, it is estimated that medication was prescribed 757,527,009, or roughly 75.1%, of the 1,008,802,005 (=251,274,996 + 757,527,009) distinct visits to a physician's office that were estimated to have occurred in the NAMCS target population during calendar year 2010. From the corollary of Equation 6.6 noted previously, we know that the weighted sum of predicted probabilities for all 31,229 visits must also equal 757,527,009.

The message in the Model Convergence Status section notifies us that the Fisher scoring algorithm was successful in finding a unique solution to the ML estimates, which are provided in the Analysis of Maximum Likelihood Estimates table. Each of the p parameters is given alongside their standard errors, and the corresponding Wald chi-square test statistic (from Equation 6.8) for the underlying hypothesis test H_0: $\beta_j = 0$ versus H_1: $\beta_j \neq 0$. Again, a large p value suggests that the true parameter is not significantly different from zero. From the output, it appears all parameters are significant at the $\alpha = 0.05$ level except for those associated with SEX, AGE, and TIMEMD.

In terms of the overall model, the section with heading "Testing Global Null Hypothesis: BETA=0" provides three test statistics for the statistical hypotheses H_0: $\beta_1 = \beta_2 = \cdots = \beta_{(p-1)} = 0$ versus H_1: $\beta_j \neq 0$ for one or more values of $j = 1, 2, \ldots, (p - 1)$. Remember that, in the presence of complex survey design features, we should only heed the results of the Wald test. In this example, however, all tests lead to the same conclusion, that the null hypothesis should be rejected.

The Odds Ratio Estimates table reports an estimated odds ratio for all variables not involved in an interaction. As an example, we can gather from the output that the odds of medication being prescribed increases 19.1% with each additional chronic condition afflicting the patient. Because the confidence limits for this estimate do not encompass 1, we can conclude this is a significant increase, albeit we could have concluded as much from the fact that the Wald chi-square test statistic for this model parameter indicates it is significantly different from 0. Whereas values reported in the Odds Ratio Estimates table for a continuous predictor variable represent the estimated odds ratio given a one-unit increase, for a categorical predictor variable, the value represents the estimated odds ratios of the given category relative to the reference category. In either case, these quantities are algebraically equivalent to exponentiating the corresponding variable's parameter. Using SEX and AGE as examples, we can verify exp(0.0183) = 1.018 and exp(0.00306) = 1.003. In fact, specifying the option EXPB after the slash in the MODEL statement

TABLE 6.4

Summary of Useful MODEL Statement Options Available after the Slash in PROC SURVEYLOGISTIC

Option	Effect
NOINT	Excludes the intercept from the model by eliminating the column of ones in the design matrix
NODUMMYPRINT	Suppresses the Class Level Information table
COVB	Outputs the model parameter covariance matrix
VADJUST=NONE	Omits the model parameter covariance matrix adjustment factor applied by default as discussed in Hidiroglou et al. (1980)
CLPARM	Reports confidence limits for model parameters
CLODDS	Produces a new section of output containing odds ratio estimates and confidence limits, which only differ from those reported by default if the UNITS statement is used
ALPHA=*decimal*	Modifies the default significance level for confidence intervals of parameters and odds ratios (default is ALPHA=.05)
EXPB	Adds a column to the Analysis of Maximum Likelihood Estimates table housing the exponentiated model parameter estimates
RSQUARE	Outputs two generalized R^2 measures similar in spirit (but not with the exact same interpretation) to the familiar R^2 from linear regression
LINK=*option*	Specifying CLOGLOG or PROBIT that will fit the complementary log-log or probit regression versions of the model, respectively; for an outcome with more than two categories, the cumulative (ordinal) version of any given model is fitted, but specifying GLOGIT explicitly requests the generalized (multinomial) logistic regression model (see Section 6.7)

is another way to obtain these odds ratios, which are then reported in a new column appearing within the Analysis of Maximum Likelihood Estimates table. Techniques to tailor odds ratios to those of particular interest to the researcher are covered in Section 6.6.

Table 6.4 summarizes some of the most useful MODEL statement options. We have already touched on several of these. We will discuss more about the LINK= option later in this chapter as well as Chapter 7. Specifying RSQUARE outputs two generalized versions of the R^2 measure we discussed in Chapter 5. The first is attributable to Cox and Snell (1989), and the other is attributable to Nagelkerke (1991). Many researchers caution against using these pseudo-R^2 measures (e.g., Hosmer et al., 2013) arguing that they can produce specious conclusions because they really do not carry the same interpretation as in the linear regression case. Granted, these cautions are generally voiced within the context of simple random sampling as opposed to complex sample designs. But since these measures are a function of likelihood ratios, whose behavior and distributional properties we have already noted change in the presence of weights, it is probably safest to avoid them.

6.4 Testing for a Reduced Model and Adequate Model Fit

We found from the output of Program 6.1 that the parameters associated with SEX, AGE, and TIMEMD were all individually insignificant. Suppose we sought to test whether the model could be reduced by simultaneously eliminating these three effects from the model. This requires us to calculate a Wald chi-square statistic, which we can do with the CONTRAST statement. Our primary responsibility is to correctly define the contrast matrix **C**. It will have three columns, each comprised of all zeros except the row associated with one of the three given parameters, which is assigned a 1. Program 6.2 demonstrates syntax to define this matrix. Remember that a comma demarcates columns of **C**, and that any effect not listed has an implied coefficient(s) of zero. One noteworthy difference relative to the CONTRAST statement in PROC SURVEYREG, however, is that it is no longer necessary to specify a coefficient of –1 for the reference level of a categorical predictor (*cf.* Program 5.2).

From the Contrast Test Results table in the output we observe a Wald chi-square value of 3.4495 with a corresponding *p* value much higher than any conventional cutoff point when referenced against a chi-square distribution with $q = 3$ degrees of freedom. A large *p* value such as this suggests that the null hypothesis postulating that the three parameters are simultaneously equal 0 cannot be rejected. Hence, it appears we can eliminate these three factors from the model.

Note that SAS does not report the modified test statistic F_{ADJ} recommended by Korn and Graubard (1990) defined in Equation 6.18, but it would not substantively change the conclusion in this case because $F_{ADJ} = (3.4495/3) \times (550/552) = 1.1457$, which has a *p* value of 0.66 based on a reference *F* distribution with $q = 3$ numerator and $df - q + 1 = 552 - 3 + 1 = 550$ denominator degrees of freedom.

Program 6.2: Testing Whether a Logistic Regression Model Can Be Reduced

```
proc surveylogistic data=NAMCS_2010;
  strata CSTRATM;
  cluster CPSUM;
  class SEX RACER MAJOR / param=ref;
  model MED (event='1')=SEX RACER MAJOR
                        AGE TOTCHRON TIMEMD;
weight PATWT;
contrast 'Test of Reduced Model' SEX 1, AGE 1, TIMEMD 1;
run;
```

Contrast Test Results			
Contrast	DF	Wald Chi-Square	Pr>ChiSq
Test of reduced model	3	3.4495	0.3274

After fine-tuning the set of predictor variables, we may wish to evaluate the overall fit of the model. At present, there is no LACKFIT option available in PROC SURVEYLOGISTIC like there is in PROC LOGISTIC, but Program 6.3 demonstrates syntax to retrieve all necessary quantities underlying the alternative test from Equation 6.19 attributable to Archer and Lemeshow (2006). To be discussed in more detail in the next section, the OUTPUT statement stores the predicted probability of the outcome in a variable named P_HAT in the data set named RESIDUALS. The actual residual is calculated in the subsequent DATA step in the variable RESIDUAL, and the RESIDUALS data set is then sorted in ascending order according to the predicted probabilities.

To assign $g = 10$ deciles, thresholds are inserted at points where the cumulative sum of weights first exceeds $(1/g)$th of the total sum of weights. Specifically, the numeric variable DECILE ranges from 1 to 10, where units having DECILE=1 are those with the smallest predicted probabilities and DECILE=10 are those with the largest, and the sum of weights for units in any particular decile is approximately one-tenth of the total sum of weights. This customized DATA step programming is imperative because the decile assignments per Graubard et al. (1997) are not the same as those provided using, for example, PROC RANK or the DECILE option in the PROC SURVEYMEANS statement.

In effect, the PROC SURVEYREG step in Program 6.3 tricks SAS into computing the term W_{A-L}/g in Equation 6.19. From the output, we observe that this value is 6.28. The adjustment factor is $(df - g + 2)/df = (552 - 10 + 2)/552 \approx 1$, so even without carrying out the computations we know we would get something very similar. The probability of seeing a test statistic this large under the null hypothesis is very small, suggesting there is room for improvement with respect to model fit.

Program 6.3: Testing for Lack of Fit in a Logistic Regression Model Accounting for Complex Survey Design Features

```
proc surveylogistic data=NAMCS_2010;
  strata CSTRATM;
  cluster CPSUM;
  class RACER MAJOR / param=ref;
  model MED (event='1') = RACER MAJOR TOTCHRON;
```

```
weight PATWT;
output out=residuals p=p_hat;
run;

data residuals;
  set residuals;
residual=MED-p_hat;
run;

proc sort data=residuals;
  by p_hat;
run;

* assign risk decile thresholds where cumulated sum of weights
exceeds (1/g)th of the total sum of weights;
data residuals;
  set residuals;
PATWT_sum=1008802005;
PATWT_running_sum+PATWT;
decile=ceil(10*(PATWT_running_sum/PATWT_sum));
run;

proc surveyreg data=residuals;
  strata CSTRATM;
  cluster CPSUM;
  class decile;
  model residual=decile / noint solution vadjust=none;
weight PATWT;
contrast 'A-L LOF Test'
  decile 1 0 0 0 0 0 0 0 0 0,
  decile 0 1 0 0 0 0 0 0 0 0,
  decile 0 0 1 0 0 0 0 0 0 0,
  decile 0 0 0 1 0 0 0 0 0 0,
  decile 0 0 0 0 1 0 0 0 0 0,
  decile 0 0 0 0 0 1 0 0 0 0,
  decile 0 0 0 0 0 0 1 0 0 0,
  decile 0 0 0 0 0 0 0 1 0 0,
  decile 0 0 0 0 0 0 0 0 1 0,
  decile 0 0 0 0 0 0 0 0 0 1;
run;
```

Analysis of Contrast			
Contrast	Num DF	F Value	Pr>F
A-L LOF Test	10	6.28	<.0001

6.5 Computing Unit-Level Statistics

The OUTPUT statement of PROC SURVEYLOGISTIC is designed to generate unit-level statistics and output them alongside the original input data
set to a separate data set named in the OUT= option. If we let x_i denote the
*i*th sampling unit's row in the design matrix X, perhaps the two most useful
statistical keywords are P=, which requests the predicted probability of the
event, or

$$\hat{p}_i = \frac{\exp(x_i\hat{B})}{1+\exp(x_i\hat{B})} \tag{6.20}$$

and XBETA=, which requests the logit, or

$$\ln\left(\frac{\hat{p}_i}{1-\hat{p}_i}\right) = x_i\hat{B} \tag{6.21}$$

Two related keywords are L= and U=, which will output the lower and upper
bound of the predicted probability, respectively. Similarly to what was outlined in Section 4.2.2, these are obtained by finding the lower and upper
bounds of the logit confidence interval and then exponentiating. Specifically,
the logit confidence interval is defined by

$$x_i\hat{B} \pm t_{df,\alpha/2}\sqrt{x_i \, \text{cov}(\hat{B})x_i^T} \tag{6.22}$$

where $t_{df,\alpha/2}$ is the $100(1-\alpha/2)$th percentile of a *t* distribution with *df* complex
survey degrees of freedom. The STDXBETA= keyword can be specified to
output the quantity within the square root in Equation 6.22.

Using the reduced model following results of Program 6.2, Program 6.4
demonstrates syntax to save in a new data set named PREDS the original
data plus the predicted probability and lower and upper bounds from a 95%
confidence interval. The three new variables housing these three statistics
are P_HAT, P_HAT_LOWER, and P_HAT_UPPER, respectively. Note that the
latter two will not necessarily be symmetric about P_HAT. And although
these bounds are constructed with respect to the default $\alpha=0.05$, you can
specify an alternative significance level using the ALPHA= option after the
slash in the OUTPUT statement.

Program 6.4: Outputting Unit-Level Statistics from a Logistic Regression Model

```
proc surveylogistic data=NAMCS_2010;
  strata CSTRATM;
  cluster CPSUM;
  class RACER MAJOR / param=ref;
  model MED (event='1')=RACER MAJOR TOTCHRON;
weight PATWT;
output out=preds p=p_hat l=p_hat_lower u=p_hat_upper;
run;
```

6.6 Customizing Odds Ratios

The purpose of this section is to demonstrate how to use results from a logistic regression model to compare two targeted covariate profiles. The basic quantity estimated by the model parameters for the first unit is $\hat{l}_1$, the estimated logit or log-odds of experiencing the event, so comparing two units boils down to finding the difference between two unit's logits, $\ln(\hat{O}) = \hat{l}_1 - \hat{l}_2$. This difference is expressed $\ln(\hat{O})$ because it actually represents a log-odds ratio since the natural logarithm of the difference between two quantities is equivalent to the natural logarithm of the ratio of the two quantities.

Before continuing, it should be mentioned that the ODDSRATIO statement is an extremely handy tool for customizing odds ratios, but at the time of this writing, it is only available in PROC LOGISTIC. In all likelihood, it will be available in PROC SURVEYLOGISTIC in an upcoming release of SAS. When this happens, the methods to be shown in this section may be less syntactically efficient, so check the documentation to determine whether it is available in your version of SAS.

The general approach to follow when customizing an odds ratio is to write out the estimated logit for the first case in terms of the estimated model parameters, do the same for the second case, and then take the difference. Once the difference is expressed in terms of model parameters, we then populate coefficients of a $p \times 1$ contrast vector **C** and utilize the CONTRAST statement to carry out the pertinent calculations.

To motivate an example, recall that the default series of odds ratios reported for a categorical variable effect are those pitting the nonreference categories against the reference category. But what if we wanted to compare two nonreference categories? For instance, we may have observed from the output of Program 6.1 that the log-odds ratios for both black patients and white patients were both positive and significant, indicating an increased likelihood that medication is prescribed relative to patients of other races, but we may wish to assess whether the difference between black and white patients is significant, all else equal. To answer this question, we can reason that the logit difference reduces to the parameter associated with black patients minus the parameter associated with white patients. Program 6.5 demonstrates how to estimate this quantity for the reduced model consisting only of patient race (RACER), the primary reason for the visit (MAJOR), and the total number of chronic conditions afflicted the patient (TOTCHRON).

In the background, PROC SURVEYLOGISTIC constructs $\mathbf{C}$ as a $p \times 1$ vector of all zeros except an entry of -1 for the RACER=1 parameter and 1 for the RACER=2 parameter. The estimate of the log-odds ratio is $\mathbf{C}^{\mathrm{T}}\hat{\mathbf{B}}$, and the standard error is $\sqrt{\mathbf{C}^{\mathrm{T}}\mathrm{cov}(\hat{\mathbf{B}})\mathbf{C}}$. The option ESTIMATE=BOTH requests values in both the logit scale (Type=PARM row) and the odds ratio scale (Type=EXP row), the latter simply being exponentiated versions of the former. The estimated odds ratio of 1.2176 reported means black patients are 21.76% more likely to have medication prescribed, all else equal. Because the 95% confidence limits on the odds ratio scale do not contain 1—or, equivalently, because the 95% confidence limits on the logit scale do not contain 0—we can conclude that this is a significant difference.

Program 6.5: Estimating a Customized Odds Ratio for Two Covariate Profiles in a Logistic Regression Model

```
proc surveylogistic data=NAMCS_2010;
  strata CSTRATM;
  cluster CPSUM;
  class RACER MAJOR / param=ref;
  model MED (event='1') = RACER MAJOR TOTCHRON;
weight PATWT;
contrast 'B vs. W Patients' RACER -1 1 / estimate=both;
run;
```

Contrast Estimation and Testing Results by Row									
Contrast	Type	Row	Estimate	Standard Error	Alpha	Confidence Limits		Wald Chi-Square	Pr>ChiSq
B vs. W Patients	FARM	1	0.1968	0.0815	0.05	0.0371	0.3566	5.8337	0.0157
B vs. W Patients	EXP	1	1.2176	0.0992	0.05	1.0378	1.4284	5.8337	0.0157

To motivate another example, recall that the odds ratios provided by PROC SURVEYLOGISTIC are given in terms of a one-unit difference in the underlying variable. This may not correspond to the most fitting comparison for continuous predictors. For example, perhaps the investigators were more interested in estimating the effect of a three-unit increase in the total number of chronic conditions afflicting the patient. Program 6.6 demonstrates CONTRAST statement syntax to estimate this effect. The log-odds ratio reported is 0.5581, which is three times the TOTCHRON parameter estimate of 0.1860. The odds ratio is approximately 1.75, meaning the likelihood of receiving a prescription is about 75% higher for a patient with three chronic conditions more than another otherwise identical patient. Note that an alternative method to procure this estimate is via a UNITS statement (e.g., UNITS TOTCHRON=3). Consult the documentation for more details on that statement.

Program 6.6: Estimating a Customized Odds Ratio for a Continuous Variable in a Logistic Regression Model

```
proc surveylogistic data=NAMCS_2010;
  strata CSTRATM;
  cluster CPSUM;
  class RACER MAJOR / param=ref;
  model MED (event='1') = RACER MAJOR
                    TOTCHRON;
weight PATWT;
contrast 'TOTCHRON + 3' TOTCHRON 3 / estimate=both;
run;
```

Analysis of Maximum Likelihood Estimates						
Parameter		DF	Estimate	Standard Error	Wald Chi-Square	Pr>ChiSq
Intercept		1	0.1827	0.1885	0.9399	0.3323
RACER	1	1	0.5013	0.1807	7.6998	0.0055
RACER	2	1	0.6982	0.1963	12.6503	0.0004
MAJOR	1	1	0.3678	0.0856	18.4523	<.0001
MAJOR	2	1	0.6156	0.1134	29.4838	<.0001
MAJOR	3	1	0.4944	0.1026	23.2282	<.0001
MAJOR	4	1	-0.7380	0.1231	35.9579	<.0001
TOTCHRON		1	0.1860	0.0196	91.1766	<.0001

Contrast Estimation and Testing Results by Row									
Contrast	Type	Row	Estimate	Standard Error	Alpha	Confidence Limits		Wald Chi-Square	Pr>ChiSq
TOTCHRON + 3	PARM	1	0.5581	0.0584	0.05	0.4436	0.6727	91.1766	<.0001
TOTCHRON + 3	EXP	1	1.7474	0.1021	0.05	1.5582	1.9595	91.1766	<.0001

We conclude this section with a somewhat more complicated example. The goal is still to estimate a customized odds ratio from a model composed of the same three predictor variables used in previous examples, but we will now assume more disparate covariate profiles and that the model has an interaction term between patient race and the patient's total number of chronic conditions. In essence, including this interaction allows the effect of TOTCHRON to vary according to a patient's race. There is no need to create additional variables in a separate DATA step to account for the interaction. By specifying RACER*TOTCHRON in the MODEL statement, PROC SURVEYLOGISTIC will augment the design matrix with the two necessary columns: one housing the product of the 0/1 indicator variable representing RACER=1 and TOTCHRON and another housing the like for the indicator variable representing RACER=2.

Let us suppose that the objective is to estimate the odds ratio for a hypothetical black patient on a visit tending to a routine chronic problem (MAJOR=2) relative to a white patient seeing the physician because of a flare up in a chronic problem (MAJOR=3). Assume that the two patients have the same number of total chronic conditions. The logit difference will only consist of parameters for which the two patients differ. The first patient gets a +1 for the following coefficients: MAJOR=2, RACER=2, and the interaction term for (RACER=2) × TOTCHRON. The second patient gets a −1 for the coefficients of MAJOR=3, RACER=1, and the interaction for (RACER=1) × TOTCHRON.

The CONTRAST statement in Program 6.7 estimates this odds ratio. Its structure parallels what we have seen previously, but take heed of how the coefficients for the two interaction parameters are specified. The option E appearing after the slash requests the contrast vector **C** to be output, which is wise to do in a more complicated scenario such as this because it offers the opportunity to verify that all coefficients were properly assigned. For instance, how are we to be sure which of the two interaction terms are linked to the first coefficient listed after RACER*TOTCHRON? From the portion of the output labeled "Coefficients of Contrast Custom Odds Ratio," we can confirm that the first term associated with the interaction is that for RACER=1 and the second for RACER=2. Indeed, the contrast coefficients have been assigned correctly.

Despite necessitating more careful thought to coax out of SAS, there is nothing unique about how this particular odds ratio is to be interpreted. The hypothetical visit of the black patient in which a routine chronic problem is addressed is about 38.11% more likely to lead to medication being prescribed than the visit of a white patient prompted by a chronic problem worsening. The 95% confidence interval of this estimate is (4.14%, 83.16%), indicating that the effect is statistically significant, albeit with a very large degree of uncertainty.

Program 6.7: Estimating a Customized Odds Ratio for Two Covariate Profiles in a More Complicated Logistic Regression Model Involving Interaction Terms

```
proc surveylogistic data=NAMCS_2010;
  strata CSTRATM;
  cluster CPSUM;
  class RACER MAJOR / param=ref;
  model MED (event='1') = RACER MAJOR
                          TOTCHRON
                          RACER*TOTCHRON;
weight PATWT;
contrast 'Custom Odds Ratio'
         MAJOR 0 1 -1 0 RACER -1 1 RACER*TOTCHRON -1 1 /
         estimate=both e;
run;
```

Coefficients of Contrast Custom Odds Ratio	
Parameter	Row1
Intercept	0
RACER1	−1
RACER2	1
MAJOR1	0
MAJOR2	1
MAJOR3	−1
MAJOR4	0
TOTCHRON	0
RACER1TOTCHRON	−1
RACER2TOTCHRON	1

Contrast Estimation and Testing Results by Row									
Contrast	Type	Row	Estimate	Standard Error	Alpha	Confidence Limits		Wald Chi-Square	Pr> ChiSq
Custom Odds Ratio	PARM	1	0.3229	0.1440	0.05	0.0406	0.6052	5.0244	0.0250
Custom Odds Ratio	EXP	1	1.3811	0.1989	0.05	1.0414	1.8316	5.0244	0.0250

6.7 Extensions for Modeling Variables with More than Two Outcomes

6.7.1 Introduction

The aim of this section is to demonstrate how the logistic regression model for a binary survey outcome variable can be extended to handle outcome variables with $K > 2$ possible categories. In Section 6.7.2, we introduce the multinomial logistic regression model, designed to handle nominally scaled outcome variables. For outcome variables measured on an ordinal scale, where there is some sort of inherent ordering, the cumulative logistic regression model offers a more parsimonious alternative. That type of model is discussed in Section 6.7.3.

6.7.2 Multinomial Logistic Regression Model for Nominal Outcomes

For an outcome variable consisting of $K > 2$ possible events, the "multinomial logistic regression model," referred to in the SAS documentation as the "generalized logistic regression model," is essentially the combination of $K - 1$ sets of binary logistic regression models. This means that if there are p parameters (i.e., design matrix columns) associated with each binary model, then we can expect the multinomial model to consist of $(K - 1) \times p$ parameters. Each of the $K - 1$ sets of parameters represents logit differences between a particular event and an omitted event called the "baseline category." It is important to realize, however, that all model parameters must be simultaneously estimated. Parameters estimated by partitioning the data and fitting $K - 1$ separate binary models will be similar in magnitude but will generally have larger standard errors—see Section 5.5 of Allison (2012) for a discussion.

Let us walk through the mechanics of the multinomial model by considering an outcome with $K = 3$ events denoted A, B, and C. If we designate event C as the baseline category, then the first of the $K - 1 = 2$ binary logistic regression models is specified by $\ln(\Pr(A|\mathbf{x}_i)/\Pr(C|\mathbf{x}_i)) = \mathbf{x}_i \mathbf{B}_{A:C}$, where $\mathbf{x}_i$ represents the $1 \times p$ covariate vector for the ith sampling unit and $\mathbf{B}_{A:C} = \begin{bmatrix} B_{0A:C} & B_{1A:C} & \cdots & B_{(p-1)A:C} \end{bmatrix}^{\mathrm{T}}$ represents a $p \times 1$ vector of parameters, an intercept and $p - 1$ terms reflecting the expected change in the log-odds of event A occurring relative to event C given a one-unit increase in the particular covariate. The second of the binary logistic regression models is specified by $\ln(\Pr(B|\mathbf{x}_i)/\Pr(C|\mathbf{x}_i)) = \mathbf{x}_i \mathbf{B}_{B:C}$. The third possible binary model is redundant because

$$\ln\left(\frac{\Pr(A|\mathbf{x}_i)}{\Pr(B|\mathbf{x}_i)}\right) = \ln\left(\frac{\Pr(A|\mathbf{x}_i)}{\Pr(C|\mathbf{x}_i)}\right) - \ln\left(\frac{\Pr(B|\mathbf{x}_i)}{\Pr(C|\mathbf{x}_i)}\right) = \mathbf{x}_i \mathbf{B}_{A:C} - \mathbf{x}_i \mathbf{B}_{B:C} = \mathbf{x}_i(\mathbf{B}_{A:C} - \mathbf{B}_{B:C})$$

which also reveals how we can compare any two *non*baseline events by simply subtracting one parameter from the other. Changing the baseline category has no impact on statistics reflecting the model's likelihood or the overall significance level of a particular effect; it merely alters how parameters are to be interpreted.

Working with the two sets of parameters, we can convert the logits into probabilities at any time by exploiting the following relationships: (1) $\Pr(A|\mathbf{x}_i)=\exp(\mathbf{x}_i\mathbf{B}_{A:C})/(1 + \exp(\mathbf{x}_i\mathbf{B}_{A:C}) + \exp(\mathbf{x}_i\mathbf{B}_{B:C}))$, (2) $\Pr(B|\mathbf{x}_i)=\exp(\mathbf{x}_i\mathbf{B}_{B:C})/(1 + \exp(\mathbf{x}_i\mathbf{B}_{A:C}) + \exp(\mathbf{x}_i\mathbf{B}_{B:C}))$, and because probabilities of the three events must sum to 1, (3) $\Pr(C|\mathbf{x}_i)=1/(1 + \exp(\mathbf{x}_i\mathbf{B}_{A:C}) + \exp(\mathbf{x}_i\mathbf{B}_{B:C}))$.

To generalize matters to any number of events indexed by $k=1, \ldots, K$, and, without loss of generality, if we suppose that the Kth event has been assigned as the baseline category, then the multinomial model is defined by

$$\ln\left(\frac{\Pr(k \mid \mathbf{x}_i)}{\Pr(K \mid \mathbf{x}_i)}\right) = \mathbf{x}_i\mathbf{B}_{k:K} \tag{6.23}$$

for $k=1,\ldots, K - 1$. Individual probabilities for the kth event can be extracted by

$$\Pr(k \mid \mathbf{x}_i) = \frac{\exp(\mathbf{x}_i\mathbf{B}_{k:K})}{1+\sum_{k=1}^{K-1}\exp(\mathbf{x}_i\mathbf{B}_{k:K})} \tag{6.24}$$

with the exception of the baseline category's probability, which is

$$\Pr(K \mid \mathbf{x}_i) = \frac{1}{1+\sum_{k=1}^{K-1}\exp(\mathbf{x}_i\mathbf{B}_{k:K})} \tag{6.25}$$

Examples in the preceding sections of this chapter demonstrated modeling the dichotomy of whether medication was prescribed during the physician visit using the 0/1 indicator variable named MED on the NAMCS_2010 data set. This is a collapsed version of the trinomial variable NUMMED, coded 0, 1, or 2. As with the variable MED, a code of 0 means no medication was prescribed, but a code of 1 means *exactly* 1 medication was prescribed, while a code of 2 indicates two or more medications were prescribed. Program 6.8 shows syntax to fit a multinomial logistic regression model for the NUMMED outcome variable using patient race (RACER), primary reason for the visit (MAJOR), and the total number of chronic conditions afflicting the patient (TOTCHRON) as covariates. These variables' coding structures were summarized in Table 6.3. The LINK=GLOGIT option after the slash in the MODEL statement instructs PROC SURVEYLOGISTIC to fit the multinomial (i.e., generalized) logistic regression model as opposed to the cumulative logistic regression model (see Section 6.7.3), which is the default when three or more nonmissing values of the dependent variable are detected. Rather than specifying the event of interest in parentheses following identification of the dependent variable as we did for a dichotomous outcome, we can specify the baseline category with the REF= option. By default, SAS assigns the last value in the sort-ordered list of values to serve as the baseline category. In the present example, that value corresponds to NUMMED=2, but it seems more fitting here to compare the two medication categories to the group having no medications prescribed.

Only select output generated is shown here. The Response Profile section tabulates the counts and weighted sums of each distinct outcome variable value. Since there are $K=3$ such values, we find $K-1=2$ sets of parameter estimates in the Analysis of Maximum Likelihood Estimates table. The NUMMED column flags which of the two nonbaseline categories the particular parameter estimate refers to. For example, there are two parameters estimated for the first category of RACER (white patients), one for NUMMED=1 and another for NUMMED=2. Because the reference category for RACER is patients of other races (nonwhite and nonblack), the first parameter value 0.3646 represents the log-odds of a single medication being prescribed for whites relative to patients of other races, whereas the second value 0.5562 represents the log-odds of two or more medications prescribed for whites relative to patients of other races. Exponentiated versions of these two log-odds measures are reported in the first two lines of the Odds Ratio Estimates table, along with 95% confidence limits.

Program 6.8: Fitting a Multinomial Logistic Regression Model Accounting for Complex Survey Design Features

```
proc surveylogistic data=NAMCS_2010;
  strata CSTRATM;
  cluster CPSUM;
  class RACER MAJOR / param=ref;
  model NUMMED (ref='0') = RACER MAJOR
                    TOTCHRON / link=GLOGIT;
weight PATWT;
run;
```

Response Profile			
Ordered Value	NUMMED	Total Frequency	Total Weight
1	0	8,248	251,274,996
2	1	6,699	226,745,103
3	2	16,282	530,781,906

Type 3 Analysis of Effects			
Effect	DF	Wald Chi-Square	Pr>ChiSq
RACER	4	17.4511	0.0016
MAJOR	8	168.6795	<.0001
TOTCHRON	2	81.3924	<.0001

Analysis of Maximum Likelihood Estimates							
Parameter		NUMMED	DF	Estimate	Standard Error	Wald Chi-Square	Pr>ChiSq
Intercept		1	1	−0.5730	0.2050	7.8128	0.0052
Intercept		2	1	−0.4764	0.1971	5.8394	0.0157
RACER	1	1	1	0.3646	0.1920	3.6078	0.0575
RACER	1	2	1	0.5562	0.1850	9.0413	0.0026
RACER	2	1	1	0.5582	0.2093	7.1096	0.0077
RACER	2	2	1	0.7499	0.2014	13.8567	0.0002
MAJOR	1	1	1	0.2737	0.0935	8.5672	0.0034
MAJOR	1	2	1	0.4020	0.0966	17.3072	<.0001
MAJOR	2	1	1	0.2497	0.1128	4.8971	0.0269
MAJOR	2	2	1	0.6915	0.1313	27.7451	<.0001
MAJOR	3	1	1	0.2277	0.1147	3.9390	0.0472
MAJOR	3	2	1	0.5519	0.1120	24.3013	<.0001
MAJOR	4	1	1	−0.9675	0.1315	54.0931	<.0001
MAJOR	4	2	1	−0.6766	0.1437	22.1657	<.0001
TOTCHRON		1	1	0.0110	0.0102	1.1566	0.2822
TOTCHRON		2	1	0.3682	0.0414	79.1068	<.0001

Odds Ratio Estimates				
Effect	NUMMED	Point Estimate	95% Wald Confidence Limits	
RACER 1 vs 3	1	1.440	0.988	2.098
RACER 1 vs 3	2	1.744	1.214	2.508
RACER 2 vs 3	1	1.747	1.159	2.634
RACER 2 vs 3	2	2.117	1.426	3.141
MAJOR 1 vs 5	1	1.315	1.095	1.579
MAJOR 1 vs 5	2	1.495	1.237	1.807
MAJOR 2 vs 5	1	1.284	1.029	1.601
MAJOR 2 vs 5	2	1.997	1.544	2.583
MAJOR 3 vs 5	1	1.256	1.003	1.572
MAJOR 3 vs 5	2	1.737	1.394	2.163
MAJOR 4 vs 5	1	0.380	0.294	0.492
MAJOR 4 vs 5	2	0.508	0.384	0.674
TOTCHRON	1	1.011	0.991	1.032
TOTCHRON	2	1.445	1.332	1.567

We can customize multiparameter hypothesis tests in a manner comparable to what has been demonstrated previously over the course of this and
the previous chapter. To motivate a test often of interest in the multinomial
outcome setting, notice how many of the parameters for NUMMED=1 and
NUMMED=2 are of similar magnitude. With respect to the notation defined
earlier, proving $\mathbf{B}_{1:0} = \mathbf{B}_{2:0}$ would offer evidence that the model could be simplified by collapsing NUMMED=2 with NUMMED=1, thereby halving the
number of parameters necessary and, in effect, reverting to the binary version of the model previously considered.

Patterned after a similar example given on page 147 of Allison (2012),
Program 6.9 demonstrates how to utilize the TEST statement to test H_0:
$\mathbf{B}_{1:0} = \mathbf{B}_{2:0}$ versus H_1: $\mathbf{B}_{1:0} \neq \mathbf{B}_{2:0}$. Because we are interested in the joint hypothesis
that all components of the two parameter vectors are equivalent, we separate
the parameter-specific tests with commas in a single TEST statement. The
parameter affixed to NUMMED=1 is specified by suffixing the associated
model parameter with "_1" and, similarly, "_2" for NUMMED=2. Note that
the suffix is based on the underlying code of the outcome variable, not generically assigned in sequence starting at 1. Moreover, any parameter linked
to a nonreference categorical variable effect must have its underlying code
appended to the effect name prior to the underscore. For example, the syntax
RACER1_1=RACER1_2 requests a comparison of the parameter linked to the
effect of RACER=1 (versus the omitted RACER=3) for NUMMED=1 relative
to the like for NUMMED=2. The Wald chi-square statistic reported is highly
significant, which implies that we lack evidence to reduce the model by collapsing NUMMED=2 with NUMMED=1.

**Program 6.9: Testing Whether Outcome Categories Can Be Collapsed in
a Multinomial Logistic Regression Model**

```
proc surveylogistic data=NAMCS_2010;
  strata CSTRATM;
  cluster CPSUM;
  class RACER MAJOR / param=ref;
  model NUMMED (ref='0') = RACER MAJOR
                       TOTCHRON / link=GLOGIT;
weight PATWT;
test
  INTERCEPT_1=INTERCEPT_2,
  RACER1_1=RACER1_2, RACER2_1=RACER2_2,
  MAJOR1_1=MAJOR1_2, MAJOR2_1=MAJOR2_2, MAJOR3_1=MAJOR3_2,
  MAJOR4_1=MAJOR4_2,
  TOTCHRON_1=TOTCHRON_2;
run;
```

Linear Hypothesis Testing Results			
Label	Wald Chi-Square	DF	Pr>ChiSq
Test 1	360.9489	8	<.0001

For the binary case, Program 6.4 showed how to export the predicted probabilities of the designated event onto an output data set using the P= option of the OUTPUT statement. The same technique works for multinomial models, except now there are *K* events of interest, each with probabilities defined by Equations 6.24 and 6.25 that sum to 1. When PROC SURVEYLOGISTIC encounters the P= option (or any other unit-level statistical keyword) for a model with $K>2$ distinct events, it defaults to a data set structure in which one record is output for each of the *K* possible events. Hence, the output data set consists of *K* times the number of observations in the input data set, which introduces some redundancy. If you are only seeking the *K* predicted probabilities, specifying PREDPROBS=I produces a more convenient structure, whereby all probabilities are appended to each record on the input data set. They are stored in variables named IP_*code*, where *code* reflects the particular outcome variable's value—the prefix "IP" stands for "individual probability."

6.7.3 Cumulative Logistic Regression Model for Ordinal Outcomes

Certain multicategory survey outcome variables abide by an ordered structure in the sequence of response options, $y = 1, 2,..., K$. A prime example is an attitudinal question posed on a Likert-type scale, such as one with five possible choices ranging from "Completely Agree" to "Completely Disagree." It is not uncommon for analysts, particularly those with a social science background, to code the responses 1 through 5 and treat as a continuous outcome variable using, say, methods discussed in Chapter 5. An alternative for those concerned that some of the assumptions inherent with linear regression may be violated (e.g., normally distributed residuals) is to fit a *cumulative logistic regression model* (McCullagh, 1980). Note, however, that this alternative carries assumptions of its own that will become clear as we discuss it further.

Some perceive the cumulative logistic regression model as more comprehensible than the multinomial version because there are fewer parameters. For an outcome variable with *K* categories, the cumulative version of the model results in $K - 1$ separate intercepts, but only one parameter for each covariate. The assumption permitting the simplification is that the log-odds ratios change proportionally as a function of the $K - 1$ intercept terms, but the effect of a one-unit increase in a particular covariate is identical transitioning from $y=1$ to $y=2$, $y=2$ to $y=3$, and so on. For $k=1, ..., K-1$, the cumulative logistic regression model can be expressed as

$$\ln\left(\frac{\Pr(y_i \le k \mid \mathbf{x_i})}{1 - \Pr(y_i \le k \mid \mathbf{x_i})}\right) = \ln\left(\frac{\Pr(y_i \le k \mid \mathbf{x_i})}{\Pr(y_i > k \mid \mathbf{x_i})}\right) = B_{0k} + B_1 x_{1i} + \cdots + B_p x_{pi} \quad (6.26)$$

where

B_{0k} represents a category-specific intercept term

$B_1, \ldots, B_p$ are the other parameters that represent the expected marginal change in the log-odds given a one-unit increase in the underlying covariate

Despite the model being expressed in terms of cumulative logits, we can still extract individual outcome category probabilities. For the first category, $\Pr(y_i = 1 \mid \mathbf{x_i}) = \exp(B_{01} + B_1 x_{1i} + \cdots + B_p x_{pi})/(1 + \exp(B_{01} + B_1 x_{1i} + \cdots + B_p x_{pi}))$, and for the second, $\Pr(y_i = 2 \mid \mathbf{x_i}) = \Pr(y_i \le 2 \mid \mathbf{x_i}) - \Pr(y_i = 1 \mid \mathbf{x_i})$, which can be found by taking

$$\frac{\exp(B_{02} + B_1 x_{1i} + \cdots + B_p x_{pi})}{1 + \exp(B_{02} + B_1 x_{1i} + \cdots + B_p x_{pi})} - \frac{\exp(B_{01} + B_1 x_{1i} + \cdots + B_p x_{pi})}{1 + \exp(B_{01} + B_1 x_{1i} + \cdots + B_p x_{pi})}$$

For the final category, because the probabilities must sum to 1, we can reason that

$$\Pr(y_i = K \mid \mathbf{x_i}) = 1 - \Pr(y_i \le K - 1 \mid \mathbf{x_i}) = 1 - \frac{\exp(B_{0(K-1)} + B_1 x_{1i} + \cdots + B_p x_{pi})}{1 + \exp(B_{0(K-1)} + B_1 x_{1i} + \cdots + B_p x_{pi})}$$

Program 6.10 uses the same outcome and predictor variables from Program 6.8, but the LINK=CLOGIT option specified after the slash in the MODEL statement requests the cumulative logistic regression version of the model. As expected, the output shows $K - 1$ intercept terms, but only one parameter for each covariate. No response variable option such as REF= or EVENT= is necessary because PROC SURVEYLOGISTIC cumulates the probabilities beginning with the first in the sort-ordered list by default, which is appropriate in this case.

Program 6.10: Fitting a Cumulative Logistic Regression Model Accounting for Complex Survey Design Features

```
proc surveylogistic data=NAMCS_2010;
  strata CSTRATM;
  cluster CPSUM;
  class RACER MAJOR / param=ref;
  model NUMMED = RACER MAJOR
                 TOTCHRON / link=CLOGIT;
weight PATWT;
run;
```

| Analysis of Maximum Likelihood Estimates | | | | | |
Parameter		DF	Estimate	Standard Error	Wald Chi-Square	Pr>ChiSq
Intercept	0	1	−0.2004	0.1572	1.6242	0.2025
Intercept	1	1	0.8809	0.1556	32.0332	<.0001
RACER	1	1	−0.4335	0.1468	8.7227	0.0031
RACER	2	1	−0.5641	0.1598	12.4638	0.0004
MAJOR	1	1	−0.3020	0.0744	16.4545	<.0001
MAJOR	2	1	−0.6021	0.1006	35.8142	<.0001
MAJOR	3	1	−0.4723	0.0890	28.1638	<.0001
MAJOR	4	1	0.5718	0.1262	20.5457	<.0001
TOTCHRON		1	−0.2880	0.0288	99.6846	<.0001

An alternative parameterization is to specify the logit function in terms of the complementary probabilities, $\ln(\Pr(y_i \geq k|\mathbf{x}_i)/(1 - \Pr(y_i \geq k|\mathbf{x}_i))) = \ln(\Pr(y_i \geq k|\mathbf{x}_i)/\Pr(y_i < k|\mathbf{x}_i))$. In essence, this is the version PROC SURVEYLOGISTIC fits when the DESC option is specified in parentheses following the dependent variable declaration in the MODEL statement. If we were to include that option in Program 6.10, we would find parameter estimates for covariates are of the same magnitude, only their signs have been reversed.

Also, remarks made in the previous section about outputting predicted probabilities apply just the same with the cumulative version of the model. One additional option available when fitting a cumulative logistic regression model is PREDPROBS=C, which will output $\Pr(y_i \leq k)$ in a series of variables prefixed CP_, which stands for "cumulative probability."

6.8 Summary

This chapter began by outlining the pitfalls to fitting a linear regression model to a dichotomous outcome variable. We argued why logistic regression should be used instead, which involves making a log-odds, or logit, transformation after designating one of the outcome variable categories the "event" and the other as a "nonevent." The logit transformation alters the interpretation of model parameters, however. Instead of the change in the predicted probability of the event for two cases one-unit apart with respect to the given covariate, a parameter represents the change in the natural logarithm of the odds ratio between those two cases. We demonstrated how to specify the EXPB option after the slash in the MODEL statement to exponentiate the model parameters, which converts them back into odds ratios, and also how to use the P= option in the OUTPUT statement to convert a logit into a predicted probability.

The chapter closed with a section extending the binary logistic regression model to cases in which the outcome is nominal or ordinal, consisting of $K > 2$ distinct events. After assigning one of the K categories as the baseline category, the nominal model is essentially a set of $K - 1$ binary logistic regression models relative to the baseline category serving as the "nonevent." The ordinal logistic regression model allows for $K - 1$ unique intercepts, but assumes the same log-odds change across all of the outcome category thresholds.

7

Survival Analysis with Complex Survey Data

7.1 Introduction

This chapter is a foray into *survival analysis*, also known as "time-to-event analysis," "duration modeling," or "event history analysis," depending on the discipline. Whatever the discipline, the goal is generally the same: to model the length of time units (often individuals) remain "at risk," defined as the time lapsing between the onset of some state and another event signaling termination of that state. Survival analysis is a broad enough statistical topic to command its own textbook. Two widely cited classics are Kalbfleisch and Prentice (2002) and Singer and Willett (2003), while Allison (2010) is an excellent resource for implementing these techniques within SAS.

The phrase "survival analysis" originates from applications in which the ultimate event of interest is death, perhaps following some treatment, surgery, or other biomedical intervention. It also makes sense in engineering and operations research applications where, for example, the objective is to model the lifetime of some product, bridge, or edifice. The terminology may sound strange or even a bit comical in certain contexts, such as "surviving" employment or marriage, but the modeling approaches can be adapted to any setting in which temporal information is available regarding the onset and termination of a clearly defined state. The transition times are not always precisely known, however. Indeed, sometimes all we know is that the transition time exceeds some value or falls within some interval. This introduces the concept of *censoring*. We will define the various forms of censoring in Section 7.2.2 after discussing a few data collection strategies producing longitudinal data structures. We will also define the various survival function specifications and, subsequently, classify modeling techniques used in practice.

Section 7.3 highlights the formulaic and interpretive differences when fitting survival analysis models to complex survey data using examples from data on first marriages (where applicable) as provided by female respondents to the National Survey of Family Growth (NSFG). Note that there is no single SAS/STAT procedure to be demonstrated like in Chapters 5 and 6. Rather, examples will be drawn from PROC LIFETEST, PROC SURVEYLOGISTIC, and PROC SURVEYPHREG. The latter is the only SURVEY procedure designed exclusively for survival analysis with complex survey data, but we will demonstrate PROC LIFETEST's handy repertoire of data visualization tools and PROC SURVEYLOGISTIC's capability to fit certain models in which time is measured in a finite number of intervals.

7.2 Foundations of Survival Analysis

7.2.1 Data Collection Strategies

Within the realm of applied survey research, capturing the time(s) at which a sampling unit enters and exits the state of interest can be done either prospectively or retrospectively. A prime example of a prospective design is a *panel survey design*, in which repeated measurements are taken on the same units over time. By comparison, most repeatedly administered surveys abide by a *cross-sectional survey design* in which an independent sample is selected for each new cycle. Another increasingly common prospective data collection protocol is to link survey responses from cross-sectional samples to administrative records maintaining supplemental data on, for example, program participation or mortality.

On the other hand, temporal information is sometimes produced in a retrospective manner via respondent recall. For example, the source of duration data on first marriages analyzed by Copen et al. (2012) is the respective memories of female respondents to the NSFG at the time of interview. Approximate dates of major life events such as marriage and the dissolution of marriage are likely retrievable without recall error, but one cannot ignore the well-documented cognitive issues impairing respondents' ability to recall the details of past events (Tourangeau et al., 2000). Hence, the potential for systematic recall error is one downside of this particular retrospective approach.

The data collection strategy employed can have a bearing on the precision with which time is measured, which, in turn, impacts the modeling approaches at one's disposal. A particularly decisive factor is whether the event times are measured on an effectively continuous scale (i.e., without ties) or whether they are known only to fall within some interval. For example, the Cox proportional hazards regression model was developed for continuous-time data, whereas the discrete-time hazards regression model can handle time known only on an interval basis.

7.2.2 Censoring

Another feature of survival analysis problems dictated by the data collection strategy is the *observation period* or timespan over which sampling units are monitored. It is rare when the observation period overlaps harmoniously with each sampling unit's *risk period* or duration between entering and exiting the at-risk population, even though measuring and modeling the risk period (i.e., survival time) is the fundamental goal of survival analysis. Generally speaking, when the timing information regarding when the sampling unit transitions into and out of the state of interest is incomplete, that unit's time measure is said to be *censored*. A closely related concept is *left truncation*, which occurs when a sampling unit's risk period has already begun at the start of the observation period.

Leung et al. (1997) review the many forms of censoring and truncation. To help frame the exposition, let us consider the censoring mechanisms of a survival analysis problem studied by Lewis et al. (2014) in which the objective was to model the time a federal employee voluntarily remains employed with the government after becoming eligible to retire with full benefits. The observation period for the study was bounded by the beginning of calendar year 2000 to the end of calendar year 2012. A personnel database maintained by the U.S. Office of Personnel Management was gleaned for any employee eligible to retired with full benefits at any point during that 12-year window. Figure 7.1 is provided to help visualize six of the most common types of employee risk periods and their relationship to the observation period. The start of the observation period is labeled A and the end labeled B. A horizontal line represents the time during which the individual remained on board after becoming fully eligible to retire. An X at the end of a risk period denotes retirement, whereas an O signifies some form of censoring.

The first pattern in Figure 7.1 is ideal. The individual enters the risk population during the observation period and experiences the event before the observation period concludes, so his/her entire history is known. The second

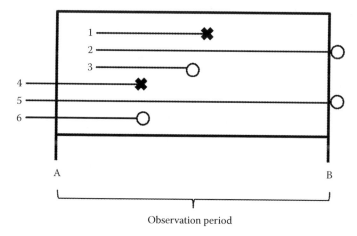

Observation period

FIGURE 7.1
Visualization of censoring and truncation in survival analysis problems.

pattern reflects data that are *right censored*. These individuals enter the risk population during the observation period, but do not experience the key event by the conclusion of the observation period. Hence, all we know about the survival time is that it exceeds some value. This is sometimes referred to as "end-of-study censoring" to distinguish it from the alternative form of right censoring represented by the third pattern, individuals whose risk period terminates because of some alternative event, not specifically the event of interest. For example, instead of retiring voluntarily, the third pattern might represent individuals who changed agencies or died. In contrast to end-of-study censoring, this is sometimes referred to as "loss-to-follow-up censoring."

The fourth pattern illustrates *left truncation*, the case for any individual already fully eligible to retire at the start of 2000. Whenever the origin of the risk period is known, adjustments to handle left truncation are straightforward, but it is a trickier situation to deal with otherwise. The fifth pattern represents individuals who were fully eligible to retire in 2000 and were still employed with the government at the end of 2012. This is an example of *doubly censored* data. Another example of doubly censored data is the sixth pattern, which portrays left-truncated individuals who were also censored at some point during the observation period.

Implicitly assumed in the definitions outlined earlier is that the precise time of censoring is known. This is referred to by Leung et al. (1997) as "point censoring." There are occasions, however, when censoring is known only to occur between two points in time, which is referred to as "interval censoring." For example, in the retirement study, the precise day on which the individual was censored could be ascertained from the personnel database by mining a "dynamics" file capturing the various ways individuals separate from the federal workforce. But suppose all we had available was a series of rosters identifying employees on board at periodic intervals. In such a setting, we would be able to identify those who departed the workforce by cross-referencing two chronologically adjacent files, yet the exact date that occurred would be unknown. Instead, all we would know is that censoring occurred at some point between the two boundaries.

The observation period end point in Figure 7.1 exemplifies *Type I censoring*. Type I means that the censoring time is under the control of the researcher. Relatively speaking, *Type II censoring* is much less common. It refers to situations when the observation period is terminated after a prespecified number of events have occurred. The complement to Type I and II censoring is *random censoring*, so named because it happens outside of the researcher's control. This is represented by the third and sixth patterns in Figure 7.1.

As Allison (2010) notes, standard survival analysis methods do not distinguish between Type I, Type II, and random censoring. They are all generically treated as right-censored observations. But the distinction is important with respect to the underlying theory. Types I and II censoring do not induce bias into the estimation process, but the same can only be said for random censoring if we assume it is *noninformative*, which is to say that the probability the observation is censored is independent of the survival time, conditional on any covariates included in the model. This is conceptually similar to the *missing at random* (MAR) assumption (Little and Rubin, 2002) underlying many of the techniques to compensate for missing data (see Section 10.2).

7.2.3 Definitions

The purpose of this section is to lay out the statistical notation and theory underpinning survival analysis. The survival time is typically denoted T. Without loss of generality, we can conceptualize the survival time as the time lapsing before some key event occurs. Naturally, it follows that $T > 0$. All approaches posit that T is a random variable abiding by some probability distribution. What distinguishes the particular methods is the probability distribution assumed and whether T is measured on a continuous scale or in discrete intervals.

For continuous-time models, we can denote the *probability density function* (PDF) of the event time as $f(t)$, interpreted as the probability that the occurs exactly at time t. For discrete-time models, we can denote the analogous PDF as $f(m)$, interpreted as the probability that the event occurs within the interval $(m, m + 1)$.

A related function is the *cumulative distribution function* (CDF) of the event time. For continuous-time models, this is specified by

$$F(t) = \int_0^t f(t)dt \tag{7.1}$$

and for discrete-time models this is specified by

$$F(m) = \sum_{j \le m} f(j) \tag{7.2}$$

The CDF is interpreted as the cumulative probability that the event occurs before or at time t for continuous-time models or before the end of the mth interval for discrete-time models. The complement to the CDF is the *survivor function*, denoted $S(t)$, which is the probability that the event occurs *after* time t, since

$$S(t) = P(T > t) = 1 - P(T \le t) = 1 - F(t) \tag{7.3}$$

For discrete-time models, the survivor function is

$$S(m) = \sum_{j > m} f(j) = 1 - \sum_{j \le m} f(j) = 1 - F(m) \tag{7.4}$$

Finally, another important component of survival analysis is the *hazard function*, denoted $h(t)$, which is defined as the instantaneous risk that the event occurs at time t, conditional upon not yet occurring up to that point. Formulaically, the hazard function is defined as the ratio of the PDF to the survivor function, or

$$h(t) = \frac{f(t)}{S(t)} \tag{7.5}$$

for continuous-time models and

$$h(m) = \frac{f(m)}{S(m-1)} \tag{7.6}$$

for discrete-time models. Note that specifying the hazard function over an interval such as this typically results in a terminology shift. Namely, the hazard function is alternatively termed the "hazard rate" or "hazard ratio" because it can be interpreted as the expected number of events to occur over the given interval (e.g., a day, a month, a year).

7.2.4 Classification of Survival Analysis Models

Survival analysis models can be classified as one of four possible types: (1) parametric, (2) nonparametric, (3) semiparametric, or (4) discrete-time. In this section, we will briefly describe each class of model, comment on the specific SAS/STAT procedures you can use to fit them, and refer the reader to the subsections of Section 7.3 in which they are fleshed out in more detail.

Parametric models assume T follows some formal probability distribution. Popular choices include the exponential, Weibull, Gompertz, log-normal, or gamma distributions. PROC LIFEREG is capable of fitting these types of models in the simple random sampling setting, but at the time of this writing there is little offered in SAS and other competing software to fit these models properly accounting for the complex sample design. As such, we will not discuss them further in this book.

On the other end of the spectrum are nonparametric models, which make no assumptions regarding the probability distribution of T. Instead, these methods aim to simply describe the survivor function, often visually. For example, a prudent first step in survival analysis modeling is to plot the estimated survivor function for the entire data set as well as subsets defined by one or more categorical covariates. PROC LIFETEST includes a suite of options to do so using, for instance, the very popular Kaplan–Meier (KM) estimator. Nonparametric modeling approaches and their adaptations to the complex survey sample design setting are discussed in Section 7.3.1.

Semiparametric models do not make the same stringent assumptions of parametric models, but they are not assumption-free. The most widely used semiparametric approach is *Cox proportional hazards regression*, named after its inventor who introduced the technique in a seminal paper (Cox, 1972). In Section 7.3.2, we will demonstrate how the plausibility of the "proportional hazards" qualifier can be assessed visually and by way of a formal statistical test. The SAS/STAT procedure devoted exclusively to fitting this type of model is PROC PHREG. PROC SURVEYPHREG, the most recent addition to the SURVEY family of procedures (Mukhopadhyay, 2010), is its counterpart to properly account for complex survey data.

The fourth and final class of survival analysis models pertains to those in which time is measured in discrete intervals. These models are also a potential solution when time is measured on an effectively continuous scale but with a lot of ties or instances where two or more sampling units share the exact same survival time since handling ties during the estimation process of parametric or semiparametric models can be a nuisance. A distinctive feature of these models is the need to transform the data set into a *person-period data set* in which there are multiple records per sampling unit, one for each period the given unit was part of the at-risk population. For example, a unit surviving until some point in the fourth time interval would be in the data set four times: The first three records correspond to the intervals during which the key event did not occur, and the fourth to the interval during which it did. Because the key outcome variable is just a dichotomous indicator of event occurrence, standard logistic regression methods available in PROC LOGISTIC are applicable, and complex survey design features are straightforward to accommodate by employing PROC SURVEYLOGISTIC instead. Discrete-time models are the focus of Section 7.3.3.

7.3 Survival Analysis with Complex Survey Data

7.3.1 Visualizing the Data Using PROC LIFETEST

Prior to fitting any kind of model, a wise preliminary step in any survival analysis problem is to visualize the data by plotting the estimated survivor function. PROC LIFETEST contains an array of graphical capabilities to facilitate this task. After defining one of the most popular estimators of the survivor function, we will highlight the formulaic modification introduced by unequal weights and then motivate an example analyzing marriage duration data in the 2006–2010 NSFG. Be advised that PROC LIFETEST should *only* be used for data visualization purposes. It is not designed to estimate measures of variability properly reflecting a complex survey design. Procedures with that capability will be discussed in subsequent sections.

Suppose that the data consist of E distinct event times $t_1 < t_2 < \cdots < t_E$. Let n_j be the number of sampling units at risk at time t_j, and let d_j be the number of sampling units who experience the event at time t_j. A sampling unit censored at time t_j is also considered to be at risk at that time. Then, for $t_1 \le t \le t_E$, the KM estimator of the survival function $S(t)$, also known as the "product-limit estimator," is defined by

$$\hat{S}(t) = \prod_{j:t_j \le t} \left(1 - \frac{d_j}{n_j}\right) \tag{7.7}$$

which is just the probability that the unit has not experienced the event by time t_j, conditional upon not experiencing the event at times t_1, t_2,..., or t_{j-1}. A few exceptions are as follows: (1) for $t < t_1$, times prior to the first event time observed in the data set, $\hat{S}(t) = 1$, and (2) for $t > t_E$, times surpassing the largest event time observed in the data set, $\hat{S}(t) = 0$ if there are no censored event times greater than t_E and $\hat{S}(t)$ is undefined otherwise. While there are no covariates explicitly stated in this formulation, we can partition the data into, say, K categories based on one or more covariates and estimate the survival function independently for each. We will see an example of this in Program 7.1.

For complex survey data sets with an analysis weight w_i affixed to the ith sampling unit's event time, the setup is generally the same, except that a weighted variant of the KM estimator in Equation 7.7 is produced as follows:

$$\hat{S}_w(t) = \prod_{j:t_j \le t} \left(1 - \frac{\hat{D}_j}{\hat{N}_j}\right) \tag{7.8}$$

where

$\hat{D}_j = \sum_{i=1}^{n} w_i I(t_i = t_j)$ can be interpreted as the estimated number of finite population units experiencing the event at time t_j

$\hat{N}_j = \sum_{i=1}^{n} w_i I(t_i \ge t_j)$ interpreted as the estimated number of finite population units still at risk at that time

To motivate an example, let us consider select NSFG marriage duration analyses presented in Copen et al. (2012). As we discovered in Chapter 4 when analyzing the 0/1 indicator variable EVRMARRY, 5,534 of the 12,279 female respondents reported wedding at least once. Although data on these respondents' full marriage histories are collected, for purposes of the forthcoming examples, we will restrict the focus to the duration of first marriages. The ultimate status of the respondent's first marriage can be ascertained from the variable MAREND01, which is missing if the marriage in still intact at time of interview, but coded as follows otherwise: 1 = divorce or annulment; 2 = separation but not officially divorced; or 3 = death of spouse. In fact, we can think of data for respondents who never married as *completely right censored* following the terminology of Leung et al. (1997). In reference to the visualization in Figure 7.1, these are individuals whose risk periods commence and terminate to the right of the end of the observation period. Because we lack data on these completely right-censored individuals, we will subset the data for only those records where EVRMARRY=1.

We will treat the "event" of interest as divorce or annulment, so separations and spousal deaths do not qualify. The DATA step syntax in Program 7.1 creates a 0/1 variable named CENSOR flagging these instances, as well as those in which the marriage is still intact at the time of interview, the other possible right-censoring mechanism. (This differs slightly from analyses in Copen et al. (2012), which treat the event of interest as a voluntary dissolution of marriage, a definition that also includes separations.) For respondents having EVRMARRY=1, the prepopulated variable MAR1DISS on the public-use file contains the number of months elapsed between the respondent's age at the onset of her first marriage and one of the MAREND01 events or the respondent's age at the time of interview, whichever came first.

Suppose that we were interested in examining the survivor function of marriage duration broken out by the four classifications of religious affiliation maintained in the variable RELIGION. Program 7.1 demonstrates PROC LIFETEST syntax to accomplish this task. Placing the variable RELIGION in the STRATA statement causes all pertinent statistics to be generated independently for each distinct value of that variable. Hence, it operates more like a BY statement than the STRATA statement in the SURVEY family of procedures.

The TIME statement is used to point PROC LIFETEST to the numeric variable housing the time until either the key event occurs or the time at which the observation is censored. To flag a censored observation, you separate the event time variable and the supplemental censoring indicator with an asterisk and identify the distinct code(s) in parentheses that indicate the event time was censored. The DATA step preceding PROC LIFETEST consolidates this information into a 0/1 indicator variable named CENSOR.

The WEIGHT statement first became available in PROC LIFETEST in Version 9.4. In earlier versions, the only way to account for the complex survey data's variable degrees of representation was to identify the weight variable in the FREQ statement. A minor wrinkle with that tactic is that PROC LIFETEST truncates the weight to the nearest integer. So long as the decimal portion truncated is trivial relative to the magnitude of the integer portion, the two approaches should provide more or less equivalent results, but

Berglund (2011) comments on a simple work-around to rescale the weights so that the truncation is no longer an issue. The NOTRUNCATE option has since been added to the FREQ statement, thereby dispelling any need for the work-around.

The KM estimator is the default, so the METHOD=KM option appearing in the PROC LIFETEST statement is technically optional. Though not shown here, a tabular summarization of pertinent statistics is sent to the listing alongside several test statistics assessing the null hypothesis of survival function (and related function) equality. We will not discuss those at present as the focus here is on visualizing the survival data. The PLOTS=(SURVIVAL(NOCENSOR)) option requests that SAS overlay the four KM estimators in one plot, synonymously labeled "Product-Limit" estimators. To reduce clutter, the NOCENSOR option is specified, which suppresses the default inclusion of a plus sign embedded along the survival curve to represent each censored observation. The MAXTIME=240 option sets the *x*-axis maximum at 240 months of marriage because the survival functions become rather noisy past that point since there are so few marriages in the data set exceeding 20 years.

Program 7.1: Plotting Weighted Survival Functions with PROC LIFETEST

```
data NSFG_F_MAR;
  set NSFG_0610_F;

* maintain only those respondents who have ever been married;
if EVRMARRY=1;

* assign an indicator of being censored observations if
individual is still married, separated (but not divorced), or
if the spouse died;
censor=(MAREND01 in(. 2 3));
run;

proc format;
  value RELIGION
    1='NO RELIGION'
    2='CATHOLIC'
    3='PROTESTANT'
    4='OTHER RELIGIONS';
run;

proc lifetest data=NSFG_F_MAR method=KM
plots=(survival(nocensor)) maxtime=240;
  time MAR1DISS*censor(1);
  strata RELIGION;
weight WGTQ1Q16;
format RELIGION RELIGION.;
run;
```

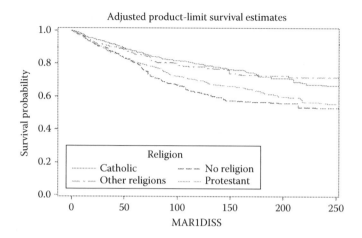

Though the estimated survivor functions plotted from Program 7.1 are not dramatically different from one another, we see that marriages in which the female is Catholic or affiliated with another non-Protestant religion have a higher survival probability than those in which the female indicates being Protestant or not religious. The gap tends to widen over time. For example, the respective probabilities of marriage lasting at least 50 months hardly differ; all fall roughly between 0.8 and 0.9. By the 150-month mark, however, the probabilities range from approximately 0.6 for nonreligious respondents to 0.8 for Catholic respondents.

Some practitioners prefer to interpret the survival data with respect to the *failure function*, found by taking 1 minus the survival function, or simply reversing the y-axis range, causing the curves to increase over time instead of decrease. You can specify PLOTS=(SURVIVAL(FAILURE)) to have PROC LIFETEST plot this transformation. Program 7.2 illustrates this alternative view for another covariate of interest in the STRATA statement, COHEVER, which is coded as a 1 if the respondent ever cohabited with an individual prior to marriage and 2 if she did not. The plot reveals how females cohabiting prior to their first marriage are much more likely to have that marriage end in divorce than are females who never cohabited prior to marriage. Specifically, at the 20-year mark, about one-half of those who cohabited beforehand experienced a divorce, whereas that figure is only about one-fifth for those who never cohabited.

Program 7.2: Plotting Weighted Failure Functions with PROC LIFETEST

```
proc format;
  value COHEVER
    1='YES'
    2='NO';
run;
```

```
proc lifetest data=NSFG_F_MAR method=KM
  plots=(survival(failure nocensor)) maxtime=240;
  time MAR1DISS*censor(1);
  strata COHEVER;
weight WGTQ1Q16;
format COHEVER COHEVER.;
run;
```

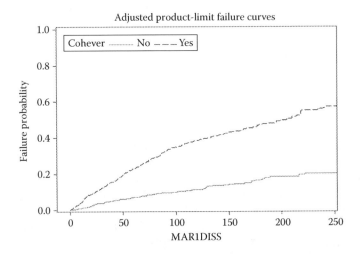

7.3.2 Fitting Cox Proportional Hazards Regression Models Using PROC SURVEYPHREG

In this section, we flesh out some of the basic concepts of survival time modeling and introduce the ability to account for covariates by way of the extraordinarily popular Cox proportional hazards regression approach (Cox, 1972). Despite the fact that there are model coefficients (i.e., betas) to be estimated, the Cox regression technique is labeled "semiparametric" because the survival time is not assumed to conform to any formal probability distribution. Following the order of exposition in the two previous chapters on modeling survey data, we will begin by covering the fundamentals within the context of simple random sampling and then introduce modifications necessary when adapted to the complex survey data setting.

The notion behind Cox regression is to model $h(t)$, the hazard function, as a function of p covariates. In particular, the assumption is that the hazard for the ith unit can be expressed as

$$h(t \mid \mathbf{x}_i) = h_0(t)\exp(\beta_1 x_1 + \beta_2 x_2 + \cdots + \beta_p x_p) \tag{7.9}$$

where $\mathbf{x}_i$ is a $1 \times p$ vector of covariates. In words, this means that the hazard equals a *baseline hazard* $h_0(t)$, a value common to all units, which is adjusted in proportion to an exponential function of the unit's covariates.

Evident from Equation 7.9 is that the hazard is not a *linear* function of covariates, but similarly to what we do in logistic regression, a logarithmic transformation is applied to convert it into one. Specifically, taking the natural logarithm of both sides, we get

$$\ln[h(t \mid \mathbf{x_i})] = \ln[h_0(t)] + \beta_1 x_1 + \beta_2 x_2 + \cdots + \beta_p x_p \tag{7.10}$$

Notice how there is no intercept term. Indeed, we will not see an intercept output by PROC SURVEYPHREG. The intercept, or value for a unit with all p covariates equal to 0, is absorbed into the $\ln[h_0(t)]$ term. Interestingly, Cox showed in his paper how that term need not be explicitly specified during the estimation process. We will see why shortly.

To understand how model parameters are to be interpreted, consider two units subscripted A and B identical except for one particular covariate indexed by j that is one unit greater for unit A. On the natural logarithm scale, the hazard difference between these two units can be written as

$$\ln[h(t \mid \mathbf{x_A}) - h(t \mid \mathbf{x_B})] = \beta_j x_{jA} - \beta_j x_{jB} = \frac{\beta_j x_{jA}}{\beta_j x_{jB}} = \frac{\beta_j(x_{jB} + 1)}{\beta_j(x_{jB})} = \beta_j \tag{7.11}$$

This is because of the property that $\ln(x - y) = \ln(x/y)$ and because all other terms (including $\ln[h_0(t)]$) cancel out in the difference. Hence, each model parameter represents the change in the log-hazard ratio resulting from a one-unit increase in the covariate. When employing the reference parameterization for a categorical covariate with K distinct values, the $K - 1$ nonzero parameters are interpreted as log-hazard changes relative to the omitted category. In either case, exponentiating the parameter returns a hazard ratio, which is easier to interpret. For example, $\exp(\beta_j) = 1.2$ implies that a one-unit increase in the corresponding predictor variable results in a 20% increase with respect to the hazard, which is basically the same interpretation as the relative risk statistic (see Section 4.3.3). SAS reports the exponentiated versions of the estimated betas by default under the "Hazard Ratio" column in the Analysis of Maximum Likelihood Estimates table.

The technique used to estimate the Cox regression model parameters is *partial likelihood* (Cox, 1972, 1975). Compared to maximum likelihood (e.g., the method used in Chapter 6 to estimate logistic regression parameters), which involves writing the likelihood function as the product of likelihoods for sampling units, partial likelihood calls for writing the likelihood function as the product of the likelihood of *events*. For a sample data set with E distinct events, if we denote the partial likelihood PL and the likelihood of the jth event L_j, this implies

$$PL = \prod_{j=1}^{E} L_j \tag{7.12}$$

The L_js, or event likelihoods, are formed by taking the ratio of the hazard for the unit experiencing that particular event to the sum of the hazards for units still in the risk set at that particular time. For the first event, this looks something like

$$L_1 = \frac{h_1(t_j)}{\sum_{i=1}^{n} h_i(t_j)} = \frac{h_0(t_j)\exp(\mathbf{x}_1\boldsymbol{\beta})}{\sum_{i=1}^{n} h_0(t_j)\exp(\mathbf{x}_i\boldsymbol{\beta})} = \frac{h_0(t_j)\exp(\mathbf{x}_1\boldsymbol{\beta})}{h_0(t_j)\exp(\mathbf{x}_1\boldsymbol{\beta}) + \cdots + h_0(t_j)\exp(\mathbf{x}_n\boldsymbol{\beta})}$$

(7.13)

where $\mathbf{x}_1$ denotes the covariate vector for the unit experiencing the event first. Notice how the same baseline hazard is a leading factor in the numerator and all terms in the denominator. Therefore, it cancels out of the likelihood, which explains why there is no need to explicitly specify it.

After writing down the L_js for all E distinct events at times $t_1 < t_2 < \cdots < t_E$, terms can be rearranged to express the partial likelihood as a function of the sample size as follows:

$$PL = \prod_{i=1}^{n} \left(\frac{\exp(\mathbf{x}_i\boldsymbol{\beta})}{\sum_{i'=1}^{n} I(t_{i'} \le t_i)\exp(\mathbf{x}_{i'}\boldsymbol{\beta})} \right)^{\delta_i}$$

(7.14)

where
  δ_i is 1 if the unit experienced the event and 0 if the unit was censored
  $I(t_{i'} \le t_i)$ is an indicator of whether the unit i' is still in the risk set at the time unit i experiences the event

Although the outer product encompasses the entire sample, note that the power term $\delta_i = 0$ for censored units effectively nullifies their contribution to the partial likelihood.

Just as was done in Chapter 6 for the logistic regression model, maximizing the natural logarithm of the likelihood makes things easier, and so Equation 7.14 becomes

$$\ln PL = \sum_{i=1}^{n} \delta_i \left(\mathbf{x}_i\boldsymbol{\beta} - \ln \sum_{i'=1}^{n} I(t_{i'} \le t_i)\exp(\mathbf{x}_{i'}\boldsymbol{\beta}) \right)$$

(7.15)

From here, just as was remarked about estimating logistic regression model parameters in Chapter 6, one of several available optimization techniques can be used, all of which should arrive at the same approximate solution. By default, PROC SURVEYPHREG uses the Newton–Raphson algorithm, but alternatives can be requested via the TECHNIQUE= option in the NLOPTIONS statement.

One nuance worth mentioning is that the likelihood in Equation 7.15 is only valid when there are no tied data, meaning no instances of two or more units sharing the exact same event time. When there are tied data, an approximation must be made. Allison (2010) discusses at length the mathematical reasoning and details of several approximations used in practice. SAS and most other competing software default to a technique proposed by Breslow (1974) that Allison (2010) asserts works well when there are few ties. When ties are abundant, he argues one of the alternatives you can specify in the TIES= option after the slash in the MODEL statement would be preferred, such as TIES=EFRON based on a technique discussed in Efron (1977). Of course, another way to handle this situation is to fit a discrete-time hazards model, which is the topic of the next section.

Now let us consider the impact a complex sample design has on fitting the Cox proportional hazards regression model. As we have done in previous modeling chapters, we will distinguish finite population model parameters with Roman letters instead of Greek letters. The setup of the partial likelihood function is comparable except that we must account for the weights. This is accomplished by amending Equation 7.14 to

$$PL = \prod_{i=1}^{n} \left(\frac{\exp(\mathbf{x}_i \mathbf{B})}{\sum_{i'=1}^{n} I(t_{i'} \leq t_i) \exp(\mathbf{x}_{i'} \mathbf{B})} \right)^{\delta_i w_i} \tag{7.16}$$

where w_i represents the ith sampling unit's weight. The quantity in Equation 7.16 can be thought of as a *pseudo-partial likelihood*, and we take the natural logarithm as before to facilitate the maximization process, which produces the following log-pseudo-partial likelihood function:

$$\ln PL = \sum_{i=1}^{n} \delta_i w_i \left(\mathbf{x}_i \mathbf{B} - \ln \sum_{i'=1}^{n} I(t_{i'} \leq t_i) \exp(\mathbf{x}_{i'} \mathbf{B}) \right) \tag{7.17}$$

If we denote the $p \times 1$ vector of model parameters $\hat{\mathbf{B}}$, the next step is to estimate the $p \times p$ covariance matrix $\text{cov}(\hat{\mathbf{B}})$. The Taylor series linearization approach to estimating $\text{cov}(\hat{\mathbf{B}})$ follows the same principles of Binder (1983) used previously for the logistic regression model, but we will skip the exceedingly intricate step-by-step derivation. The reader seeking more details is referred to the documentation, the Appendix of Mukhopadhyay (2010), and Binder (1990). Once we have $\hat{\mathbf{B}}$ and $\text{cov}(\hat{\mathbf{B}})$, we are able to test hypotheses of particular model coefficients and whether the model can be reduced in a procedurally equivalent manner to the methods demonstrated in Chapters 5 and 6.

Returning to the NSFG marriage duration analysis, suppose we sought to fit a Cox regression model of first marriage duration simultaneously

TABLE 7.1

Names and Coding Structure of Variables Used in the NSFG Marriage Duration
Cox Proportional Hazards Regression Model

Variable Name	Description	Coding Structure
MAR1DISS	Months between first marriage onset and dissolution (or respondent age at interview if still intact)	Integer
CENSOR	Indicator of MAR1DISS time being right censored	0 = Not censored 1 = Censored
RELIGION	Religious affiliation of female respondent	1 = No religion 2 = Catholic 3 = Protestant 4 = Other religions
HISPRACE	Race/ethnicity	1 = Hispanic 2 = Non-Hispanic White 3 = Non-Hispanic Black 4 = Other non-Hispanic races
COHEVER	Indicator of ever cohabiting prior to marriage	1 = Yes 2 = No
FMAR1AGE	Female respondent age at first marriage	Integer

controlling for religious affiliation, race/ethnicity, cohabitation prior to marriage, and the female's age at the time of marriage. Table 7.1 summarizes the naming conventions and coding structures of these four covariates, as well as the key outcome variable (MAR1DISS) and the censoring indicator (CENSOR).

Program 7.3 shows the PROC SURVEYPHREG syntax to fit the model. The dependent variable specification in the MODEL statement shares the same structure as the TIME statement in the PROC LIFETEST syntax from Programs 7.1 and 7.2. The outcome variable and censoring indicator variable are separated by an asterisk, and the particular code(s) of the latter flagging a censored observation are itemized in parentheses immediately thereafter. The PARAM=REF option after the slash in the CLASS statement invokes the reference parameterization. Reference categories are specified for RELIGION and HISPRACE in parentheses. For instance, the "no religion" category was chosen as the reference, permitting the exponentiated parameter estimates of the other three categorizations to be interpreted as the hazard ratio relative to those respondents who do not affiliate with a religion. For brevity, the output reported here is restricted to only model parameter estimates. As an example of how they are to be interpreted, we see that the hazard ratio for the parameter representing RELIGION=2 is 0.747, which means that the probability of a marriage ending in divorce for Catholics is $(1 - 0.747) \times 100 = 25.3\%$ less likely than that of respondents not affiliated with any kind of religion.

Program 7.3: Cox Proportional Hazards Regression with Complex Survey Data

```
proc surveyphreg data=NSFG_F_MAR;
  strata SEST;
  cluster SECU;
  class RELIGION (ref='1') HISPRACE (ref='2')
        COHEVER / param=ref;
  model MAR1DISS*censor(1) = RELIGION HISPRACE COHEVER
                             FMAR1AGE;
weight WGTQ1Q16;
run;
```

Analysis of Maximum Likelihood Estimates						
Parameter	DF	Estimate	Standard Error	t Value	Pr > |t|	Hazard Ratio
RELIGION 2	96	−0.291849	0.131720	−2.22	0.0291	0.747
RELIGION 3	96	−0.123318	0.100331	−1.23	0.2220	0.884
RELIGION 4	96	−0.065012	0.170721	−0.38	0.7042	0.937
HISPRACE 1	96	−0.361939	0.113645	−3.18	0.0020	0.596
HISPRACE 3	96	0.197634	0.109314	1.81	0.0737	1.219
HISPRACE 4	96	−0.054304	0.137458	−0.40	0.5937	0.947
COHEVER 1	96	1.400467	0.112406	12.46	<.0001	4.057
FMAR1AGE	96	−0.140112	0.011308	−12.39	<.0001	0.869

The Analysis of Maximum Likelihood Estimates table summarizes the results of parameter-specific hypothesis tests of the form $H_0: B_j = 0$ versus H_1: $B_j \neq 0$ for the $j = 1, 2,..., p$ parameters. The "t Value" column houses the test statistic $t = \hat{B}_j / \text{se}(\hat{B}_j)$ and the corresponding p value is given in the next column to the right, as determined from a student t distribution with complex survey degrees of freedom (calculated via the usual rule of thumb: the number of PSUs minus the number of strata). Recall that despite the parameter estimates being quantified on the natural logarithm scale, it still makes sense to test whether they are significantly different from 0. This is because of their interpretation as log-hazard ratios, which implies they can be exponentiated to convert into hazard ratios, and $\exp(B_j = 0) = 1$ indicates no difference in the two hazards constituting the ratio.

Similarly in spirit to what was done for the other modeling procedures, Table 7.2 is a concise summary of useful MODEL statement options that can be specified after the slash to modify or supplement the default statistics reported by PROC SURVEYPHREG.

You can use the ESTIMATE statement to construct customized or multi-parameter hypotheses. For example, notice how only one of the parameters associated with HISPRACE was significant at the $\alpha = 0.05$ level. Perhaps we sought to test whether this effect, on the whole, was truly an influential predictor variable. To answer this question, we can carry out an F test or a chi-square test in the same way we did for the linear and logistic

TABLE 7.2

Summary of Useful MODEL Statement Options Available after the Slash in PROC SURVEYPHREG

Option	Effect
ALPHA= *decimal*	Modifies default significance level for confidence intervals of model parameters and hazard ratios
CLPARM	Reports confidence limits for the model parameters at the α significance level specified in the ALPHA= option (α = 0.05 by default)
RISKLIMITS	Reports confidence limits for the hazard ratios at the α significance level specified in the ALPHA= option (α = 0.05 by default)
COVB	Outputs the model parameter covariance matrix
VADJUST= NONE	Omits the model parameter covariance matrix adjustment factor applied by default as discussed in Hidiroglou et al. (1980)
TIES=	Specifies the method to handle tied event times in partial likelihood computations—default is BRESLOW (Breslow, 1974) but an alternative is EFRON (Efron, 1977)

regression models. SAS will carry out all the pertinent computations; our main concern is to properly define the contrast vector **C**. Program 7.4 illustrates how to conduct an effect-level test for RELIGION. The JOINT option is needed to have **C** populated with the three columns demarcated by commas in the ESTIMATE statement. Otherwise, they will be treated as three separate $p \times 1$ vectors, resulting in three independent tests. The test statistic here is an F statistic with numerator degrees of freedom equaling the number of columns of **C** and denominator degrees of freedom equaling the complex survey degrees of freedom. You can request a chi-square version of the test, defined as the F statistic divided by the numerator degrees of freedom, by specifying JOINT(CHISQ).

From the output, we observe that the F statistic of 1.74 is insignificant at the α = 0.05 level, which leaves little doubt that all terms representing the effect of religion can be dropped from the model. Finally, be advised that the adjustment recommended by Korn and Graubard (1990) applies for test statistics generated via the ESTIMATE statement in PROC SURVEYPHREG. From Equation 5.17, the adjustment factor would be $(96 - 3 + 1)/96 \approx 1$ and so is inconsequential in the present case, but be advised that the same may not be true for alternative designs with fewer complex survey degrees of freedom.

Program 7.4: Testing Whether a Cox Proportional Hazards Regression Model Can Be Reduced

```
proc surveyphreg data=NSFG_F_MAR;
  strata SEST;
  cluster SECU;
  class RELIGION (ref='1') HISPRACE (ref='2')
      COHEVER / param=ref;
```

```
model MAR1DISS*censor(1) = RELIGION HISPRACE COHEVER
                          FMAR1AGE;
weight WGTQ1Q16;
estimate 'Test of Reduced Model - Dropping Religion'
         RELIGION 1 0 0, RELIGION 0 1 0, RELIGION 0 0 1 / joint;
run;
```

F Test for Estimates				
Label	Num DF	Den DF	*F* Value	Pr > *F*
Test of reduced model—dropping religion	3	96	1.74	0.1635

The proportional hazards assumption underlying the Cox regression modeling approach implies that the log-hazards are constant across the range of event times. As an example from the present model, this is to say that the effect of cohabiting is the same 1 year into a marriage as it is 10 years into a marriage. The plausibility of this assumption can be assessed visually or statistically. The visual method involves plotting $\ln(-\ln(S_w(t)))$ for the respective "levels" of the covariate as a function of $\ln(t)$. If the curves appear more or less parallel, then there is evidence that the proportional hazards assumption holds. On the other hand, clear departures from parallelism provide evidence that the effect of the covariate varies as a function of time.

The easiest way to generate these plots is by using the PLOTS=(LOGLOGS) option in the PROC LIFETEST statement while naming the covariate of interest the STRATA statement. Program 7.5 demonstrates syntax to investigate COHEVER, the indicator of the female ever having cohabited prior to her first marriage. The two lines are generally parallel throughout, so there is no evidence of a proportional hazards assumption violation for this particular covariate.

Program 7.5: Visually Inspecting Whether the Proportional Hazards Assumption Holds for a Predictor Variable Used in a Cox Regression Model

```
proc format;
  value COHEVER
    1='YES'
    2='NO';
run;

proc lifetest data=NSFG_F_MAR method=KM plots=(loglogs)
  maxtime=240;
  time MAR1DISS*censor(1);
  strata COHEVER;
weight WGTQ1Q16;
format COHEVER COHEVER.;
run;
```

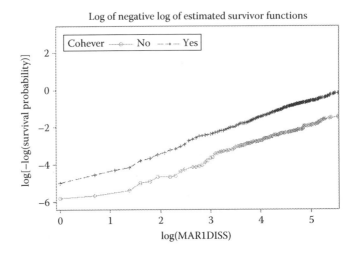

Log of negative log of estimated survivor functions

A more formal statistical approach for assessing the proportional hazards assumption is to create an additional model covariate defined as the product of event time and the covariate of interest, and then test whether its associated parameter(s) is significantly different from zero. If the parameter(s) is not significantly different from zero, there is no evidence against the proportional hazards assumption. Program 7.6 shows how to carry out this evaluation for the effect of cohabiting prior to one's first marriage. PROC SURVEYPHREG will not allow us to add a term such as COHEVER*MAR1DISS to the right of the equals sign in the MODEL statement because MAR1DISS has already been designated as the outcome variable. So, the onus is on us to create the appropriate interaction variables. Interestingly, however, we can do this within PROC SURVEYPHREG as it is one of the few SAS procedures permitting DATA step–type programming statements to create new variables on the fly.

Since COHEVER is dichotomous and we are exploiting the reference parameterization, there is only one nonzero parameter incorporating its effect into the model. Therefore, only one parameter is needed to introduce an interaction with the event time variable MAR1DISS. The new variable created is named COHEVERT, and from the output, we observe its parameter is not significantly different from zero, agreeing with what the visual investigation suggested, that the proportional hazards assumption appears to hold for this particular covariate.

Program 7.6: A Formal Statistical Test to Assess Whether the Proportional Hazards Assumption Holds for a Predictor Variable Used in a Cox Regression Model

```
proc surveyphreg data=NSFG_F_MAR;
  strata SEST;
  cluster SECU;
```

```
class HISPRACE (ref='2')
      COHEVER / param=ref;
model MAR1DISS*censor(1) = HISPRACE FMAR1AGE COHEVER
                          COHEVERT;
weight WGTQ1Q16;
COHEVERT=(COHEVER=1)*MAR1DISS;
run;
```

Analysis of Maximum Likelihood Estimates						
Parameter	DF	Estimate	Standard Error	*t* Value	Pr > |*t*|	Hazard Ratio
HISPRACE 1	96	−0.442960	0.098044	−4.52	<.0001	0.642
HISPRACE 3	96	0.197783	0.105743	1.87	0.0645	1.219
HISPRACE 4	96	−0.057598	0.134670	−0.43	0.6698	0.944
FMAR1AGE	96	−0.140940	0.011491	−12.27	<.0001	0.869
COHEVER 1	96	1.438096	0.197053	7.30	<.0001	4.213
COHEVERT	96	−0.000434	0.002186	−0.20	0.8430	1.000

There is a tendency for analysts to conclude that whenever the proportional hazards assumption fails to hold, the Cox regression approach is altogether invalidated. This is not exactly true. Rather, the implication is that some form of interaction with time needs to be factored into the model, although the downside in doing so is that the model becomes more complex and harder to interpret, especially when interaction terms are imperative for a large number of covariates.

7.3.3 Fitting Discrete-Time Hazards Regression Models Using PROC SURVEYLOGISTIC

In this section, we turn our attention to discrete-time hazards models, which can handle event times known only to have occurred within one of $m = 1,...,$ M distinct intervals. Even when time is measured on an effectively continuous scale, it can prove advantageous to coarsen, or discretize, time to permit one of these alternative models to be fitted. This is especially true in the presence of time-varying covariates. The other key benefit is that these models are naturally amenable to tied data. Recall that an approximation to the partial likelihood approach to fitting a Cox regression model is required when two or more event times are exactly the same. No such approximation is needed for the maximum likelihood approach used when fitting discrete-time hazards models.

The logistic form of the discrete-time hazards model is defined by

$$\ln\left(\frac{h_{mi}}{1-h_{mi}}\right) = \alpha_m + \beta_1 x_{1mi} + \cdots + \beta_p x_{pmi} = \alpha_m + \mathbf{x}_{\mathbf{mi}}\beta \qquad (7.18)$$

where h_{mi} is the hazard (i.e., probability) of the event occurring during the mth interval for the ith unit, conditional on the event not previously occurring. Under the parameterization in Equation 7.18, α_m represents a time-specific intercept term or the expected hazard during the mth interval for a sampling unit with all covariates equal to zero. Hence, there are M unique intercept terms to be estimated. When M is large, alternative parameterizations are possible to produce a more parsimonious model. For example, one can assume a linear trend by substituting $\alpha_0 + \alpha_1 m$ for the α_ms or a quadratic trend by substituting $\alpha_0 + \alpha_1 m + \alpha_2 m^2$. A visual technique to assess the plausibility of either of these two simplifying assumptions is to fit the "full" model, the one with the α_ms, and plot these estimated values on the y-axis as a function of m. If a linear trend is decipherable, the linear parameterization is probably sufficient. If the trend exhibits some curvature, the quadratic (or a higher-order) parameterization might be more appropriate.

We have already explored models of the type specified in Equation 7.18 in Chapter 6. To be sure, we will draw upon those same principles and the PROC SURVEYLOGISTIC syntax demonstrated previously. The major difference in the survival analysis context is that the survey data set must be reconfigured. The x_{mi} term in Equation 7.18 represents a time-specific covariate vector for the ith unit, the components of which need not be constant for each time period measured. For example, a few of the time-varying covariates incorporated into the discrete-time hazards model of federal employee retirements discussed in Lewis et al. (2014) were supervisory status and inflation-adjusted salary, both of which can change over an individual's employment history. Examples of time-invariant covariates accounted for in that study include the employee's retirement plan and gender.

The first task in fitting a discrete-time hazards model is to construct what is often referred to as a "person-period data set," a data set with one row for each period in which each unique sampling unit was at risk. It is easiest to envision in the context of a specific survival analysis problem. For example, periods in the Lewis et al. (2014) study were calendar years, and so an employee who retired in, say, the fifth year after becoming fully eligible would appear in the input data set five times. A dichotomous outcome variable included on the data set would indicate a nonevent in the first four periods and then an event in the fifth period. Right-censored individuals who do not retire by the end of the observation period are assigned a nonevent for all periods. As previously noted, certain covariates in the study were time invariant, while others were time varying, but either type was assigned to correspond to that individual's status at the start of the risk period (i.e., the calendar year).

Once the data are oriented into the person-period structure, standard methods of fitting a logistic regression model are applicable. Analysts unfamiliar with this technique are quick to voice their concerns over the seemingly unjust increase in sample size and the potential dependence introduced by taking repeated measures on the same sampling unit. If ignored, one might suspect that model parameter measures of uncertainty will be understated. Allison (1982) proves why this is not the case. Rather than working through the mathematical details, let us see an example by converting the NSFG_F_MAR data set into the person-period structure and fitting a model very similar to the Cox proportional hazards model fitted in the previous section.

The first DATA step in Program 7.7 creates a counter variable named INTERVALS defined as the number of 1-year periods a female respondent was married. Because there are a few instances of MAR1DISS, the number of months married prior to dissolution or being right censored, equaling 0, adding the fudge factor .000001 prior to executing the CEIL function, which rounds up the value in parentheses to the next integer, ensures each female respondent appears in the person-period data set at least once. The subsequent DO loop effectively appends $m - 1$ copies of each respondent's data in the output data set NSFG_F_MAR_PP. Whereas the original data set NSFG_F_MAR was comprised of 5,534 records, the person-period data set NSFG_F_MAR_PP is comprised of 44,188 records. After sorting this data set by CASEID, the unique respondent identifier on the publically available data set, the second DATA step in Program 7.7 assigns the binary outcome variable DISS to be 1 if the marriage ended in divorce in the respondent's last measured interval and 0 otherwise.

Figure 7.2 is provided to help visualize the person-period data set for three example respondents. Only a subset of the variables is shown. Note how all three variables reflecting complex survey features—SEST, SECU, and WGTQ1Q16—are identical for each record. This will generally be the case unless, for example, the longitudinal data were collected via a panel survey design in which two or more independent weight adjustment procedures (see Chapter 10) were employed to compensate for respondent attrition. We can reason that the respondent identified by CASEID=26165 experienced her marriage end in the fifth year and the respondent identified by CASEID=26166 experienced her marriage end in the second year. On the other hand, the respondent identified by CASEID=26167 was right censored during her 11th year of marriage. Therefore, the event for this individual never occurred, which is why all eleven of her records have DISS=0.

The PROC SURVEYPHREG run fits the same model from Program 7.3 using the original data set NSFG_F_MAR, whereas the PROC SURVEYLOGISTIC run fits a comparable discrete-time hazards model on the person-period data set NSFG_F_MAR_PP. Notice how the standard errors of parameters affixed are approximately equal in either case despite the person-period data set being roughly eight times as large. Though both sets of parameters are on the natural logarithm scale, recall that exponentiating a parameter yields a different interpretation in the SURVEYPHREG version (a hazard ratio) as compared to the SURVEYLOGISTIC version (odds ratio of a hazard). The fact that the parameters themselves are so close to one another is an artifact of the period-specific hazards being small in magnitude, in which case an odds ratio and a hazard ratio approximate one another. Note that the

CASEID	WGTQ1Q16	SECU	SEST	Censor	m	Diss
26165	4629.7533127	1	151	0	1	0
26165	4629.7533127	1	151	0	2	0
26165	4629.7533127	1	151	0	3	0
26165	4629.7533127	1	151	0	4	0
26165	4629.7533127	1	151	0	5	1
26166	921.41657056	1	156	0	1	0
26166	921.41657056	1	156	0	2	1
26167	11993.957144	2	114	1	1	0
26167	11993.957144	2	114	1	2	0
26167	11993.957144	2	114	1	3	0
26167	11993.957144	2	114	1	4	0
26167	11993.957144	2	114	1	5	0
26167	11993.957144	2	114	1	6	0
26167	11993.957144	2	114	1	7	0
26167	11993.957144	2	114	1	8	0
26167	11993.957144	2	114	1	9	0
26167	11993.957144	2	114	1	10	0
26167	11993.957144	2	114	1	11	0

FIGURE 7.2
Partial view of person-period data set NSFG_F_MAR_PP created to fit a discrete-time hazards model of NSFG first marriage duration data.

complementary log-log model is a variant of the logistic version in which an exponentiated model parameter *can* be interpreted as a hazard ratio. For brevity, and because the logistic transformation is arguably more popular, we will not explore that particular variant here. But note that you can request the complementary log-log model version using the LINK=CLOGLOG option after the slash in the MODEL statement of PROC SURVEYLOGISTIC. See Chapter 7 of Allison (2010) for more discussion.

Program 7.7: Comparing a Cox Proportional Hazards Model with a Discrete-Time Hazards Model

```
data NSFG_F_MAR_PP;
  set NSFG_F_MAR (keep=CASEID SEST SECU WGTQ1Q16
                       MAR1DISS censor
                       RELIGION HISPRACE COHEVER FMAR1AGE);
intervals=ceil(MAR1DISS/12 + .000001);
do m=1 to intervals;
   output;
end;
run;
```

```
proc sort data=NSFG_F_MAR_PP;
  by CASEID;
run;

data NSFG_F_MAR_PP;
  set NSFG_F_MAR_PP;
  by CASEID;
diss=0;
if last.CASEID and censor ne 1 then diss=1;
run;

title 'Cox Proportional Hazards Version';
proc surveyphreg data=NSFG_F_MAR;
  strata SEST;
  cluster SECU;
  class RELIGION (ref='1') HISPRACE (ref='2')
       COHEVER / param=ref;
  model MAR1DISS*censor(1) = RELIGION HISPRACE COHEVER
                            FMAR1AGE;
weight WGTQ1Q16;
run;

title 'Discrete-Time Hazards Version';
proc surveylogistic data=NSFG_F_MAR_PP;
  strata SEST;
  cluster SECU;
  class RELIGION (ref='1') HISPRACE (ref='2')
       COHEVER / param=ref;
  model diss (event='1') = m RELIGION HISPRACE COHEVER
                            FMAR1AGE / expb;
weight WGTQ1Q16;
run;
```

Cox Proportional Hazards Version

SURVEYPHREG Procedure

Analysis of Maximum Likelihood Estimates						
Parameter	DF	Estimate	Standard Error	*t* Value	Pr > |*t*|	Hazard Ratio
RELIGION 2	96	−0.291849	0.131720	−2.22	0.0291	0.747
RELIGION 3	96	−0.123318	0.100331	−1.23	0.2220	0.884
RELIGION 4	96	−0.065012	0.170721	−0.38	0.7042	0.937
HISPRACE 1	96	−0.361939	0.113645	−3.18	0.0020	0.696
HISPRACE 3	96	0.197634	0.109314	1.81	0.0737	1.219
HISPRACE 4	96	−0.054304	0.137458	−0.40	0.6937	0.947
COHEVER1	96	1.400467	0.112406	12.46	<.0001	4.057
FMAR1AGE	96	−0.140112	0.011308	−12.39	<.0001	0.869

Discrete-Time Hazards Version

SURVEYLOGISTIC Procedure

Analysis of Maximum Likelihood Estimates							
Parameter		DF	Estimate	Standard Error	Wald Chi-Square	Pr > ChiSq	Exp(Est)
Intercept		1	−0.7952	0.2896	7.5423	0.0060	0.451
m		1	−0.0506	0.0101	25.2296	<.0001	0.951
RELIGION	2	1	−0.2923	0.1340	4.7587	0.0291	0.747
RELIGION	3	1	−0.1211	0.1031	1.3809	0.2400	0.886
RELIGION	4	1	−0.0650	0.1737	0.1400	0.7083	0.937
HISPRACE	1	1	−0.3706	0.1163	10.1435	0.0014	0.690
HISPRACE	3	1	0.2018	0.1136	3.1571	0.0756	1.224
HISPRACE	4	1	−0.0512	0.1375	0.1385	0.7098	0.950
COHEVER	1	1	1.4174	0.1148	152.4943	<.0001	4.127
FMAR1AGE		1	−0.1451	0.0116	156.9893	<.0001	0.865

One final comment about Program 7.7: the reader may be surprised to find the two sets of parameter estimates so similar considering MAR1DISS is measured in months, whereas the discrete-time hazards approach measures time in 1-year intervals. Allison (2010, p. 129) discusses an interesting feature of the partial likelihood approach to estimating a Cox regression model intimating why this is to be expected:

> Another interesting property of the partial likelihood estimates is that they depend only on the *ranks* of the event times, not their numerical values. This implies any monotonic transformation of the event times will leave the coefficient estimates unchanged. For example, we could add a constant to everyone's time, multiply the result by a constant, take the logarithm, and then take the square root—all without producing the slightest change in the coefficients or their standard errors.

As we alluded to earlier in this section, a poignant advantage of the discrete-time hazards modeling approach is the ability to include time-varying predictor variables. To illustrate an example, suppose we sought to expand the model to account for the effect children have on the hazard of divorce. A female could enter into her first marriage with a child born previously out of wedlock or, more commonly, conceive a child with the spouse at some point during the marriage. Either way, a simple method to incorporate this factor is to insert a 0/1 indicator variable in each distinct record of the person-period data set to reflect the presence of a child at any point during the time interval.

Program 7.8 builds upon the code in Program 7.7 by creating an indicator variable named CHILDREN reflecting whether one or more child was in the picture during the given risk period. The variable is derived from chronological information stored in DATBABY1 and MARDAT01, the date of first live birth (missing if not applicable), and the marriage date of the female, respectively. The format of these two variables is "century month," which the NSFG codebook states is defined as the number of months that have passed

CASEID	MARDAT01	DATBABY1	m	Children	Diss
26180	1088	1141	1	0	1
26182	1125	1157	1	0	0
26182	1125	1157	2	0	0
26182	1125	1157	3	1	0
26182	1125	1157	4	1	0
26182	1125	1157	5	1	1
26183	1104	1153	1	0	0
26183	1104	1153	2	0	0
26183	1104	1153	3	0	0
26183	1104	1153	4	0	0
26183	1104	1153	5	1	0
26183	1104	1153	6	1	0
26183	1104	1153	7	1	0
26183	1104	1153	8	1	0
26183	1104	1153	9	1	0
26183	1104	1153	10	1	0

FIGURE 7.3
Partial view of person-period data set NSFG_F_MAR_PP2 created to fit a discrete-time hazards model of NSFG first marriage duration data with a time-varying covariate.

since January 1, 1900. It is unnecessary to convert them into SAS dates, however, because we can reference DATBABY1 against MARDAT01 plus 12 times the marriage interval number.

Figure 7.3 is a partial screenshot of the NSFG_F_MAR_PP2 data set, provided to help visualize these variables for a few example respondents. The first (CASEID=26180) respondent's marriage ended in divorce within the first year. No children were born in the first marriage, although we can gather the respondent did have a child at a later point in life since DATBABY1 is not missing and is greater than MARDAT01. The second (CASEID=26182) respondent had her first child during the third year of marriage, so her first two records have CHILDREN=0, but every record thereafter has CHILDREN=1.

After including the CHILDREN variable as a predictor variable in the model, we find its coefficient is significant and negative, which implies that the risk of divorce decreases when children are involved. (Because it was assigned as a 0/1 numeric variable, we did not need to include it in the CLASS statement.) The exponentiated coefficient is 0.755, which can be interpreted as meaning the odds are almost 25% lower.

Program 7.8: Fitting a Discrete-Time Hazards Model with a Time-Varying Covariate

```
data NSFG_F_MAR_PP2;
  set NSFG_F_MAR (keep=CASEID SEST SECU WGTQ1Q16
                  MAR1DISS censor
                  RELIGION HISPRACE COHEVER FMAR1AGE
                  DATBABY1 MARDAT01);
```

```
intervals=ceil(MAR1DISS/12 + .000001);
do m=1 to intervals;
   output;
end;
run;

proc sort data=NSFG_F_MAR_PP2;
  by CASEID;
run;

data NSFG_F_MAR_PP2;
  set NSFG_F_MAR_PP2;
  by CASEID;
* insert indicator of whether child is present during marriage
interval;
children=(DATBABY1 ne . and DATBABY1 le MARDAT01+(m*12));
diss=0;
if last.CASEID and censor ne 1 then diss=1;
run;

proc surveylogistic data=NSFG_F_MAR_PP2;
  strata SEST;
  cluster SECU;
  class RELIGION (ref='1') HISPRACE (ref='2')
        COHEVER / param=ref;
  model diss (event='1') = m RELIGION HISPRACE COHEVER
                           FMAR1AGE children / expb;
weight WGTQ1Q16;
run;
```

Analysis of Maximum Likelihood Estimates						
Parameter		DF	Estimate	Standard Error	Wald Chi-Square	Pr > ChiSq	Exp(Est)
Intercept		1	−0.6139	0.2976	4.2560	0.0391	0.541
m		1	−0.0421	0.0102	16.9472	<.0001	0.959
RELIGION	2	1	−0.2885	0.1343	4.6156	0.0317	0.749
RELIGION	3	1	−0.1139	0.1027	1.2298	0.2675	0.892
RELIGION	4	1	−0.0552	0.1738	0.1009	0.7508	0.946
HISPRACE	1	1	−0.3310	0.1170	7.9975	0.0047	0.718
HISPRACE	3	1	0.2424	0.1172	4.2757	0.0387	1.274
HISPRACE	4	1	−0.0150	0.1404	0.0115	0.9148	0.985
COHEVER	1	1	1.4445	0.1163	154.2196	<.0001	4.240
FMAR1AGE		1	−0.1482	0.0118	156.6396	<.0001	0.862
Children		1	−0.2815	0.0900	9.7859	0.0018	0.755

This concludes our discussion of discrete-time hazards models, but there are a few possible extensions worth mentioning before parting with the topic. One is that the time-varying covariates can even be contextual in nature.

For instance, in the retirement study discussed in Lewis et al. (2014), macro-economic indicators such as stock market performance and the unemployment rate were merged onto the person-period data by calendar year of the associated risk period. Another is that the underlying logistic regression model could be extended to handle an ordinal or multinomial outcome variable as was shown in Section 6.7.

The only downside to these flexibilities is that the person-period data set continues to grow and can become unwieldy with an exorbitant number of predictor variables or excessively minute intervals of time. Indeed, the KEEP statements in Programs 7.7 and 7.8 were included to manage the size of the data set because failing to do so was resulting in a significantly longer execution time.

7.4 Summary

This chapter began by motivating the need for a statistical model to explain T, the survival time of a sampling unit or, more generally, the duration between two clearly defined events. We discussed a few of the common data collection strategies that produce data amenable to one of these modeling approaches and defined the various types of censoring mechanisms introduced whenever the observation period does not cover the complete event histories of all sampling units. We also classified the modeling approaches used in practice, noting how the one chosen is often contingent upon the precision with which time is measured and whether the analyst is comfortable assuming T conforms to a formal probability distribution.

Examples were drawn from first marriage duration data collected as part of the NSFG. We demonstrated the graphical capabilities of PROC LIFETEST and the modeling capabilities of PROC SURVEYPHREG. For survival times known only to fall within one of a finite number of time intervals, we discussed the discrete-time hazards model, which can be fitted using PROC SURVEYLOGISTIC. Flexibilities of the discrete-time hazards model were noted, particularly its ability to account for time-varying covariates, or predictor variables such as salary, height, or weight, that can change over the course of a sampling unit's observation period.

As a structural aside, this chapter marks the end of the three-chapter sequence on modeling complex survey data. It also marks the end of the sequence of chapters (Chapters 2 through 7) that more or less highlights each of the six SURVEY procedures in turn. The two next two chapters deal with cross-cutting topics of survey data analysis: domain estimation and replication methods for variance estimation. As we will see, these topics are not confined to one particular SURVEY procedure.

8

Domain Estimation

8.1 Introduction

Most examples presented in previous chapters involved estimating quantities at the population level. For example, in Program 3.1, we estimated the total square footage of buildings in the commercial buildings energy consumption (CBECS) target population, and in Program 4.1, we estimated the proportion of females in the survey, national survey of family (NSFG) target population that have ever been married. This chapter covers *domain estimation*, a phrase referring to estimation and inference on a subpopulation, or portion, of the target population. To provide a few domain examples extending the scopes of the analyses in Programs 3.1 and 4.1, we might instead wish to estimate the total square footage of buildings by region of the country or tease apart the proportion of women who have ever married by some categorization of age.

Despite sounding like a clear-cut task, domain estimation is often carried out incorrectly in practice. Intuition leads the naïve analyst to use a BY statement or otherwise subset, or filter, the survey data for only those cases residing within the domain of interest. This chapter begins by discussing why this can be risky. The safer method is to employ the DOMAIN statement, which is available in every analytic SURVEY procedure except PROC SURVEYFREQ, in which the TABLES statement operates analogously to the DOMAIN statement (see Program 8.4). Depending on your version of SAS, there may be certain analyses for which the DOMAIN statement is not yet accommodated. Section 8.4 demonstrates a general workaround suggested by Graubard and Korn (1996) you can use in these circumstances.

In Section 8.5, we revisit a few examples from previous chapters, narrowing the focus to one or more specific domains. The purpose is to show how the domain estimation techniques generalize across a variety of contexts, finite population quantities, and SURVEY procedures. In addition to estimating domain statistics and their associated measures of uncertainty, we often want to carry out hypothesis tests comparing population parameters against one another. This is especially true for means. For instance, we may observe differences in the estimated mean of some quantity among two or more domains (e.g., males vs. females). A natural follow-up task is to test whether the differences observed are statistically significant. Section 8.6 is devoted exclusively to this analytic objective.

Finally, many researchers argue that you should reduce the degrees of freedom when estimating a quantity from a sparse domain, one not represented in all primary sampling units (PSUs) appearing on the complex survey data set. The chapter concludes with a few considerations and methods to apply these adjustments.

8.2 Definitions and an Example Data Set

To facilitate the exposition of domain estimation concepts, let us consider a hypothetical mathematics aptitude survey of students from a particular high school. Suppose the high school consists of four classes—grades 9 through 12—and that the students within each class are assigned to 1 of 10 homerooms, not necessarily of equal size. The objective is to select a sample of students on which to administer a mathematics examination at the start and end of school year. Supplemental information on these students will also be collected, such as whether or not they received any kind of mathematics tutoring.

Assume that a complex sample was designed exploiting class standing as the stratification factor. To simplify the data collection process, a sample of two homerooms was selected within each class, and ultimately, two students were selected within each sampled homeroom. Hence, a total of 16 students from the high school were sampled. A visualization of this population structure and corresponding sample design is provided in Figure 8.1. The population of clusters is represented by the collection of boxes, separated into four rows delineating the four strata. A colored box indicates the given cluster (i.e., homeroom) was sampled.

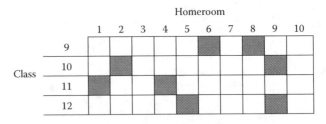

FIGURE 8.1
Visual representation of the stratified, clustered population and sample for the hypothetical mathematics aptitude survey.

Computing the mean test scores for the entire data set allows us to make inferences on the finite population of interest, the high school. Perhaps this is the primary analytic objective of the survey effort, but we might also wish to formulate estimates for each of the four classes or whether or not the student received tutoring. It is tempting to tackle these more concentrated analyses by simply filtering the survey data set for cases residing within the domain boundaries, or putting the domain identifier(s) in the BY statement of the SURVEY procedure. This is risky; depending upon the nature of the domain, doing so can lead one to make erroneous inferences. SAS will warn you when it detects potentially faulty subsetting in a SURVEY procedure. For example, in the presence of a WHERE statement, the following note appears in the log:

```
NOTE: The input data set is subset by WHERE, OBS, or
FIRSTOBS. This provides a completely separate analy-
sis of the subset. It does not provide a statisti-
cally valid subpopulation or domain analysis, where
the total number of units in the subpopulation is not
known with certainty. If you want a domain analysis,
you should include the domain variables in a DOMAIN
statement.
```

Generally speaking, the only harmless occasion to subset complex survey data is when the domain of interest is a *proper subset*, or a subset comprised of *all* sampling units from one or more strata. In terms of our hypothetical survey, this means we are free to subset the data set for students in a particular grade or any combination thereof. Figure 8.2 depicts two examples. The top figure portrays an analysis restricted to 10th grade students and the bottom figure portrays an analysis restricted to upperclassmen, those in either 11th or 12th grade.

Although proper subsets do arise in practice, more commonly the domain is represented to varying degrees among the PSUs and strata and not controlled for by the sample design. As an example, the number of students selected into the sample who received mathematics tutoring likely varies from one sample to the next, in which case the domain sample size is itself a random variable. Estimating a mean where the denominator is a random variable introduces the same computational nuances that led us to invoke Taylor series linearization as discussed in Sections 3.3 and 3.4. Simply subsetting the data set is tantamount to ignoring the domain sample size uncertainty, assuming it to be fixed over all possible samples. As suggested by the note SAS sends to the log, the built-in method to properly account for that variability is to specify the domain identifier variable(s) in the DOMAIN statement. Point estimates generated by the DOMAIN

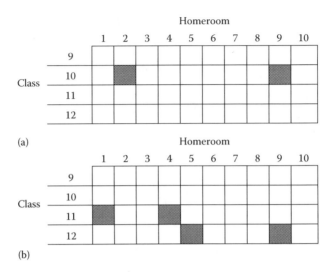

FIGURE 8.2
Visual representation of two proper subset examples in the stratified, cluster sample design employed in the hypothetical mathematics aptitude survey: Estimation of students in 10th grade (a) and Estimation of upperclassmen (b).

statement always match point estimates generated from subsetting, but measures of variability and degrees of freedom may differ.

8.3 The Risk in Subsetting a Complex Survey Data Set

Program 8.1 begins by reading in data from the hypothetical mathematics aptitude survey into a temporary SAS data set named TEST. Features of the complex sample design are maintained in the following variables:

- CLASS—the stratification factor, the high school grade of the student, ranging from 9 to 12
- HOMEROOM—the cluster identifier, a student's homeroom, coded as either a 1 or 2 that, in combination with codes of variable CLASS, defines the four distinct PSUs
- WEIGHT—the base weight, or inverse of the student's selection probability

The numeric variables GRADE1 and GRADE2 contain the student's mathematics examination results at the start and end of the school year, respectively. The maximum grade on the examination is 100. The Y/N character variable TUTOR is an indicator of whether the student received mathematics tutoring of any kind during the school year.

After reading in the raw survey data, Program 8.1 demonstrates the analysis of two proper subsets: year-end exam scores for underclassmen and upperclassmen. The analysis is conducted in two ways. The first places the variable CLASS_TYPE, a dichotomization of the stratum identifier CLASS, in the BY statement of PROC SURVEYMEANS, after first sorting the data set by this variable. Using a BY statement with a variable (or combination of variables) consisting of K distinct categories is akin to submitting K completely separate runs of PROC SURVEYMEANS, each filtered beforehand for cases in the kth category, in turn. This is evident from the Data Summary table, which reports that only one-half of the known strata and cluster counts were used for each of the two iterations. The estimated mean of the GRADE2 variable for underclassmen is 85.58 with standard error 1.5137, and for upperclassmen, these quantities are 92.50 and 1.6532, respectively.

The second PROC SURVEYMEANS run demonstrates the advocated approach, placing the domain identifier variable in the DOMAIN statement. The full-sample analysis is reported in the output first, followed by the same analysis for each distinct value of the domain variable. An identical process occurs for all SURVEY procedures, not just PROC SURVEYMEANS (see Section 8.5 for a few examples). Note how the point estimates and standard errors appearing in the Domain Analysis portion of the output match those generated from the first PROC SURVEYMEANS run. Again, this is to be expected considering that the two domains are proper subsets of the complex survey data set.

A quick aside before proceeding: to maintain tractability, examples in this chapter specify only a single variable in the DOMAIN statement, but you are free to specify more than one. When separated by spaces, domain estimation is conducted at the marginal level of each variable listed (i.e., based on each distinct value of the given variable). When separated by asterisks, such as DOMAIN VAR1*VAR2, domain estimation is conducted for cases defined by the cross-tabulation of values from VAR1 and VAR2. You can also use parentheses like you can in, say, the TABLES statement of PROC FREQ.

Program 8.1: Analysis of a Domain Characterized as a Proper Subset

```
data test;
  input class class_type $ homeroom grade1 tutor $ grade2
weight;
datalines;
12 upper 1 87 Y 94 49.5
12 upper 1 89 N 89 49.5
12 upper 2 91 Y 94 48
12 upper 2 84 Y 92 48
```

```
11 upper 1 82 N 84 47.5
11 upper 1 94 N 95 47.5
11 upper 2 93 N 95 48
11 upper 2 94 Y 97 48
10 under 1 78 N 81 39
10 under 1 84 N 84 39
10 under 2 90 N 87 37.5
10 under 2 82 N 85 37.5
 9 under 1 88 N 88 40
 9 under 1 93 Y 91 40
 9 under 2 77 Y 85 48
 9 under 2 81 N 84 48
;
run;

title '1) Analyzing Proper Subsets with the BY statement';
proc sort data=test;
  by class_type;
run;
proc surveymeans data=test mean stderr;
  by class_type;
  stratum class;
  cluster homeroom;
  var grade2;
weight weight;
run;

title '2) Analyzing Proper Subsets with the DOMAIN statement';
proc surveymeans data=test mean stderr;
  stratum class;
  cluster homeroom;
  var grade2;
weight weight;
domain class_type;
run;
```

1. Analyzing Proper Subsets with the BY Statement

SURVEYMEANS Procedure

class_type=under

Data Summary	
Number of strata	2
Number of clusters	4
Number of observations	8
Sum of weights	329

Statistics		
Variable	Mean	Std Error of Mean
grade2	85.583587	1.513723

1. Analyzing Proper Subsets with the BY Statement

SURVEYMEANS Procedure

class_type=upper

Data Summary	
Number of strata	2
Number of clusters	4
Number of observations	8
Sum of weights	386

Statistics		
Variable	Mean	Std Error of Mean
grade2	92.500000	1.653268

2. Analyzing Proper Subsets with the DOMAIN Statement

SURVEYMEANS Procedure

Domain Statistics in class_type			
class_type	Variable	Mean	Std Error of Mean
Under	grade2	85.583587	1.513723
Upper	grade2	92.500000	1.653268

Program 8.2 illustrates the impact on measures of variability in a nonproper subset setting. Here, the objective is to estimate the mean year-end exam score for the domain of students who received any kind of mathematical tutoring. We can do so by restricting analysis to records in the data set TEST where TUTOR='Y'. The first PROC SURVEYMEANS run filters those records via a WHERE statement. The second run does not subset the input data set in any way; instead, the TUTOR variable is specified in the DOMAIN statement as prescribed.

Program 8.2: Analysis of a Domain Characterized by a Random Sample Size

```
title '1) Improperly Analyzing a Domain by Subsetting the Data
Set';
proc surveymeans data=test mean stderr;
  where tutor='Y';
  stratum class;
  cluster homeroom;
  var grade2;
weight weight;
run;


title '2) Correctly Analyzing a Domain with the DOMAIN
Statement';
```

```
proc surveymeans data=test mean stderr;
  stratum class;
  cluster homeroom;
  var grade2;
weight weight;
domain tutor;
run;
```

1. Improperly Analyzing a Domain by Subsetting the Data Set

SURVEYMEANS Procedure

Data Summary	
Number of strata	3
Number of clusters	5
Number of observations	6
Sum of weights	281.5

Statistics		
Variable	Mean	Std Error of Mean
grade2	92.209591	1.058435

2. Correctly Analyzing a Domain with the DOMAIN Statement

SURVEYMEANS Procedure

Domain Statistics in Tutor			
Tutor	Variable	Mean	Std Error of Mean
N	grade2	87.439446	0.881354
Y	grade2	92.209591	1.336977

The point estimate is the same in either case, but the standard errors differ. Subsetting produces a standard error of 1.0584, whereas the standard error is 1.3370 when the DOMAIN statement is employed. The discrepancy is caused by the fact that the TUTOR='Y' condition does not occur in all PSUs. The SURVEY procedure is only aware of the stratum and PSU codes for cases meeting the WHERE statement condition, which excludes the 10th grade stratum altogether and one of the two PSUs in the 11th grade stratum. For variance estimation purposes, at least under the default Taylor series linearization method (alternatives are discussed in Chapter 9), there must be at least two PSUs in a stratum. When only one is detected, the procedure still runs, but the following message is sent to the log:

```
NOTE: Only one cluster in a stratum for variable(s)
grade2. The estimate of variance for grade2 will omit
this stratum.
```

SAS is alerting us that the overall variance estimate is ignoring one or more strata. The overall variance associated with a domain statistic is found by summing stratum-specific variances from strata where the domain is present in at least one PSU. Because these stratum-specific variances are generally greater than zero, ignoring them will lead to an underestimated variance. They are not ignored when the DOMAIN statement is used despite the same ominous note sent to the log in more recent versions of SAS. We will see why when we delve into the underlying computational details in the next section.

8.4 Domain Estimation Using Domain-Specific Weights

Although the DOMAIN statement is available in almost every SURVEY procedure, analysts are bound to encounter estimation tasks for which the DOMAIN statement is not available. One historical example is quantile estimation, which first became available in SAS in Version 9.2, but domain estimation of quantiles was not accommodated until Version 9.4. Prior to that time, if any statistical keyword related to a quantile such as MEDIAN or QUARTILES was specified in the PROC SURVEYMEANS statement as part of a step that also included a DOMAIN statement, only the population-level analyses were output, and the following warning message was sent to the log: "Quantiles are not available for domain analysis." We will demonstrate shortly a simple solution suggested by Graubard and Korn (1996) to circumvent this kind of barrier.

Before we work through a specific example, however, let us reinforce some of the concepts outlined in the previous section with more theoretical details pertaining to variance estimation in the domain estimation context. Following the notation from Section 3.3, let w_{hij} denote the weight for the jth unit in the ith PSU from the hth stratum, and let y_{hij} denote the outcome variable (e.g., mathematics examination grade). The weighted mean for the dth domain (e.g., students who received tutoring) can be expressed as

$$\hat{\bar{y}}_d = \frac{\hat{Y}_d}{\hat{N}_d} = \frac{\sum_{h=1}^{H} \sum_{i=1}^{n_h} \sum_{j=1}^{m_{hi}} w_{hij} d_{hij} y_{hij}}{\sum_{h=1}^{H} \sum_{i=1}^{n_h} \sum_{j=1}^{m_{hi}} w_{hij} d_{hij}} \qquad (8.1)$$

where
H is the number of strata
n_h is the number of PSUs selected from the hth stratum
m_{hi} is the number of units selected from the n_hth PSU
d_{hij} is a 0/1 indicator variable of the given unit falling within the particular domain

We can conceptualize this weighted mean as the ratio of an estimated domain total, $\hat{Y}_d$, to the estimated total number of cases in that domain in the population, $\hat{N}_d$. To obtain a variance estimate, we can follow the same principles of Taylor series linearization detailed in Chapter 3. Specifically, the variance is estimated by

$$\text{var}\left(\frac{\hat{Y}_d}{\hat{N}_d}\right) \approx \frac{1}{\hat{N}_d^2}\left[\text{var}(\hat{Y}_d) + \left(\frac{\hat{Y}_d}{\hat{N}_d}\right)^2 \text{var}(\hat{N}_d) - 2\left(\frac{\hat{Y}_d}{\hat{N}_d}\right)\text{cov}(\hat{Y}_d, \hat{N}_d)\right] \quad (8.2)$$

where, ignoring the finite population correction (FPCs) for the moment,

$$\text{var}(\hat{Y}_d) = \sum_{h=1}^{H}\frac{n_h}{(n_h-1)}\sum_{i=1}^{n_h}\left(\sum_{j=1}^{m_{hi}}w_{hij}d_{hij}y_{hij} - \frac{\sum_{i=1}^{n_h}\sum_{j=1}^{m_{hi}}w_{hij}d_{hij}y_{hij}}{n_h}\right)^2 \quad (8.3)$$

and

$$\text{var}(\hat{N}_d) = \sum_{h=1}^{H}\frac{n_h}{(n_h-1)}\sum_{i=1}^{n_h}\left(\sum_{j=1}^{m_{hi}}w_{hij}d_{hij} - \frac{\sum_{i=1}^{n_h}\sum_{j=1}^{m_{hi}}w_{hij}d_{hij}}{n_h}\right)^2 \quad (8.4)$$

and

$$\text{cov}(\hat{Y}_d, \hat{N}_d) = \sum_{h=1}^{H}\frac{n_h}{(n_h-1)}\sum_{i=1}^{n_h}\left(\sum_{j=1}^{m_{hi}}w_{hij}d_{hij}y_{hij} - \frac{\sum_{i=1}^{n_h}\sum_{j=1}^{m_{hi}}w_{hij}d_{hij}y_{hij}}{n_h}\right)$$

$$\times\left(\sum_{j=1}^{m_{hi}}w_{hij}d_{hij} - \frac{\sum_{i=1}^{n_h}\sum_{j=1}^{m_{hi}}w_{hij}d_{hij}}{n_h}\right) \quad (8.5)$$

Scrutinizing these formulas for a moment, we can extract theoretical support for the risks asserted about subsetting. First, if the domain appears in only one PSU within a stratum, the SURVEY procedure sees $n_h = 1$, which introduces a multiplicative factor of zero in the denominators of Equations 8.3 through 8.5. Therefore, all three terms are undefined. Second, we can reason that the stratum-specific contribution to the overall variance estimate of the domain statistic will be nonzero even if the domain is represented by a single PSU in the stratum. To see why, imagine a simple example where $n_h = 2$, each with $m_{hi} = 1$, but the domain characteristic is present in only one of the two PSUs. Except in rare circumstances, the squared deviations in, say, Equation 8.3, are nonzero for *both* PSUs, even the one that does not reside in the domain, so we do not want to omit its contribution. Comparable arguments can be made for both Equations 8.4 and 8.5. Lastly, even though we omitted them in Equations 8.3 through 8.5, subsetting the data could disturb any FPC terms $(1 - n_h/N_h)$ when they are applicable. Depending on whether the N_hs were assigned via the RATE= option or the TOTAL= option in the PROC statement, they could be either smaller or larger than they technically should. As a general rule, FPCs are based on the overall stratum sample size relative to the overall stratum population size, not the domain-specific stratum sample size relative to the domain-specific stratum population size.

Notice how the terms w_{hij} and d_{hij} are always bundled together as $w_{hij}d_{hij}$ throughout the three equations. We can interpret this as if a domain-specific weight $w_{dhij} = w_{hij}d_{hij}$ is created for purposes of estimating the dth domain's statistics and measures of uncertainty, a weight equaling the original weight if the unit is in the domain and zero otherwise. From there, the traditional formulas from Chapter 3 would apply, as would those for any other statistics. Therefore, if we were interested in estimating a particular domain mean, it would seem feasible to create the domain-specific weight in a DATA step and utilize that in the WEIGHT statement of the SURVEY procedure, foregoing the DOMAIN statement altogether. This will not work, however, because any SURVEY procedure eliminates all observations with a weight less than or equal to 0 at the outset. The workaround suggested by Graubard and Korn (1996) is to assign units outside the domain a weight that is miniscule, yet strictly greater than zero.

Recall how prior to Version 9.4 the DOMAIN statement was not available when estimating quantiles. Program 8.3 demonstrates a workaround one could use to estimate the median year-end mathematics examination grade for the domain of students who received any kind of tutoring. A new data set named TEST_D is created from the original survey data set TEST and populated with a domain-specific weight WEIGHT_D. Note that placing a condition in parentheses returns 0 if false and 1 if true. The net effect is that WEIGHT_D is assigned the full-sample weight for cases where TUTOR='Y' and a trivial, though positive, weight otherwise. The keywords MEAN and STDERR are specified such that the output can be cross-referenced with that generated from the PROC SURVEYMEANS run in Program 8.2 where TUTOR was specified in the DOMAIN statement. We can see how the standard errors in either run are identical (approximately 1.3370).

Program 8.3: Creating and Utilizing a Domain-Specific Weight for a Domain Analysis

```
data test_d;
  set test;
weight_d=weight*(tutor='Y') + .0000000000001;
run;

proc surveymeans data=test_d mean stderr median;
  stratum class;
  cluster homeroom;
  var grade2;
weight weight_d;
run;
```

Data Summary	
Number of strata	4
Number of clusters	8
Number of observations	16
Sum of weights	281.5

Statistics		
Variable	Mean	Std Error of Mean
grade2	92.209591	1.336977

Quantiles					
Variable	Percentile		Estimate	Std Error	95% Confidence Limits
grade2	50%	Median	92.097436	1.648637	87.5200856	96.6747862

8.5 Domain Estimation for Alternative Statistics

In this section, we revisit a few examples from prior chapters, refining population-level estimation tasks down to domain estimation tasks. The intent is to demonstrate how the fundamental concepts and syntax rules carry over intact to any finite population quantity, not just sample means and quantiles discussed thus far in the chapter.

In Program 3.2, we estimated the total number of buildings in the CBECS 2003 target population using district steam. Perhaps there was an interest in estimating that total by U.S. region. Program 8.4 accomplishes this task by amending the original PROC SURVEYMEANS syntax with a DOMAIN statement in which the region identifier variable REGION8 is specified. A format is applied to the region identifier such that a more informative label is printed in the output.

Notice how the sum of the four domain-specific estimated totals equals the overall estimated total, or 14,439 + 8,286 + 14,030 + 10,351 = 47,106. The second part of Program 8.5 shows how the same results can be obtained from PROC SURVEYFREQ despite the fact that a DOMAIN statement is not available in that procedure. Essentially, any variable (or combination of variables) specified to the left of an asterisk in a TABLES statement is treated as if it had appeared in a DOMAIN statement, so rest assured that PROC SURVEYFREQ will properly estimate the variances of domain totals.

Program 8.4: Estimating Totals within Domains

```
proc format;
  value YESNO
    1='Yes'
    2='No';

  value REGION
    1='Northeast'
    2='Midwest'
    3='South'
    4='West';
run;

proc surveymeans data=CBECS_2003 nobs sum;
  strata STRATUM8;
  cluster PAIR8;
  class STUSED8;
  var STUSED8;
weight ADJWT8;
domain REGION8;
format STUSED8 YESNO. REGION8 REGION.;
run;

* an equivalent analysis using PROC SURVEYFREQ;
proc surveyfreq data=CBECS_2003;
  strata STRATUM8;
  cluster PAIR8;
  tables REGION8*STUSED8 / nopercent;
weight ADJWT8;
format STUSED8 YESNO. REGION8 REGION.;
run;
```

SURVEYMEANS Procedure

Data Summary	
Number of strata	44
Number of clusters	88
Number of observations	5,215
Sum of weights	4,858,749.82

Complex Survey Data Analysis with SAS®

Class Level Information		
CLASS Variable	Levels	Values
STUSED8	2	Yes No

Statistics				
Variable	Level	N	Sum	Std Dev
STUSED8	Yes	265	47,106	7,602.878471
	No	4950	4,811,644	239,130

Domain Statistics in REGION8					
REGIONS	Variable	Level	N	Sum	Std Dev
Northeast	STUSED8	Yes	82	14,439	3,441.342497
		No	810	746,707	85,320
Midwest	STUSED8	Yes	59	8,286.370000	3,108.224143
		No	1322	1,296,453	131,788
South	STUSED8	Yes	82	14,030	4,537.608964
		No	1887	1,859,165	172,366
West	STUSED8	Yes	42	10,351	4,322.839074
		No	931	909,319	74,058

SURVEYFREQ Procedure

Data Summary	
Number of strata	44
Number of clusters	88
Number of observations	5,215
Sum of weights	4,858,749.82

Table of REGIONS by STUSED8				
REGIONS	STUSED8	Frequency	Weighted Frequency	Std Dev of Wgt Freq
Northeast	Yes	82	14,439	3,441
	No	810	746,707	85,320
	Total	892	761,145	86,534
Midwest	Yes	59	8,286	3,108
	No	1322	1,296,453	131,788
	Total	1381	1,304,739	131,990
South	Yes	82	14,030	4,538
	No	1887	1,859,165	172,366
	Total	1969	1,873,195	171,492
West	Yes	42	10,351	4,323
	No	931	909,319	74,058
	Total	973	919,670	74,197
Total	Yes	265	47,106	7,603
	No	4950	4,811,644	239,130
	Total	5215	4,858,750	239,288

Let us now return to the model of time spent with the physician (TIMEMD) from Chapter 5, the reduced model after eliminating insignificant predictor variables. Remaining in the model were the following factors: whether the doctor is the lone physician in the practice (SOLO), the primary reason for the visit (MAJOR), patient age (AGE), and the total number of chronic conditions afflicting the patient (TOTCHRON). It turns out that for a small subset of visits on the NAMCS 2010 public-use data set—854 out of 31,229—the physician and patient did not actually see one another. These visits are flagged by the 0/1 indicator variable PHYS, where a 1 signifies the physician met with the patient. For records where PHYS=0, the TIMEMD variable is 0. Hence, we may wish to refit the model from Chapter 5 controlling for the domain defined by PHYS=1.

Program 8.5 demonstrates how to do so using the DOMAIN statement in PROC SURVEYREG. Just like other procedures, PROC SURVEYREG first conducts an overall analysis and then repeats for each distinct level of the DOMAIN identifier variable(s). Only the output components housing the estimated model parameters are provided here. Both the parameters and standard errors for the domain defined by PHYS=0 are all zero because TIMEMD is always 0 in that instance, but notice how the parameters have shifted somewhat with the 854 records of all 0s excluded. The second part of Program 8.5 reproduces these results following the domain-specific weight approach suggested by Graubard and Korn (1996). This is illustrated to reinforce that their approach generalizes to other analyses.

Program 8.5: Fitting a Linear Model within Domains

```
title1 '1) Illustrating the DOMAIN Statement in PROC
SURVEYREG';
proc surveyreg data=NAMCS_2010;
  strata CSTRATM;
  cluster CPSUM;
  class SOLO MAJOR;
  model TIMEMD = SOLO MAJOR
                 AGE TOTCHRON / solution;
weight PATWT;
domain PHYS;
run;

title1 '2) Illustrating the Domain-Specific Weight Approach
for Estimating Regression Coefficients';
data NAMCS_2010_d;
  set NAMCS_2010;
PATWT_d=PHYS*PATWT + .0000000000001;
run;
```

```
proc surveyreg data=NAMCS_2010_d;
   strata CSTRATM;
   cluster CPSUM;
   class SOLO MAJOR;
   model TIMEMD = SOLO MAJOR
                  AGE TOTCHRON / solution;
weight PATWT_d;
run;
```

1. Illustrating the DOMAIN Statement in PROC SURVEYREG

PHYS=0

Tests of Model Effects			
Effect	Num DF	F Value	Pr > F
Model	1	0.00	1.0000
Intercept	0		
SOLO	0		
MAJOR	0		
AGE	0		
TOTCHRON	0		
Note: The denominator degrees of freedom for the F tests is 552.			

Estimated Regression Coefficients				
Parameter	Estimate	Standard Error	t Value	Pr > |t|
Intercept	0	0		
SOLO 1	0	0		
SOLO 2	0	0		
MAJOR 1	0	0		
MAJOR 2	0	0		
MAJOR 3	0	0		
MAJOR 4	0	0		
MAJOR 5	0	0		
AGE	0	0		
TOTCHRON	0	0		

PHYS=1

Tests of Model Effects			
Effect	Num DF	F Value	Pr > F
Model	7	7.33	<.0001
Intercept	1	1361.31	<.0001

(Continued)

PHYS=1

Tests of Model Effects			
Effect	Num DF	F Value	Pr > F
SOLO	1	3.80	0.0518
MAJOR	4	8.62	<.0001
AGE	1	3.76	0.0531
TOTCHRON	1	10.27	0.0014

Note: The denominator degrees of freedom for the F tests is 552.

Estimated Regression Coefficients				
Parameter	Estimate	Standard Error	t Value	Pr > |t|
Intercept	19.7897450	0.56642195	34.94	<.0001
SOLO 1	1.5219434	0.78102878	1.95	0.0518
SOLO 2	0.0000000	0.00000000		
MAJOR 1	−1.0414505	0.60738851	−1.71	0.0870
MAJOR 2	−0.7595231	0.83130481	−0.91	0.3613
MAJOR 3	0.3745995	0.84882931	0.44	0.6592
MAJOR 4	−3.7441933	0.86567683	−4.33	<.0001
MAJOR 5	0.0000000	0.00000000		
AGE	0.0233785	0.01206302	1.94	0.0531
TOTCHRON	0.3244247	0.10124397	3.20	0.0014

2. Illustrating the Domain-Specific Weight Approach for Estimating Regression Coefficients

Tests of Model Effects			
Effect	Num DF	F Value	Pr > F
Model	7	7.33	<.0001
Intercept	1	1361.31	<.0001
SOLO	1	3.80	0.0518
MAJOR	4	8.62	<.0001
AGE	1	3.76	0.0531
TOTCHRON	1	10.27	0.0014

Note: The denominator degrees of freedom for the F tests is 552.

Estimated Regression Coefficients				
Parameter	Estimate	Standard Error	t Value	Pr > |t|
Intercept	19.7897450	0.56642196	34.94	<.0001
SOLO 1	1.5219434	0.78102878	1.95	0.0518
SOLO 2	0.0000000	0.00000000		
MAJOR 1	−1.0414505	0.60738851	−1.71	0.0870
MAJOR 2	−0.7595231	0.83130481	−0.91	0.3613

(Continued)

Estimated Regression Coefficients				
Parameter	Estimate	Standard Error	t Value	Pr > $|t|$
MAJOR 3	0.3745995	0.84882931	0.44	0.6592
MAJOR 4	−3.7441933	0.86567683	−4.33	<.0001
MAJOR 5	0.0000000	0.00000000		
AGE	0.0233785	0.01206302	1.94	0.0531
TOTCHRON	0.3244247	0.10124397	3.20	0.0014

8.6 Significance Testing for Domain Mean Differences

After obtaining domain point estimates and measures of uncertainty, a commonly ensuing objective is to seek to determine whether the differences observed are statistically significant. This is especially true when the point estimates at hand are sample means. To that end, this section of the chapter is devoted exclusively to illustrating a few of the tools and methods you can employ to conduct significance tests on means.

Let us begin by defining some notations and reminding ourselves of few fundamental formulas. Suppose we have estimated two domain means $\hat{\bar{y}}_1$ and $\hat{\bar{y}}_2$, and that we wish to conduct the hypothesis test H_0: $\bar{y}_1 = \bar{y}_2$ versus H_1: $\bar{y}_1 \neq \bar{y}_2$. To assess significance, we can construct a t statistic:

$$t = \frac{\hat{\bar{y}}_1 - \hat{\bar{y}}_2}{se(\hat{\bar{y}}_1 - \hat{\bar{y}}_2)} \tag{8.6}$$

where $se(\hat{\bar{y}}_1 - \hat{\bar{y}}_2) = \sqrt{var(\hat{\bar{y}}_1) + var(\hat{\bar{y}}_2) - 2\,cov(\hat{\bar{y}}_1, \hat{\bar{y}}_2)}$, and compare to the appropriate reference distribution. The terms $var(\hat{\bar{y}}_1)$ and $var(\hat{\bar{y}}_2)$ can be obtained from PROC SURVEYMEANS by specifying the keyword VAR in the PROC SURVEYMEANS statement. You can also square the standard errors of the domain means, which are typically reported by default. But there is no keyword available in PROC SURVEYMEANS to obtain $cov(\hat{\bar{y}}_1, \hat{\bar{y}}_2)$, the estimated covariance of the two domain means. Technically speaking, we could write freestanding code to obtain it, but that is unnecessary. Instead, we will demonstrate a method to acquire $se(\hat{\bar{y}}_1 - \hat{\bar{y}}_2)$ directly without having to put together the pieces independently.

Before we move on to the illustration, note that there is one scenario in which the covariance term is zero and can be ignored: when the two domains are drawn from disjoint sets of PSUs. Whenever cases from the two domains appear in one or more of the same PSUs, a nonzero (and often positive) covariance typically results. Consideration of whether a covariance is in play is so heavily emphasized because it is the decisive factor as to whether output from PROC SURVEYMEANS can be used directly to conduct a domain mean significance test. Generally speaking, PROC SURVEYMEANS provides all necessary ingredients only when the two domains emanate from a disjoint sets of PSUs (i.e., when the covariance term is zero). It is not equipped to account for nonzero covariances among domain mean estimates, nor is the CONTRAST= parameter of the %SMSUB macro developed several years back for added PROC SURVEYMEANS functionality. Regardless of whether a covariance exists, however, we can exploit a few of the relatively new tools built into PROC SURVEYREG for efficient domain mean significance testing.

Why PROC SURVEYREG? Interestingly, it is because a linear regression model can be used to test whether two domain means are significantly different from one another. The general approach is to fit a model with an intercept and a single predictor variable, a dichotomous 0/1 indicator variable distinguishing the two domains. From there, a two-sided t test assessing whether the indicator variable's parameter is significantly different from zero is equivalent to the t statistic specified in Equation 8.6.

To elucidate the idea, let us consider two example domain mean significance tests from the hypothetical mathematics aptitude survey. The first is the difference in the year-end test scores between upperclassman and underclassmen. The first PROC SURVEYREG run in Program 8.6 shows how to conduct this test using the indicator variable regression method. As was stated in earlier chapters, recall that all three SURVEY procedures with modeling capabilities (PROC SURVEYREG, PROC SURVEYLOGISTIC, and PROC SURVEYPHREG) include a factor $(n - 1)/(n - p)$ in the calculation of the model parameter covariance matrix by default based on a recommendation by Hidiroglou et al. (1980). To omit this adjustment factor, you can specify VADJUST=NONE after the slash in the MODEL statement.

From the Estimated Regression Coefficients table in the output, we can infer that the dummy variable associated with upperclassmen has been omitted. Hence, the intercept serves as an estimate of this domain's mean and the parameter affixed to the underclassmen domain represents the difference between the two domain means—that is, transitioning from upperclassmen to underclassmen. (Granted, a more intuitive interpretation might follow from reversing the parameterization, but this is no matter in a two-sided test.) We can observe the t statistic is $-6.9164/2.2416 = -3.09$, which is significant at the $\alpha = 0.05$ significance level, meaning we are safe to conclude a significant difference exists in test scores between underclassmen and upperclassmen. Moreover, referring back to the output following Program 8.1, we can verify the absence of a covariance between the two means since the denominator of this test is equal to the square root of the sum of the two domain mean variances, $\sqrt{1.5137^2 + 1.6533^2} \approx 2.2416$. Again, this is something we would anticipate considering the two domains consist of mutually exclusive sets of PSUs.

The second PROC SURVEYREG run in Program 8.6 illustrates how to determine significance for the year-end mean test difference between those who received tutoring and those who did not. There are multiple occurrences of a PSU containing students from both domains, rendering a nonzero covariance. The parameterization scheme PROC SURVEYREG utilizes omits the indicator variable associated with TUTOR='Y', which causes the estimate associated with TUTOR='N' to signify the estimated change transitioning from TUTOR='Y' to TUTOR='N'. Therefore, we can reason that students who did not receive any kind of mathematics tutoring score 4.77 points lower on the test, on average, than those who received tutoring. The *t* test is highly significant, providing strong evidence that the true population means for these two domains are different from one another.

Finally, we can verify the presence of a covariance in the second PROC SURVEYREG run since the standard error of the test, 1.0017, does not equal the square root of the sum of the squared domain mean standard errors reported in Program 8.2 ($\sqrt{0.8814^2 + 1.3370^2} = 1.6014$). The precipitous reduction in variance is a by-product of the positive covariance caused by correlated test scores within PSUs. This is commonly encountered for many outcome variables measured in practice (see discussion in Section 1.4.4).

Program 8.6: Testing Whether Domain Mean Differences Are Significant

```
title '1) Testing Year-End Mean Score for Underclassmen vs.
Upperclassmen';
proc surveyreg data=test;
   stratum class;
   cluster homeroom;
   class class_type;
   model grade2 = class_type / solution vadjust=none;
weight weight;
run;

title '2) Testing Year-End Mean Score for Tutored Students vs.
Non-Tutored Students';
proc surveyreg data=test;
   stratum class;
   cluster homeroom;
   class tutor;
   model grade2 = tutor / solution vadjust=none;
weight weight;
run;
```

1. Testing Year-End Mean Score for Underclassmen versus Upperclassmen

Estimated Regression Coefficients				
Parameter	Estimate	Standard Error	t Value	Pr> $|t|$
Intercept	92.5000000	1.65326793	55.95	<.0001
class_type under	−6.9164134	2.24157393	−3.09	0.0367
class_type upper	0.0000000	0.00000000		

2. Testing Year-End Mean Score for Tutored Students versus Nontutored Students

Estimated Regression Coefficients				
Parameter	Estimate	Standard Error	t Value	Pr> $|t|$
Intercept	92.2095915	1.33697702	68.97	<.0001
tutor N	−4.7701451	1.00169127	−4.76	0.0089
tutor Y	0.0000000	0.00000000		

Domain examples appearing in the remainder of this section will be drawn from NAMCS 2010, but before parting with the hypothetical mathematics aptitude survey data set, a few words are warranted regarding another relevant domain mean comparison. Given that the examinations were administered at the start and end of the school year, one may be interested in estimating the mean change in mathematics aptitude over those two points in time. This estimation problem introduces two possible causes for a covariance between the two domain means, one attributable to the sample design and another attributable to repeated measures being taken from the same set of students.

This is the scenario of *repeated measures*, which is typically addressed in introductory statistical methods textbooks (e.g., Example 7.7 of Moore and McCabe, 2006). An advocated method that evades the explicit calculation of a covariance is to create a new variable for each sampling unit defined as the difference between the two measures. The appeal is that a bivariate analysis is reduced to a univariate analysis because testing whether the mean of the difference variable is significantly different from 0 is equivalent to testing whether there was a significant change in the mean measure across the two points in time.

Korn and Graubard (1999, p. 79) note that this approach also works in the complex sample design setting. In terms of the hypothetical mathematics aptitude example, we create a new variate $d_{hij} = y_{2hij} - y_{1hij}$ for the jth student in the ith homeroom of the hth grade, where y_{2hij} is the student's year-end test score and y_{1hij} is the test score from the beginning of the school year. Using only this new variate, we can carry out the following hypothesis tests: H_0: $\bar{d} = \bar{y}_2 - \bar{y}_1 = 0$ versus H_1: $\bar{d} = \bar{y}_2 - \bar{y}_1 \neq 0$.

The first part of Program 8.7 illustrates this method. The difference variable is created in a DATA step and named DIFF. From there, we test whether the mean of this variable is significantly different from 0 by specifying the statistical keyword T in the PROC statement of PROC SURVEYMEANS. Examining the output, we see that the observed t statistic of 2.96 is significant at the $\alpha = 0.05$ significance level, suggesting that the students from the school did, in fact, experience a significant increase in mathematics aptitude over the course of the school year.

The second part of Program 8.7 demonstrates an alternative method to conduct the same t test. It is derived from a technique shown in Example 5.13 of Heeringa et al. (2010). The notion is to reorient the data set such that the indicator variable regression approach described in previous examples can be implemented. Specifically, the first step is to stack the data set with observations from the first domain (i.e., the first time point) on top of observations from the second domain (i.e., the second time point). This is accomplished in the DATA step producing the data set named TEST_STACKED. GRADE1 and GRADE2 are renamed so that they can be stored in same column of the concatenated data set. A domain indicator variable named DOMAIN assigns records from the second domain a value of 1 and 0 otherwise. The assimilated outcome variable GRADE is then modeled as a function of the variable DOMAIN while simultaneously accounting for the complex survey features. Despite the input data set TEST_STACKED having twice the number of observations as the original data set, this will not impact variance estimates because there is no increase in the number of distinct PSU or stratum codes. Indeed, we can verify that the t statistic and p value for the parameter estimate corresponding to DOMAIN match precisely with the output generated by the preceding PROC SURVEYMEANS run.

Program 8.7: Illustrating the Difference Variable Approach for Sample Mean Significance Testing When Multiple Measures Are Taken of the Same Set of Sampling Units

```
* difference variable approach;
data test;
   set test;
diff=grade2-grade1;
run;
proc surveymeans data=test mean stderr t;
   stratum class;
   cluster homeroom;
   var diff;
weight weight;
run;
```

```
* data stacking approach;
data test_stacked;
   set test (in=d1 rename=(grade1=grade))
      test (in=d2 rename=(grade2=grade));
domain=d2;
run;

proc surveyreg data=test_stacked;
   stratum class;
   cluster homeroom;
   model grade = domain / vadjust=none;
weight weight;
run;
```

SURVEYMEANS Procedure

Data Summary	
Number of strata	4
Number of clusters	8
Number of observations	16
Sum of weights	715

Statistics

Variable	Mean	Std Error of Mean	t Value	Pr > |t|
diff	2.548252	0.860450	2.96	0.0415

SURVEYREG Procedure

Estimated Regression Coefficients				
Parameter	Estimate	Standard Error	t Value	Pr > |t|
Intercept	86.7692308	1.72841398	50.20	<.0001
domain	2.5482517	0.86044990	2.96	0.0415

The data stacking technique generalizes to other settings, including entirely separate survey administrations. For example, it is not unusual for the same stratification and primary-stage clustering scheme to be used in adjacent administrations of a repeated cross-sectional survey. In fact, this is the case for the NSFG data set analyzed throughout this book, relative to its previous cycle (see Appendix 2 of the data user's guide: http://www.cdc.gov/nchs/nsfg/nsfg_2006_2010_puf.htm). The subsequent stages of clustering and ultimate units of analysis are still sampled independently, but maintaining the core structure within the ultimate cluster paradigm offers efficiencies with respect to both variances of estimates of change and the costs associated with completely revamping the sample design.

TABLE 8.1

Coding Structure of Variable MAJOR on the NAMCS 2010 Public-Use
Data Set, the Primary Reason for the Patient Visiting the Physician

Code	Meaning
1	New problem (<3 months onset)
2	Chronic problem, routine
3	Chronic problem, flare up
4	Pre-/Postsurgery
5	Preventive care (e.g., routine prenatal, well-baby, screening, insurance, general exams)

Note that you can still use the data stacking approach even if the sample design has changed substantively, but you should ensure strata and cluster identifiers are coded uniquely across the two data sets. Additionally, the approach is not limited to data derived from complex sample surveys. For instance, had a simple random sample of students from the high school been selected, a similar sequence of steps could be followed. The only additional detail one must adhere to is to create a unique student identifier and specify that in the CLUSTER statement of PROC SURVEYREG.

The techniques demonstrated thus far in this section are sufficient for targeted domain mean significance tests, but they would be monotonous to apply in a production environment where a vast number of domain comparisons are desired. For example, even a single-domain identifier variable consisting of D distinct domains would necessitate $D(D - 1)/2$ unique domain mean significance tests, requiring us to code in as many 0/1 indicator variables and as many MODEL statements. The LSMEANS statement in PROC SURVEYREG is an efficient tool to employ in these situations.

To motivate an example using NAMCS 2010, suppose we were interested in contrasting mean time spent with the physician (variable TIMEMD) among the primary reasons for the visit (variable MAJOR). Table 8.1 itemizes the $D = 5$ categorizations available in the NAMCS_2010 data set.

Program 8.8 shows syntax to conduct all possible $5(5 - 1)/2 = 10$ mean differences. The general approach is to specify the domain identifier variable in the CLASS statement and as the sole predictor variable of the outcome variable of interest, and also specify this variable in the LSMEANS statement with the DIFF option after the slash. The LSMEANS statement produces a summarization of the estimated marginal means and standard errors. It also conducts significance tests for whether the estimated means are significantly different from zero. The DIFF option generates the 10 pairwise estimated mean differences and reports results of the respective two-sided t tests.

Program 8.8: Conducting Pairwise Mean Significance Tests Using a Categorical Variable Defining Domains

```
proc surveyreg data=NAMCS_2010;
  stratum CSTRATM;
  cluster CPSUM;
  class MAJOR;
  model TIMEMD = MAJOR / vadjust=none;
weight PATWT;
lsmeans MAJOR / diff;
run;
```

MAJOR Least Squares Means					
MAJOR	Estimate	Standard Error	DF	*t* Value	Pr > |*t*|
1	19.9857	0.5793	552	34.50	<.0001
2	20.5501	0.3991	552	51.49	<.0001
3	21.9394	0.6398	552	34.29	<.0001
4	17.6291	0.6836	552	25.79	<.0001
5	20.4920	0.5478	552	37.41	<.0001

Differences of MAJOR Least-Squares Means						
MAJOR	_MAJOR	Estimate	Standard Error	DF	*t* Value	Pr > |*t*|
1	2	−0.5645	0.6291	552	−0.90	0.3700
1	3	−1.9538	0.6706	552	−2.91	0.0037
1	4	2.3565	0.8340	552	2.83	0.0049
1	5	−0.5064	0.5776	552	−0.88	0.3810
2	3	−1.3893	0.5548	552	−2.12	0.0343
2	4	2.9210	0.7524	552	3.88	0.0001
2	5	0.05808	0.6413	552	0.09	0.9279
3	4	4.3103	0.8340	552	5.17	<.0001
3	5	1.4474	0.7624	552	1.90	0.0582
4	5	−2.8629	0.7798	552	−3.67	0.0003

A few additional options available after the slash in the LSMEANS statement deserve brief mention:

- The significance level for the *t* tests can be modified using the ALPHA= option.

- Related to the aforementioned point, there are a variety of techniques to adjust *p* values using the ADJUST= option. These are aimed at controlling the overall Type I error rate, many of which also appear in PROC MULTTEST. Consult the documentation for more details.

- The PLOTS= option offers a suite of preconfigured ODS Graphics-based visualizations that could prove useful for deciphering underlying patterns.

In fact, the analysis in Program 8.8 could be extended even further by including one or more variables in a DOMAIN statement. For example, if we had specified SEX in the DOMAIN statement, the same analysis would be output, followed by comparable analyses for males and females. Another potential extension of interest would be to control for one or more covariates by adding them to the MODEL statement. For example, if we instead modified Program 8.8 by adding SEX as an independent variable in the MODEL statement, the LSMEANS differences reported would no longer be interpreted as the marginal differences. Instead, they would be interpreted as differences in mean time spent with the physician while simultaneously accounting for the effect of patient gender. In essence, that type of analysis is in the same spirit as what was presented in Program 5.6.

Another point worth making prior to moving on to the next topic is that all of the PROC SURVEYREG approaches demonstrated in this section translate to dichotomous outcome variables as long as the two categories are coded one unit apart (e.g., via a 0/1 indicator variable). In fact, we can conceive of the risk difference statistic produced in Program 4.7 as a domain mean significance test when both the outcome and domain indicator are dichotomous. Program 8.9 confirms the equivalence in an analysis contrasting the probability of medication being prescribed during the visit depending upon whether or not the physician practices alone or as part of a group. The variable MED is a 0/1 indicator of a prescription being written, and SOLO is coded 1 if the physician operates his/her own practice and 2 otherwise. As we can gather from the PROC SURVEYFREQ output, the probability that MED=1 given SOLO=1 is 0.7118, whereas the probability that MED=1 given SOLO=2 is 0.7689. The latter is the intercept parameter estimate following the PROC SURVEYREG run in Program 8.9, while the parameter affixed to SOLO=2 represents the difference in probabilities between SOLO=1 and SOLO=2 (−0.0571 = 0.7118 − 0.7689). Notice how the 95% confidence limits on this parameter estimate match the like from the row labeled "Difference" as part of the Colum 2 Risk Estimates section of the PROC SURVEYFREQ output.

Program 8.9: Demonstrating the Linear Regression Dummy Variable Approach for Testing Whether Two Domain Proportions Are Significantly Different and Its Equivalence to the Risk Difference Statistic Generated by PROC SURVEYFREQ

```
proc surveyreg data=NAMCS_2010;
   stratum CSTRATM;
   cluster CPSUM;
   class SOLO;
   model MED = SOLO / vadjust=none solution clparm;
weight PATWT;
run;
```

```
proc surveyfreq data=NAMCS_2010;
  stratum CSTRATM;
  cluster CPSUM;
  tables SOLO*MED / nowt row riskdiff;
weight PATWT;
run;
```

SURVEYREG Procedure

		Estimated Regression Coefficients					
Parameter	Estimate	Standard Error	t Value	Pr > |t|	95% Confidence Interval	
Intercept	0.7688868	0.01222044	62.92	<.0001	0.7448826	0.7928911
SOLO 1	−0.0571077	0.02083697	−2.74	0.0063	−0.0980372	−0.0161783
SOLO 2	0.0000000	0.00000000			0.0000000	0.0000000

SURVEYFREQ Procedure

		Table of SOLO by MED				
SOLO	MED	Frequency	Percent	Std Err of Percent	Row Percent	Std Err of Row Percent
1	0	2,710	9.0691	0.7759	28.8221	1.6751
	1	6,081	22.3967	1.6656	71.1779	1.6751
	Total	8,791	31.4657	2.1523	100.000	
2	0	5,538	15.8392	0.8832	23.1113	1.2220
	1	16,900	52.6951	2.0004	76.8887	1.2220
	Total	22,438	68.5343	2.1523	100.000	
Total	0	8,248	24.9083	1.0074		
	1	22,981	75.0917	1.0074		
	Total	31,229	100.000			

Column 1 Risk Estimates				
	Risk	Standard Error	95% Confidence Limits	
Row 1	0.2882	0.0168	0.2553	0.3211
Row 2	0.2311	0.0122	0.2071	0.2551
Total	0.2491	0.0101	0.2293	0.2689
Difference	0.0571	0.0208	0.0162	0.0980
Difference is (Row 1 − Row 2)				
Sample Size = 31,229				

Column 2 Risk Estimates				
	Risk	Standard Error	95% Confidence Limits	
Row 1	0.7118	0.0168	0.6789	0.7447
Row 2	0.7689	0.0122	0.7449	0.7929
Total	0.7509	0.0101	0.7311	0.7707
Difference	−0.0571	0.0208	−0.0980	−0.0162
Difference is (Row 1 − Row 2)				
Sample Size = 31,229				

This exercise should not be construed as an endorsement for fitting linear regression models to binary outcome variables; rather, logistic regression is preferred for the reasons cited in Section 6.2. The purpose was simply to demonstrate how the algebraic equivalence between the indicator variable regression approach and a two-sample t test holds when applied to proportions just as it does for means of continuous outcome variables.

8.7 Degrees of Freedom Adjustments

Most survey statisticians would agree that the traditional complex survey degrees of freedom rule of thumb—the number of distinct PSUs minus the number of strata—is too large when estimating a quantity for a sparse domain, or one not present in all PSUs and/or strata. The literature is a bit mixed regarding the ideal alternative, however. Heeringa et al. (2010) suggest reducing the degrees of freedom to the number of PSUs less the number of strata for strata where the domain appears in a least one PSU. Korn and Graubard (1999) recommend a (potentially) more conservative adjustment: the number of unique PSUs containing at least one domain case minus the number of strata in which these PSUs are situated.

Regardless of one's school of thought, be advised that the options for assigning a fixed value for the degrees of freedom vary among the SURVEY procedures. Table 8.2 summarizes the available methods. At present, there are no built-in options for PROC SURVEYMEANS and PROC SURVEYLOGISTIC. According to the documentation, you can specify the DFADJ option after the slash in the DOMAIN statement of PROC SURVEYMEANS to implement the adjustment advocated by Heeringa et al. (2010), but it contains a bug at the time of this writing—rather than being reduced, the degrees of freedom actually increase. A general workaround will be demonstrated shortly.

Remember that the degrees of freedom will not affect point estimates or their estimated variances, but can impact the p values of hypothesis tests and

TABLE 8.2

SURVEY Procedure Options to Adjust the Degrees of Freedom When Utilizing Taylor Series Linearization for Variance Estimation

Procedure	Option
PROC SURVEYMEANS	Nothing available at the time of this writing
PROC SURVEYFREQ	DF=*number* option after the slash in the TABLES statement
PROC SURVEYREG	DF=*number* option after the slash in the MODEL statement
PROC SURVEYLOGISTIC	Nothing available at the time of this writing
PROC SURVEYPHREG	DF=*number* option after the slash in the MODEL statement

confidence interval widths because the underlying test statistic's reference distribution is based on assumed degrees of freedom. Not all domain estimation tasks necessitate the adjustments discussed in this section. Practically speaking, neglecting to modify the degrees of freedom is only hazardous in scenarios where there are few complex survey degrees of freedom to begin with or when the domain is extremely rare. Conclusions are far less likely to change if the degrees of freedom are reduced from 100 to 96, say, then if they are reduced from 100 to 16.

To motivate an example, let us revisit the domain analysis of the mean year-end test score in the hypothetical mathematics aptitude survey for students who received tutoring. Instead of estimating the mean and standard error for this domain using the domain-specific weight approach as we did in Program 8.3, suppose we sought to find the endpoints of a 95% confidence interval about the mean. We may have observed from the DATA step in Program 8.1 that not all PSUs contribute to the tutoring domain— namely, neither homeroom in the 10th grade stratum nor the first homeroom in the 11th grade stratum. The first PROC SURVEYMEANS run in Program 8.10 requests a confidence interval without any kind of adjustment. The second run reduces the degrees of freedom per the guidance from Heeringa et al. (2010). This is done indirectly using PROC SQL to subset the original data set for only those strata containing at least one PSU in the domain of interest. The result is stored in a data set named TEST_D_ADJ, which is analyzed in the second PROC SURVEYMEANS run. We can verify that the adjustment has no effect on the mean and standard error, but results in a somewhat wider 95% confidence interval. This is because PROC SURVEYMEANS references t distribution with 3 degrees of freedom as opposed to 4.

Program 8.10: Adjusting the Degrees of Freedom for the Sample Mean of a Sparse Domain

```
title '1) 95% Confidence Limits without a DF Adjustment';
data test_d;
  set test;
weight_d=weight*(tutor='Y') + .0000000000001;
run;
proc surveymeans data=test_d mean stderr clm df;
  stratum class;
  cluster homeroom;
  var grade2;
weight weight_d;
run;

title '2) 95% Confidence Limits with DF Adjustment per
Heeringa et al. (2010)';
```

```
* preliminary PROC SQL step subsets the data for strata with
at least one PSU in the domain defined by TUTOR='Y';
proc sql;
  create table test_d_adj as
  select *
    from test_d
    where class in(select distinct class from test_d where
    tutor='Y');
quit;
proc surveymeans data=test_d_adj mean stderr clm df;
  stratum class;
  cluster homeroom;
  var grade2;
weight weight_d;
run;
```

1. 95% Confidence Limits without a DF Adjustment

Data Summary	
Number of strata	4
Number of clusters	8
Number of observations	16
Sum of weights	281.5

Statistics					
Variable	DF	Mean	Std Error of Mean	95% CL for Mean	
grade2	4	92.209591	1.336977	88.4975482	95.9216348

2. 95% Confidence Limits with DF Adjustment per Heeringa et al. (2010)

Data Summary	
Number of strata	3
Number of clusters	6
Number of observations	12
Sum of weights	281.5

Statistics					
Variable	DF	Mean	Std Error of Mean	95% CL for Mean	
grade2	3	92.209591	1.336977	87.9547339	96.4644491

Be advised that this workaround could also be used for other SURVEY procedures. A bit more programming work would be required to derive the alternative degrees of freedom reduction proposed by Korn and Graubard (1999). We will not explore an example of that here.

8.8 Summary

This chapter discussed the cross-cutting topic of domain estimation, analyses aimed at only a portion, or domain, of the target population. After defining some terminology, we delineated the risks introduced by subsetting the complex survey data set. A few exceptions to the guidance never to subset the complex survey data set were given; however, it is strongly recommended that you employ the DOMAIN statement whenever possible. If faced with a situation where the DOMAIN statement is not (yet) permitted for a particular analysis, you can use the domain-specific weight approach illustrated in Section 8.4.

Lastly, note that the issues raised apply only when utilizing the default technique for variance estimation, Taylor series linearization. In the next chapter, we will explore into an alternative class of variance estimation techniques collectively referred to as *replication techniques*, which abide by a different set of rules with respect to domain estimation and degrees of freedom adjustments. As will become evident, however, these alternative approaches are advantageous in many respects.

9

Replication Techniques for Variance Estimation

9.1 Introduction

Up to this point in this book, all examples estimating the variance of a statistic have utilized Taylor series linearization (TSL), the SURVEY family of procedures' default strategy. This chapter is an all-inclusive exploration into the increasingly popular alternative class of variance estimators referred to as "replication techniques" (Rust, 1985; Rust and Rao, 1996). While there are various user-written macros that can be used to conduct these techniques (e.g., Hawkes, 1997; Bienias, 2001; Berglund, 2004), the release of SAS Version 9.2 ushered in new capabilities to greatly simplify the syntax required (Mukhopadhyay et al., 2008).

The basic idea behind replication is to treat the survey data set as if it were the population and repeatedly sample from it in some systematic fashion. From each sample, or *replicate*, you estimate the quantity of interest. The variance of the full survey data set estimate is then determined as a simple function of the variability amongst the replicate-specific estimates. As we will see, the process can be efficiently implemented by appending a series of *replicate weights* to the analysis data set. One very appealing feature is that there is generally only one variance formula (per method), regardless of the point estimate at hand. Certain other advantages relative to TSL will be highlighted as well.

This chapter begins with a detailed explanation of the computational algorithm SAS employs behind the scenes as part of the TSL process. Next, three of the most commonly used replication techniques—balanced repeated replication (BRR), the jackknife, and the bootstrap—are introduced and demonstrated via examples drawn primarily from the same fictitious mathematics aptitude complex survey data set used in Chapter 8. It is this author's firm belief that a data set of manageable size is essential for the uninitiated to comprehend these techniques since it allows one to wholly visualize the replicate weights and how they are constructed. We also touch on the multivariate generalization of replication techniques, demonstrating how they can be used to estimate the covariance matrix of linear model parameters as well as variances of more complex estimators. The chapter concludes with a brief discussion regarding degrees of freedom considerations.

9.2 More Details Regarding Taylor Series Linearization

As we have now defined on several occasions, the TSL variance approxima-tion of a weighted mean $\hat{\bar{y}} = \sum_{i=1}^{n} w_i y / \sum_{i=1}^{n} w_i = \hat{Y}/\hat{N}$ is given by

$$\mathrm{var}\left(\frac{\hat{Y}}{\hat{N}}\right) \approx \frac{1}{\hat{N}^2}\left[\mathrm{var}(\hat{Y}) + \left(\frac{\hat{Y}}{\hat{N}}\right)^2 \mathrm{var}(\hat{N}) - 2\left(\frac{\hat{Y}}{\hat{N}}\right)\mathrm{cov}(\hat{Y},\hat{N})\right] \quad (9.1)$$

where the variance terms follow the structure of Equation 3.2 and the covari-ance term follows that given by Equation 3.5. Computing all of these terms in the expression is rather laborious. In actuality, SAS uses a computational maneuver proposed by Woodruff (1971), which effectively reduces to a uni-variate problem any estimator that can be expressed as a differentiable func-tion of totals. Although it may not be immediately obvious, this covers a wide range of estimators.

We will walk through the details of this process for a weighted sample mean, which is instructive to do for two reasons: (1) there may be occasions where a TSL variance estimate has not been preprogrammed into the given SURVEY procedure and you must derive it; and (2) to instill an appreciation for the simplicity behind using replicate weights to estimate the variance of *any* statistic.

The weighted sample mean is clearly a function of $p = 2$ estimated totals since $\hat{\bar{y}} = \sum_{i=1}^{n} w_i y_i / \sum_{i=1}^{n} w_i = \hat{Y}/\hat{N} = \hat{T}_1/\hat{T}_2$. The first step of the Woodruff (1971) method is to create a primary sampling unit (PSU)-level variate equaling the sum of the function's partial derivatives times the given PSU-level estimate of the total. In the present case, we can express this variate as $u_i = \sum_{j=1}^{p} (\partial\bar{y}/\partial T_j)\hat{t}_{ji}$, where $\hat{t}_{ji}$ represents the PSU-level estimate of the jth total in the function, T_j. Note that the partial derivative with respect to the numerator is $\partial\bar{y}/\partial T_1 = 1/\hat{T}_2$ and the partial derivative with respect to the denominator is $\partial\bar{y}/\partial T_2 = -\hat{T}_1/\hat{T}_2^2$. Furthermore, $\hat{t}_{1i} = w_i y_i$ and $\hat{t}_{2i} = w_i$. After a little algebra, we can write u_i as $u_i = (1/\hat{T}_2)(w_i y_i - (\hat{T}_1/\hat{T}_2)w_i)$.

We will refer to the variate u_i as a *linear substitute*, although it is some-times referred to as a *score variable*. Following Equation 3.2, the variance of the original function is then estimated by the variance of the estimated sum of the linear substitute with respect to the complex survey design. Hence, the novelty of the approach is that there is only one variance to calculate, not two variances and a covariance as prescribed by Equation 9.1.

Program 9.1 demonstrates this approach on the same TEST data set introduced in Chapter 8, which consists of data from a hypothetical complex sample survey of 16 high school students given a mathematics aptitude examination on two occasions—the variables GRADE1 and GRADE2 are the students' examination scores (out of 100) at the beginning and end of the school year, respectively. Recall that the student body was stratified by class standing (variable CLASS) and a cluster sample of two homerooms (variable HOMEROOM) were selected within each of the four class standings, with two students ultimately selected within each homeroom. The variable WEIGHT reflects the inverse of each student's selection probability, representing the number of students in the school the sampled student represents.

Let us consider the year-end math examination score for an example weighted mean. The PROC SQL step in Program 9.1 creates the linear substitute variate for each of the four distinct PSUs and stores in a data set called LIN_SUBS. The first PROC SURVEYMEANS run requests the variance and standard error of the estimated sum of the u_is accounting for the complex survey design. Since the data set LIN_SUBS has only one observation per PSU, the CLUSTER statement is technically unnecessary. Note from the output how the variance matches what is output from the second PROC SURVEYMEANS run, the standard approach.

Program 9.1: Demonstrating Woodruff's Method of Taylor Series Linearization Using the Math Aptitude Examination Complex Survey Data Set

```
data test;
  input class class_type $ homeroom grade1 tutor $ grade2
  weight;
datalines;
12 upper 1 87 Y 94 49.5
12 upper 1 89 N 89 49.5
12 upper 2 91 Y 94 48
12 upper 2 84 Y 92 48
11 upper 1 82 N 84 47.5
11 upper 1 94 N 95 47.5
11 upper 2 93 N 95 48
11 upper 2 94 Y 97 48
10 under 1 78 N 81 39
10 under 1 84 N 84 39
10 under 2 90 N 87 37.5
10 under 2 82 N 85 37.5
```

```
  9 under 1 88 N 88 40
  9 under 1 93 Y 91 40
  9 under 2 77 Y 85 48
  9 under 2 81 N 84 48
;
run;

title '1) Demonstrating the Woodruff Algorithm for Taylor
Series Linearization';
* compute the PSU-level linear substitute term;
proc sql;
   select sum(weight)         into :N_hat from test;
   select sum(weight*grade2) into :Y_hat from test;

   create table lin_subs as
   select class,homeroom,
         (1/&N_hat)*(sum(weight*grade2) -
         (&Y_hat/&N_hat)*sum(weight)) as u_i
    from test
    group by class,homeroom;
quit;

* estimate variance of the sum of the linear substitute with
respect to the design;
proc surveymeans data=lin_subs varsum std;
   stratum class;
   cluster homeroom;
   var u_i;
run;

title '2) Standard PROC SURVEYMEANS Syntax for a Weighted
Mean';
proc surveymeans data=test mean var stderr;
   stratum class;
   cluster homeroom;
   var grade2;
weight weight;
run;
```

1. Demonstrating the Woodruff Algorithm for Taylor Series Linearization

SURVEYMEANS Procedure

Data Summary	
Number of strata	4
Number of clusters	8
Number of observations	8

Statistics		
Variable	Std Dev	Var of Sum
u_i	1.183587	1.400879

2. Standard PROC SURVEYMEANS Syntax for a Weighted Mean

SURVEYMEANS Procedure

Data Summary	
Number of strata	4
Number of clusters	8
Number of observations	16
Sum of weights	715

Statistics			
Variable	Mean	Std Error of Mean	Var of Mean
grade2	89.317483	1.183587	1.400879

To reiterate, the purpose of this illustration was to convey that while SAS may already have these linear substitutes preprogrammed for commonly requested statistics, an understanding of the process is valuable because you may encounter a situation in which you need to derive the variance of a complex estimator not readily available. We will consider one such example later in the chapter (see Program 9.14).

9.3 Balanced Repeated Replication

The first replication technique we will consider is *balanced repeated replication* (McCarthy, 1966; Chapter 3 of Wolter, 2007), which was originally developed to estimate variances in a two-PSU-per-stratum sample design. In BRR, we select one of the two PSUs from each of H strata and double the weights of all units therein while setting the weights of all units in the unselected PSU to 0. This new weight is referred to as a "replicate weight." The idea is to perform this process a number of times, creating a series of replicate weights that get appended to the analysis file. The variability among estimates calculated using each distinct replicate weight serves as the estimate of variability for the full-sample point estimate. As will be shown shortly, SAS will carry out this routine when you specify the VARMETHOD=BRR option in the PROC statement of a SURVEY procedure.

Before we continue, two questions immediately arise: (1) How many replicates (or replicate weights) are sufficient? (2) How do we know which PSUs to select? To achieve a desirable property called "full orthogonal balance," the number of replicates necessary is the first multiple of four strictly greater than the number of strata. That is, if we denote R to be the number of replicates (or replicate weights), we require $H < R \leq H + 4$. To answer the second question, we make use of Hadamard matrices from the field of experimental design. A Hadamard matrix, typically denoted **H**, is a square matrix of +1s and −1s where columns correspond to strata and rows replicates. (Columns and rows are technically interchangeable.) After randomly numbering the two PSUs (i.e., homerooms) within a stratum, let a +1 indicate selecting the first PSU and a −1 indicate selecting the second PSU. The "orthogonal" qualifier comes from the fact that the sum of the product of entries along any two rows or columns is 0. What follows is one example Hadamard matrix that could be used for the hypothetical mathematics aptitude survey data set:

$$\mathbf{H} = \begin{bmatrix} +1 & +1 & +1 & +1 & +1 & +1 & +1 & +1 \\ +1 & -1 & +1 & -1 & +1 & -1 & +1 & -1 \\ +1 & +1 & -1 & -1 & +1 & +1 & -1 & -1 \\ +1 & -1 & -1 & +1 & +1 & -1 & -1 & +1 \\ +1 & +1 & +1 & +1 & -1 & -1 & -1 & -1 \\ +1 & -1 & +1 & -1 & -1 & +1 & -1 & +1 \\ +1 & +1 & -1 & -1 & -1 & -1 & +1 & +1 \\ +1 & -1 & -1 & +1 & -1 & +1 & +1 & -1 \end{bmatrix} \tag{9.2}$$

There are eight columns, but we only need four, and any will suffice, so we can take the first four. We can think of the leftmost column as representing the ninth grade stratum and the fourth column as representing the twelfth grade stratum. The first replicate will consist of the first homeroom from each stratum, while the second replicate will consist of the first homeroom for the 9th and 11th grade strata, but the second homeroom from the 10th and 12th grade strata, and so on. SAS has Hadamard matrices stored internally to accommodate; however, many strata are detected in the input data set, but you can input your own with the VARMETHOD=BRR(HADAMARD=*data-set-name*) option in the PROC statement if desired. See the documentation for more details.

Once the replicates have been selected, the replicate-specific estimates are computed, which we can denote by $\hat{\theta}_r$ ($r = 1, ..., R$). If we symbolize the full-sample estimate $\hat{\theta}$, the BRR variance estimate is the mean squared error of the replicate-specific estimates about $\hat{\theta}$, or

$$\text{var}_{BRR}(\hat{\theta}) = \frac{1}{R} \sum_{r=1}^{R} (\hat{\theta}_r - \hat{\theta})^2 \tag{9.3}$$

where generic theta notation is used here to emphasize that this is a universal formula applicable to *any* finite population quantity.

Program 9.2 requests an estimated variance via BRR for the mean of GRADE2 in the TEST data set. Relative to the second PROC SURVEYMEANS run in Program 9.1, the only difference is that the option VARMETHOD=BRR is specified in the PROC statement. Comparing the output to that generated from Program 9.1, we note the point estimate is the same, but measures of variability are slightly different. Neither is superior; both are unbiased estimates of the sampling error in the estimated mean associated with the sample design.

Program 9.2: BRR Variance Estimation

```
proc surveymeans data=test varmethod=BRR mean var stderr;
  stratum class;
  cluster homeroom;
  var grade2;
weight weight;
run;
```

SURVEYMEANS Procedure

Data Summary	
Number of strata	4
Number of clusters	8
Number of observations	16
Sum of weights	715

Variance Estimation	
Method	BRR
Number of replicates	8

Statistics			
Variable	Mean	Std Error of Mean	Var of Mean
grade2	89.317483	1.201095	1.442630

While SAS is fully capable of proceeding through all steps of the BRR process independently for each call to the given SURVEY procedure, it is more efficient to merge the replicate weights onto the analysis file for subsequent BRR variance estimation requests. We can do this by specifying OUTWEIGHTS=*data-set-name* in parentheses immediately after VARMETHOD=BRR. Program 9.3 shows how to do so, storing the result in a data set called TEST_BRR. Figure 9.1 illustrates how TEST_BRR contains all input data set variables and observations, but also the replicate weights. There are eight of them, but only the first four are shown. SAS affixes a common label of "Replicate Weight," but the variables are actually named REPWT_1, REPWT_2,…, REPWT_8.

Program 9.3: Creating an Analysis Data Set with BRR Replicate Weights Appended

```
proc surveymeans data=test varmethod=BRR (outweights=test_BRR)
  mean var stderr;
  stratum class;
```

	Class	Homeroom	grade2	Weight	Replicate Weight	Replicate Weight	Replicate Weight	Replicate Weight
1	12	1	94	49.5	99	99	99	99
2	12	1	89	49.5	99	99	99	99
3	12	2	94	48	0	0	0	0
4	12	2	92	48	0	0	0	0
5	11	1	84	47.5	95	0	0	95
6	11	1	95	47.5	95	0	0	95
7	11	2	95	48	0	96	96	0
8	11	2	97	48	0	96	96	0
9	10	1	81	39	78	78	0	0
10	10	1	84	39	78	78	0	0
11	10	2	87	37.5	0	0	75	75
12	10	2	85	37.5	0	0	75	75
13	9	1	88	40	80	0	80	0
14	9	1	91	40	80	0	80	0
15	9	2	85	48	0	96	0	96
16	9	2	84	48	0	96	0	96

FIGURE 9.1
Partial view of the BRR replicate weights on data set TEST_BRR created in Program 9.3.

```
   cluster homeroom;
   var grade2;
weight weight;
run;
```

Inspecting the replicate weights, we can infer that the first replicate was formed by selecting the first PSU (HOMEROOM=1) from all strata as the weights for those cases have doubled, while the weights for students in the other homeroom (HOMEROOM=2) have been set to zero. A similar line of reasoning applies to the other three replicate weights shown. If needed, you can have SAS output the particular Hadamard matrix used by specifying the PRINTH option in parentheses immediately after VARMETHOD=BRR.

Another appealing feature of the replicate weights is that they provide all of the complex design information necessary to properly calculate a variance. This means you can use them in later calls to any SURVEY procedure via a REPWEIGHTS statement without the STRATUM or CLUSTER statements. In fact, when those statements appear in combination with the REPWEIGHTS statement, they are ignored. For survey sponsors concerned that releasing stratum and PSU codes could pose a data confidentiality risk, this is viewed as a distinct advantage over TSL.

Program 9.4 uses the data set with the BRR replicate weights appended to produce the same results (output not shown) as Program 9.3. Note that we still must specify the VARMETHOD=BRR in the PROC statement. When the REPWEIGHTS statement appears without the VARMETHOD= option, the replicate weights are assumed to be derived from the jackknife replication technique, which we will discuss in Section 9.5. Be advised that variance estimates will generally be incorrect if the technique specified in the VARMETHOD= option is not that which was used to construct the replicate weights.

Program 9.4: BRR Variance Estimation Using a Data Set with Replicate Weights Appended

```
proc surveymeans data=test_BRR varmethod=BRR mean var stderr;
  var grade2;
weight weight;
repweights RepWt_1-RepWt_8;
run;
```

9.4 Fay's Variant to BRR

Judkins (1990) discusses a variant to the traditional BRR method whereby instead of doubling the weights of units in the selected PSU, weights are inflated by a factor of $2-\varepsilon$, where $0 \le \varepsilon < 1$, and weights of units in the "unselected" PSU are inflated by a factor of ε. Named after its inventor, Robert Fay, this method is typically called "Fay's BRR" and the epsilon term "Fay's coefficient." Note that when $\varepsilon = 0$, the method defaults to traditional BRR; otherwise, there is a slight modification to Formula 9.3 in that we divide through by a factor of $(1 - \varepsilon)^2$ as follows:

$$\text{var}_{BRR-Fay}(\hat{\theta}) = \frac{1}{R(1-\varepsilon)^2} \sum_{r=1}^{R} (\hat{\theta}_r - \hat{\theta})^2 \tag{9.4}$$

The main advantage of Fay's BRR is that each PSU is represented in every replicate, which is believed to foster more stability in variance estimates. This is particularly important for surveys with fewer degrees of freedom or analyses of rare population domains.

Fay's BRR can be implemented by specifying VARMETHOD=BRR(FAY<= *Fay coefficient>*) in the PROC statement of any SURVEY procedure. As with Program 9.4, we can have SAS do the legwork to create and append the Fay BRR replicate weights or provide the replicate weights directly. Declaring the Fay coefficient is technically optional; if nothing is specified, $\varepsilon = 0.5$ is assigned by default. There is no universally optimal ε, although Lee and Forthofer (2006) cite simulation results suggesting $\varepsilon = 0.3$ yields some desirable properties.

	Class	Homeroom	grade2	Weight	Replicate Weight	Replicate Weight	Replicate Weight	Replicate Weight
1	12	1	94	49.5	84.15	84.15	84.15	84.15
2	12	1	89	49.5	84.15	84.15	84.15	84.15
3	12	2	94	48	14.4	14.4	14.4	14.4
4	12	2	92	48	14.4	14.4	14.4	14.4
5	11	1	84	47.5	80.75	14.25	14.25	80.75
6	11	1	95	47.5	80.75	14.25	14.25	80.75
7	11	2	95	48	14.4	81.6	81.6	14.4
8	11	2	97	48	14.4	81.6	81.6	14.4
9	10	1	81	39	66.3	66.3	11.7	11.7
10	10	1	84	39	66.3	66.3	11.7	11.7
11	10	2	87	37.5	11.25	11.25	63.75	63.75
12	10	2	85	37.5	11.25	11.25	63.75	63.75
13	9	1	88	40	68	12	68	12
14	9	1	91	40	68	12	68	12
15	9	2	85	48	14.4	81.6	14.4	81.6
16	9	2	84	48	14.4	81.6	14.4	81.6

FIGURE 9.2
Partial view of the Fay BRR replicate weights on data set TEST_BRR_FAY created in Program 9.5.

Program 9.5 requests Fay's BRR with $\varepsilon = 0.3$ to estimate the variance of the mean of GRADE2. The OUTWEIGHTS=*data-set-name* code in parentheses instructs SAS to attach the replicate weights to the original input data set and save as a new data set named TEST_BRR_FAY. The standard error (1.1956) is slightly smaller than the traditional BRR standard error from Program 9.2 (1.2011). Figure 9.2 is a partial view of the first four replicate weights appended to the output data set TEST_BRR_FAY. Notice how all replicate weights are nonzero. For example, in the first replicate, weights of observations in the first PSU in both strata are multiplied by 1.7 while the "unselected" PSU's observations have their weights multiplied by 0.3.

Program 9.5: Fay's BRR Method for Variance Estimation

```
proc surveymeans data=test varmethod=BRR
    (outweights=test_BRR_Fay fay=.3)
  mean stderr;
  stratum class;
  cluster homeroom;
  var grade2;
weight weight;
run;
```

SURVEYMEANS Procedure

Data Summary	
Number of strata	4
Number of clusters	8
Number of observations	16
Sum of weights	715

Variance Estimation	
Method	BRR
Number of replicates	8
Fay coefficient	0.3

Statistics		
Variable	Mean	Std Error of Mean
grade2	89.317483	1.195644

Program 9.6 compares BRR to Fay's BRR for two domain estimates. Specifically, the mean year-end exam score is estimated for students who received any kind mathematics tutoring (TUTOR='Y') as well as those that did not (TUTOR='N') by simply specifying the variable TUTOR in the DOMAIN statement. All means are identical, as we would expect, but the standard errors under Fay's BRR are somewhat smaller than those calculated using traditional BRR. Although this is a small example for illustrative purposes, it is possible this finding is attributable to the added stability garnered from having all sampled PSUs appearing in every replicate.

This example brings up another advantage of replication: since the replicate weights contain all pertinent sample design information, subsetting the data for observations in the domain of interest is technically permitted. For example, if we were to subset cases in the data set TEST_BRR where TUTOR='Y' prior to running the PROC SURVEYMEANS step in Program 9.6, we would obtain the same estimated standard error for that domain reported via the DOMAIN statement. Note that this is in stark contrast to what transpired under the TSL variance estimation approach. As we saw in Program 8.2, subsetting for cases where TUTOR='Y' did not yield the same standard error as that reported via the DOMAIN statement. Although it was stressed previously that, unless the domain constitutes a proper subset, we must use a DOMAIN statement or create a domain-specific weight, technically speaking, that directive only applies to TSL variance estimation.

Program 9.6: Comparing BRR and Fay's BRR for Domain Estimation

```
title '1) Domain Estimation with BRR';
proc surveymeans data=test_BRR varmethod=BRR mean stderr;
  var grade2;
```

```
weight weight;
repweights RepWt_1-RepWt_8;
domain tutor;
run;

title "2) Domain Estimation with Fay's BRR";
proc surveymeans data=test_BRR_Fay varmethod=BRR (fay=.3)
mean stderr;
   var grade2;
weight weight;
repweights RepWt_1-RepWt_8;
domain tutor;
run;
```

1. Domain Estimation with BRR

Domain Statistics in Tutor			
Tutor	Variable	Mean	Std Error of Mean
N	grade2	87.439446	0.996418
Y	grade2	92.209591	1.441920

2. Domain Estimation with Fay's BRR

Domain Statistics in Tutor			
Tutor	Variable	Mean	Std Error of Mean
N	grade2	87.439446	0.953260
Y	grade2	92.209591	1.385047

9.5 The Jackknife

Another popular replication procedure is the *jackknife* (Chapter 4 of Wolter, 2007), the origins of which can be traced to Quenouille (1949), Tukey (1958), and Durbin (1959). There are actually several closely related forms of the jackknife used in practice, but the one we will focus on in this section is the traditional method, what Valliant et al. (2008) refer to as the "delete-one" version. The scheme is to delete each PSU, in turn, and weight up the remaining PSUs to form a replicate-specific estimate. As with BRR, the variance of any full-sample estimate is found by a universal function of variation among the replicate-specific estimates.

Since each PSU is deleted in one replicate, the number of replicate weights equals the number of PSUs. Although there tends to be more replicate weights to deal with, the jackknife is not restricted to two-PSU-per-stratum design as is BRR. In general, the replicate weights are constructed as follows:

1. For units in the dropped PSU, set all weights to 0.
2. For units in the same stratum as the dropped PSU, what SAS refers to as the "donor stratum," inflate the weights by a factor of $n_h/(n_h - 1)$, where n_h is the number of PSUs in the donor stratum.
3. For units outside the donor stratum, the replicate weight takes on the same value as the original weight. (If there is no stratification in the sample design, you can skip this step.)

After computing the full-sample point estimate, $\hat{\theta}$, and the like using all replicate weights, $\hat{\theta}_r$ ($r = 1,...,R$), the jackknife variance estimate is

$$\text{var}_{JK}(\hat{\theta}) = \sum_{r=1}^{R} \frac{n_h - 1}{n_h}(\hat{\theta}_r - \hat{\theta})^2 \tag{9.5}$$

Observe how the formula is no longer the mean squared error of the replicate-specific estimates, but instead a multiplicative factor of $(n_h - 1)/n_h$ is applied to each replicate estimate's squared deviation from the full-sample estimate. This stratum-specific factor is called the "jackknife coefficient."

Program 9.7 requests a standard error for the mean of GRADE2 via the jackknife. Specifying VARMETHOD=JK is all that is needed to have SAS carry out the three-step process sketched out above. The optional syntax in parentheses immediately thereafter tells SAS to append the jackknife replicate weights to the input data set and store the result in a data set named TEST_JK, and also to store the jackknife coefficients in a data set named TEST_JK_COEFS. As we will see in Program 9.8, we need to specify this supplemental data set as part of the REPWEIGHTS statement when VARMETHOD=JK. We find in the output the same point estimate is produced, but the standard error (1.1839) is slightly different from the techniques employed previously in this chapter, but unbiased nonetheless.

Program 9.7: Jackknife Variance Estimation

```
proc surveymeans data=test varmethod=JK
  (outweights=test_JK outjkcoefs=test_JK_coefs) mean stderr;
  stratum class;
  cluster homeroom;
  var grade2;
weight weight;
run;
```

SURVEYMEANS Procedure

Data Summary	
Number of strata	4
Number of clusters	8
Number of observations	16
Sum of weights	715

Variance Estimation	
Method	Jackknife
Number of replicates	8

Statistics		
Variable	Mean	Std Error of Mean
grade2	89.317483	1.183876

Figures 9.3 and 9.4 offer a visualization of the output data sets housing the jackknife replicate weights and jackknife coefficients, respectively. There are eight replicate weights generated, one for as many PSUs in the sample, although only the first four are shown in Figure 9.3. Underneath the label "Replicate Weight," they are named REPWT_1, REPWT_2,..., REPWT_8,

	Class	Homeroom	grade2	Weight	Replicate Weight	Replicate Weight	Replicate Weight	Replicate Weight
1	12	1	94	49.5	49.5	49.5	49.5	49.5
2	12	1	89	49.5	49.5	49.5	49.5	49.5
3	12	2	94	48	48	48	48	48
4	12	2	92	48	48	48	48	48
5	11	1	84	47.5	47.5	47.5	47.5	47.5
6	11	1	95	47.5	47.5	47.5	47.5	47.5
7	11	2	95	48	48	48	48	48
8	11	2	97	48	48	48	48	48
9	10	1	81	39	39	39	0	78
10	10	1	84	39	39	39	0	78
11	10	2	87	37.5	37.5	37.5	75	0
12	10	2	85	37.5	37.5	37.5	75	0
13	9	1	88	40	0	80	40	40
14	9	1	91	40	0	80	40	40
15	9	2	85	48	96	0	48	48
16	9	2	84	48	96	0	48	48

FIGURE 9.3
Partial view of the jackknife replicate weights on data set TEST_JK created in Program 9.7.

Replicate Number	Donor Stratum	Jackknife Coefficient
1	1	1	0.5
2	2	1	0.5
3	3	2	0.5
4	4	2	0.5
5	5	3	0.5
6	6	3	0.5
7	7	4	0.5
8	8	4	0.5

FIGURE 9.4
Data set view of the TEST_JK_COEFS data set containing the jackknife coefficients created in Program 9.7.

the same nomenclature SAS uses when appending BRR replicate weights. SAS proceeds through the algorithm of dropping PSUs according to the sort order of their codes. This is why the PSU associated HOMEROOM=1 in CLASS=9 was dropped in the first replicate. We can also observe in the first replicate how units in the second PSU within the same donor stratum had their weights multiplied by $n_h/(n_h - 1) = 2$, whereas the weights of all units in other strata were not changed. In Figure 9.4, we can observe that the data set TEST_JK_COEFS contains key variables REPLICATE and JKCOEFFICIENT that the SURVEY PROC will be looking for in subsequent analyses.

Program 9.8 reproduces the analysis in Program 9.7 by supplying PROC SURVEYMEANS with the data set containing the jackknife replicate weights. With the REPWEIGHTS statement and VARMETHOD=JK appearing in the PROC statement, we no longer need to include the STRATUM or CLUSTER statements. The JKCOEF=TEST_JK_COEFS option after the slash in the REPWEIGHTS statement points to the supplemental data set housing the jackknife coefficients, the data set shown in Figure 9.4. You can specify a scalar if, as in the present case, all coefficients are the same. Hence, alternative syntax for this particular example would be JKCOEF=0.5.

Program 9.8: Jackknife Variance Estimation Using a Data Set with Replicate Weights Appended

```
proc surveymeans data=test_JK varmethod=JK mean stderr;
   var grade2;
weight weight;
repweights RepWt_1-RepWt_8 / jkcoefs=test_JK_coefs;
run;
```

SURVEYMEANS Procedure

Data Summary	
Number of observations	16
Sum of weights	715

Variance Estimation	
Method	Jackknife
Replicate weights	TEST_JK
Number of replicates	8

Statistics		
Variable	Mean	Std Error of Mean
grade2	89.317483	1.183876

In the presence of a two-PSU-per-stratum sample design, a noteworthy simplification to the jackknife procedure can be employed (cf. Section 3.6.3.1 of Heeringa et al., 2010). Westat (2007, Appendix A) refers to this as the *JK*2 approach. For linear estimators such as totals, it can be shown that retaining only one of the two replicates from each stratum and setting the jackknife coefficient to 1 is algebraically equivalent to the variance estimate obtained via the full jackknife procedure. For nonlinear estimates such as weighted means, the variance is not necessarily the same, but still unbiased. Thus, we can cut our workload in half by randomly selecting one of the two replicates from all H strata and employing the following modified formula:

$$\mathrm{var}_{JK2}(\hat{\theta}) = \sum_{\tilde{r}=1}^{R/2} (\hat{\theta}_{\tilde{r}} - \hat{\theta})^2 \qquad (9.6)$$

where $\tilde{r}$ indexes the replicate randomly chosen within each stratum.

Suppose we did this for our TEST_JK data set and selected the second, third, fifth, and eighth jackknife replicates. Although we want to set the jackknife coefficient to 1, SAS requires a value strictly between 0 and 1. (If we omit the JKCOEF= option altogether, the SURVEY procedure assumes it to be $(R - 1)/R$, where R is the number of replicate weights in the REPWEIGHTS statement.) The work-around is to specify a number that is inconsequentially less than 1, like 0.99999, as is done in Program 9.9. The corresponding output shows how the standard error (1.1918) is in the neighborhood of the standard error obtained from implementing the full jackknife procedure (1.1839).

Program 9.9: Illustrating the JK2 Variance Estimation Method, a Simplified Jackknife Procedure for Two-PSU-per-Stratum Designs

```
proc surveymeans data=test_JK varmethod=JK mean stderr;
   var grade2;
weight weight;
repweights RepWt_2 RepWt_3 RepWt_5 RepWt_8 / jkcoefs=0.99999;
run;
```

SURVEYMEANS Procedure

Data Summary	
Number of observations	16
Sum of weights	715

Variance Estimation	
Method	Jackknife
Replicate weights	TEST_JK
Number of replicates	4

Statistics		
Variable	Mean	Std Error of Mean
grade2	89.317483	1.191791

9.6 The Bootstrap

The third replication technique considered in this chapter is the *bootstrap* (Efron and Tibshirani, 1993) and its application to variance estimation of complex survey data (McCarthy and Snowden, 1985; Rao and Wu, 1988; Lahiri, 2003). Although the technique is not explicitly offered in the SURVEY procedures at the time of this writing (i.e., there is no VARMETHOD=BOOT), techniques shown in Lohr (2012) can be used as a work-around.

Perhaps the most popular form of the bootstrap in complex survey statistics is the *nonparametric bootstrap*. The first step is to independently select $n_h - 1$ PSUs with replacement from each stratum. If we let n_{hib}^* denote the number of times ith PSU from the hth stratum was selected in the bth bootstrap sample, we can define the bth set of replicate weights as $w_{hijb} = w_{hij}\left(n_h/(n_h-1)\right)n_{hib}^*$, where j indexes units within the PSU. In words, this means we multiply the weights of all units in the ith PSU from the hth stratum by $n_h/(n_h - 1)$ times the number of times that PSU was selected. Using each replicate weight, we calculate $\hat{\theta}_b$, an estimate of the quantity of interest. The idea is to repeat the process independently *B* times, where *B* is often 200 or more, and then use the following formula to estimate the full-sample variance:

$$\text{var}_{boot}(\hat{\theta}) = \frac{1}{B-1} \sum_{b=1}^{B} (\hat{\theta}_b - \hat{\theta})^2 \tag{9.7}$$

where, as before, $\hat{\theta}$ denotes the estimate calculated from the full sample.

To use this technique with our example complex survey data set TEST, we must first create the bootstrap replicate weights. For this task, we can use the %BOOTWT macro developed by Lohr (2012). The first macro parameter names the input data set, and the second identifies the full-sample weight variable. The third and fourth identify the stratum and cluster identifiers, respectively. The fifth parameter is the number of bootstrap replicates requested. The sixth parameter names the output data set to consist of the input data set with replicate weights appended, and the seventh parameter is a prefix to be used in naming the replicate weights. The eighth and final parameter is a seed the user can specify to ensure the replicate weight creation process is reproducible—an addition this author made to the original macro definition. The macro call after the compilation step results in 200 bootstrap replicate weights named REPWT_1, REPWT_2,..., REPWT_200 being tacked onto the output data set called TEST_BOOT.

Program 9.10: Creating Bootstrap Replicate Weights Using the %BOOTWT Macro

```
%macro bootwt (fulldata,wt,stratvar,psuvar,numboot,fullrep,
               rep wt,seed);
proc sort data=&fulldata out=fulldata;
  by &stratvar &psuvar;
run;

proc sql stimer;
  create table psulist as
  select distinct &stratvar, &psuvar
    from fulldata
   order by &stratvar, &psuvar;

  create table numpsu as
  select distinct &stratvar, count(*)-1 as _nsize_
    from psulist
   group by &stratvar
   order by &stratvar;
quit;
```

```
data fulldata (drop=_nsize_);
  merge fulldata (in=inf)
        numpsu    (in=inn);
  by &stratvar;
if inf & inn;
wtmult = (_nsize_ + 1) / _nsize_;
run;

proc surveyselect data=psulist method=urs sampsize=numpsu
                  out=repout outall reps=&numboot seed=&seed;
  strata &stratvar;
  id &psuvar;
run;

proc sort data=repout (keep=&stratvar &psuvar replicate
numberhits) out=repout_sorted;
  by &stratvar &psuvar replicate;
run;

proc transpose data=repout_sorted (keep=&stratvar &psuvar
                                   replicate numberhits)
     out=repout_tr (keep=&stratvar &psuvar repmult:)
     prefix=repmult;
  by &stratvar &psuvar;
  id replicate;
  var numberhits;
run;

data &fullrep (drop=i repmult1-repmult&numboot wtmult);
array &repwt   (&numboot);
array repmult (&numboot);
  merge fulldata   (in=inf)
        repout_tr (in=inr);
  by &stratvar &psuvar;
  do i = 1 to &numboot;
    &repwt(i) = &wt * repmult(i) * wtmult;
  end;
run;

%mend bootwt;

%bootwt (test,weight,class,homeroom,200,test_boot,RepWt_,
    399448);
```

The replicate weights shown in Figure 9.5 might strike you as closely resembling the BRR replicate weights shown in Figure 9.1. Indeed, BRR can be characterized as a kind of "smart" bootstrap requiring far fewer replications in the two-PSU-per-stratum setting. An advantage of the bootstrap technique,

	RepWt_1	RepWt_2	RepWt_3	RepWt_4	RepWt_5
1	0	80	0	80	0
2	0	80	0	80	0
3	96	0	96	0	96
4	96	0	96	0	96
5	78	78	78	0	0
6	78	78	78	0	0
7	0	0	0	75	75
8	0	0	0	75	75
9	0	0	95	95	95
10	0	0	95	95	95
11	96	96	0	0	0
12	96	96	0	0	0
13	99	99	99	99	0
14	99	99	99	99	0
15	0	0	0	0	96
16	0	0	0	0	96

FIGURE 9.5
View of first 5 bootstrap replicate weights on data set TEST_BOOT produced from Program 9.10.

however, just like the jackknife, is that it is directly amenable to variable PSU counts per stratum.

Once the replicate weights have been appended to the survey data set, we can utilize the REPWEIGHTS statement to do the heavy lifting in calculating all $B = 200$ bootstrap replicate estimates. Looking at Equation 9.7, Lohr (2012) notes how you can move the $1/(B - 1)$ term inside the summation and arrive at essentially the same structure given by Equation 9.5. This implies we could specify VARMETHOD=JK in the PROC statement and insert $1/(B - 1)$ as the JKCOEF= option. Acknowledging $1/(200 - 1) = 0.005025$ in the present case, Program 9.11 shows how this can be done for an estimate of the mean of GRADE2. From the output, we observe yet another valid standard error estimate (1.1561).

Program 9.11: Bootstrap Variance Estimation Using a Data Set with Replicate Weights Appended

```
proc surveymeans data=test_boot varmethod=JK mean stderr;
  var grade2;
weight weight;
repweights RepWt_1-RepWt_200 / jkcoef=0.005025;
run;
```

SURVEYMEANS Procedure

Data Summary	
Number of observations	16
Sum of weights	715

Variance Estimation	
Method	Jackknife
Replicate weights	TEST_BOOT
Number of replicates	200

Statistics		
Variable	Mean	Std Error of Mean
grade2	89.317483	1.156067

9.7 Replication with Linear Models

While the exposition thus far has focused on a single descriptive statistic, a weighted mean, replication techniques can also be used to estimate variances for multivariate statistics such as a vector of estimated linear model parameters $\hat{\mathbf{B}}$. These are alternatives that can be used in lieu of whichever TSL approach has been derived to estimate $\text{cov}(\hat{\mathbf{B}})$ (e.g., Fuller, 1975; Binder, 1983). For instance, the multivariate generalization to BRR is

$$\text{cov}_{BRR}(\hat{\mathbf{B}}) = \frac{1}{R} \sum_{r=1}^{R} (\hat{\mathbf{B}}_r - \hat{\mathbf{B}})(\hat{\mathbf{B}}_r - \hat{\mathbf{B}})^{\mathsf{T}} \qquad (9.8)$$

where

$\hat{\mathbf{B}}_r$ denotes the rth replicate-weight estimate of the given model parameters

$\hat{\mathbf{B}}$ symbolizes the parameters estimated using the full-sample weight

Suppose that we wanted to fit a simple linear regression model predicting GRADE2 based on a 0/1 indicator variable of whether the student regularly receives math tutoring. We can use the same VARMETHOD=BRR and REPWEIGHTS syntax on the data set TEST_BRR to estimate $\text{cov}_{BRR}(\hat{\mathbf{B}})$. In Program 9.12, we first create the indicator variable TUTOR_Y in a DATA step and then run PROC SURVEYREG to fit the model of GRADE based on an intercept and TUTOR_Y. The COVB option is specified after the slash in the MODEL statement to output $\text{cov}_{BRR}(\hat{\mathbf{B}})$. Since we have already discussed in

Chapter 5 various components of the PROC SURVEYREG output, only that pertaining to the covariance matrix is shown here.

Program 9.12: BRR Variance Estimation for a Linear Model

```
data test_brr;
  set test_brr;
tutor_Y=(tutor='Y');
run;

proc surveyreg data=test_brr varmethod=BRR;
  model grade2 = tutor_Y / covb;
weight weight;
repweights RepWt_1-RepWt_8;
run;
```

Covariance of Estimated Regression Coefficients		
	Intercept	tutor_Y
Intercept	0.9928488886	−0.201326971
tutor_Y	−0.201326971	1.4889379508

To get a better handle on the underlying calculations prescribed by Equation 9.8, let us briefly walk through how PROC SURVEYREG arrives at these numbers. Essentially, the weighted least-squares parameters are estimated using the variable WEIGHT and then for REPWT_1, REPWT_2,…, and REPWT_8. The estimates of the intercept and slope using the variable WEIGHT are 87.4394 and 4.7701, respectively, and Table 9.1 summarizes the eight sets of estimates found using each of the BRR replicate weights.

TABLE 9.1

Summary of BRR Replicate-Specific Parameter Estimates for Simple Linear Regression Model Fitted in Program 9.12

Replicate	$\hat{B}_0$	$\hat{B}_1$
1	87.0971	5.5621
2	86.9463	5.0743
3	89.1082	5.0663
4	87.4393	2.1300
5	86.6549	5.7568
6	86.3621	5.6379
7	89.1411	4.4676
8	87.0849	3.2485

The figures in Table 9.1 are coalesced to produce the following matrix generated by Program 9.12:

$$\text{cov}_{BRR}(\hat{\mathbf{B}}) = \begin{bmatrix} \text{var}_{BRR}(\hat{B}_0) & \text{cov}_{BRR}(\hat{B}_0, \hat{B}_1) \\ \text{cov}_{BRR}(\hat{B}_0, \hat{B}_1) & \text{var}_{BRR}(\hat{B}_1) \end{bmatrix} \tag{9.9}$$

where, for example

$$\text{var}_{BRR}(\hat{B}_0) = \frac{1}{8}\left[(87.0971 - 87.4394)^2 + \cdots + (87.0849 - 87.4394)^2 \right]$$

$$\text{cov}_{BRR}(\hat{B}_0, \hat{B}_1) = \frac{1}{8}\left[\begin{array}{l} (87.0971 - 87.4394)(5.5621 - 4.7701) + \cdots \\ + (87.0849 - 87.4394)(3.2485 - 4.7701) \end{array} \right].$$

Other replication techniques can be employed analogously. For example, let us consider applying the jackknife replication technique to the reduced logistic regression model from Chapter 6 in which we used the NAMCS 2010 complex survey data set to model the log-odds of medication being prescribed (MED=1) during the visit as a function of patient race (RACER), primary reason for the visit (MAJOR), and the total number of chronic conditions afflicting the patient (TOTCHRON). Program 9.13 refits this model and estimates the model parameter covariance matrix using the default TSL generalization attributable to Binder (1983) as well as via the jackknife. Notice from the output how the estimated model parameters produced from either run are identical, but the standard errors vary somewhat. As a consequence, the Wald chi-square statistics and associated *p* values change somewhat, yet none of the discrepancies are large enough to substantively modify any conclusions.

Program 9.13: Comparing Taylor Series Linearization and Jackknife Variance Estimation for a Linear Model

```
title '1) Taylor Series Linearization';
proc surveylogistic data=NAMCS_2010 varmethod=Taylor;
   strata CSTRATM;
   cluster CPSUM;
   class RACER MAJOR / param=ref;
   model MED (event='1') = RACER MAJOR TOTCHRON;
weight PATWT;
run;

title '2) Jackknife Repeated Replication';
proc surveylogistic data=NAMCS_2010 varmethod=JK;
   strata CSTRATM;
   cluster CPSUM;
```

```
class RACER MAJOR / param=ref;
model MED (event='1') = RACER MAJOR
                        TOTCHRON;
weight PATWT;
run;
```

1. Taylor Series Linearization

SURVEYLOGISTIC Procedure

Analysis of Maximum Likelihood Estimates						
Parameter		DF	Estimate	Standard Error	Wald Chi-Square	Pr > ChiSq
Intercept		1	0.1827	0.1885	0.9399	0.3323
RACER	1	1	0.5013	0.1807	7.6998	0.0055
RACER	2	1	0.6982	0.1963	12.6503	0.0004
MAJOR	1	1	0.3678	0.0856	18.4523	<.0001
MAJOR	2	1	0.6156	0.1134	29.4838	<.0001
MAJOR	3	1	0.4944	0.1026	23.2282	<.0001
MAJOR	4	1	−0.7380	0.1231	35.9579	<.0001
TOTCHRON		1	0.1860	0.0195	91.1766	<.0001

2. Jackknife Repeated Replication

SURVEYLOGISTIC Procedure

Analysis of Maximum Likelihood Estimates						
Parameter		DF	Estimate	Standard Error	Wald Chi-Square	Pr > ChiSq
Intercept		1	0.1827	0.1958	0.8707	0.3508
RACER	1	1	0.5013	0.1886	7.0650	0.0079
RACER	2	1	0.6982	0.2040	11.7122	0.0006
MAJOR	1	1	0.3678	0.0860	18.2751	<.0001
MAJOR	2	1	0.6156	0.1144	28.9435	<.0001
MAJOR	3	1	0.4944	0.1039	22.6361	<.0001
MAJOR	4	1	−0.7380	0.1249	34.9048	<.0001
TOTCHRON		1	0.1860	0.0199	87.7793	<.0001

Be advised that there is not yet a SURVEY procedure to accommodate all possible linear models. One example is Poisson regression models, which are warranted when the outcome variable is a count of some kind. As stated previously, reading complex survey data into a procedure like PROC GLM with a WEIGHT statement will produce valid point estimates for model parameters, but the standard errors will not be correct. If faced with this situation, one plausible work-around would be to use a SURVEY procedure or the %BOOTWT macro to append replicate weights to the analysis file, and then tweak code in one of the user-written macros developed to perform replication for linear models—for instance, Bienias (2001) or Berglund (2004). Another potential work-around is to use the %SASMOD module of IVEware, a free set of SAS-callable macros developed by researchers at the University of Michigan (Raghunathan et al., 2002).

9.8 Replication as a Tool for Estimating Variances of Complex Point Estimates

In Program 3.6, we estimated the overall electricity intensity of commercial buildings in the CBECS 2003 target population, where intensity was defined as the ratio of annualized kilowatt hours of electricity used per square foot. Perhaps we wanted to test whether the intensities of two domains were significantly different from one another. Specifically, suppose the two domains were defined by primary building activity (variable PBA8) inpatient health care (PBA8=16) and outpatient health care (PBA8=8). We can express this domain comparison as the hypothesis test $H_0: R_1 = R_2$ versus $H_1: R_1 \neq R_2$, where R_1 denotes the intensity for buildings primarily devoted to inpatient health-care activities and R_2 is the like for outpatient health-care activities, and then proceed by computing the ratio difference and dividing through by the square root of the variance of that difference. This quotient would then be referenced against a t distribution with complex survey degrees of freedom. The variance is quite complex and involves numerous covariance terms, but using the skills we have honed over the course of this chapter, we have two options at our disposal: (1) use a replication approach such as BRR or (2) derive a linear substitute via the Woodruff method. We will demonstrate both in Program 9.14.

The first part of Program 9.14 estimates this ratio difference and estimates its variance via BRR. The initial PROC SURVEYMEANS run forms the BRR replicate weights and stores on an output data set called CBECS_2003_BRR. There are 48 replicate weights, the first multiple of four greater than the 44 strata in the 2003 CBECS sample design. The sole parameter of the macro %BRR_RATIODIFF is the number of replicates. While it is not essential to discuss every detail of the macro, the gist is that the ratio difference is computed once overall using the full-sample weight variable (ADJWT8) and again using each of the 48 replicate weights (REPWT_1, REPWT_2,..., REPWT_48). The macro produces a summary data set BRR_EST_ALL in which rows represent replicates. There are only two columns: (1) EST, the replicate-specific ratio difference, and (2) VAR, the squared deviation of this difference from the full-sample ratio difference, divided through by $1/R = 1/48$. See Figure 9.6 for a partial view of this summary data set. From here, values in the second column can be summed to arrive at the BRR variance estimate.

The second part of Program 9.14 demonstrates an application of the Woodruff (1971) algorithm introduced in Section 9.2 to obtain a TSL variance estimate of the same quantity. While the syntax may seem simpler at first glance, deriving the linear substitute is the hard part. If we express the domain ratio difference as $\hat{R}_1 - \hat{R}_2 = \hat{T}_1/\hat{T}_2 - \hat{T}_3/\hat{T}_4$, we see how the estimate is comprised of four estimated totals. From Section 9.2, the procedure is to multiply each of the four totals by the partial derivative of the function with

	est	var
1	9.1677114776	0.0971692694
2	14.3026411	0.1844216275
3	6.2244607955	0.5424933823
4	14.56286731	0.2180925593
5	14.145572976	0.1654639169
6	12.319765006	0.0205176635
7	10.470324141	0.0153026584
8	9.239917285	0.0907803924
9	12.136818959	0.0136501587
10	13.213540197	0.0741174585

FIGURE 9.6
Partial view of data set BRR_EST_ALL created from the %BRR_RATIODIFF macro in
Program 9.14.

respect to that total, and sum all four resulting terms. After some algebra, we
arrive at the following:

$$u_{hij} = d_{1hij}\left(\frac{1}{\hat{T}_2}\right)\left(w_{hij}x_{hij} - \left(\frac{\hat{T}_1}{\hat{T}_2}\right)w_{hij}y_{hij}\right) - d_{2hij}\left(\frac{1}{\hat{T}_4}\right)\left(w_{hij}x_{hij} - \left(\frac{\hat{T}_3}{\hat{T}_4}\right)w_{hij}y_{hij}\right) \quad (9.10)$$

where
  d_{1hij} is a 0/1 indicator variable of the jth building in the ith PSU in the hth
    stratum residing in the first domain
  d_{2hij} is the like for the second domain
  w_{hij} is the analysis weight
  x_{hij} is the annualized kilowatt hours of electricity used by the building
  y_{hij} is the square footage of the building

The PROC SQL step in Program 9.14 amalgamates all of these terms. The
result is provided to PROC SURVEYMEANS, which is used to obtain an
estimated total and associated variance accounting for the stratification and
clustering in the CBECS sample design.

**Program 9.14: BRR Variance Estimate of the Difference between Two
Domain Ratios**

```
* generate BRR replicate weights;
proc surveymeans data=CBECS_2003 varmethod=BRR
  (outweights=CBECS_2003_brr);
  stratum STRATUM8;
  cluster PAIR8;
```

```
   var ELCNS8 SQFT8;
   ratio ELCNS8 / SQFT8;
weight ADJWT8;
domain PBA8;
run;

%macro BRR_ratiodiff(replicates);
* housecleaning;
proc sql; drop table BRR_est_all; quit;

* store full-sample estimates in macro variables;
proc sql noprint;
select sum(ADJWT8*ELCNS8)/sum(ADJWT8*SQFT8) into :ratio1
   from CBECS_2003_brr where PBA8=16;
select sum(ADJWT8*ELCNS8)/sum(ADJWT8*SQFT8) into :ratio2
   from CBECS_2003_brr where PBA8=8;
select sum((PBA8=16)*ADJWT8*ELCNS8)/
sum((PBA8=16)*ADJWT8*SQFT8)  -
      sum((PBA8=8 )*ADJWT8*ELCNS8)/sum((PBA8=8)
      *ADJWT8*SQFT8)
         into :ratio_diff_FS
   from CBECS_2003_brr;
quit;

* loop through all replicate weights and get the same
estimate;
%do r=1 %to &replicates;
proc sql;
   create table BRR_est_&r as
   select ratio1 - ratio2 as est from
  (select sum((PBA8=16)*RepWt_&r*ELCNS8)/
         sum((PBA8=16)*RepWt_&r*SQFT8) as
      ratio1
    from CBECS_2003_brr),
  (select sum((PBA8=8)*RepWt_&r*ELCNS8)/
         sum((PBA8=8)*RepWt_&r*SQFT8) as
      ratio2
    from CBECS_2003_brr);
quit;
proc datasets nolist;
   append data=BRR_est_&r base=BRR_est_all FORCE;
   delete BRR_est_&r;
run;
quit;
%end;

* consolidate replicate-specific estimates and find squared
deviations from
   full-sample estimate;
data BRR_est_all;
```

```
  set BRR_est_all;
var=(1/&replicates)*(est - &ratio_diff_FS)**2;
run;

proc sql;
   select &ratio1 as Ratio1,&ratio2 as Ratio2,
          &ratio1 - &ratio2 as Difference,
          sqrt(sum(var)) as SE_brr,
          sum(var) as var_brr
     from BRR_est_all;
quit;
%mend;

%BRR_ratiodiff(48);

* Woodruff TSL approach to estimate the variance of the same
quantity;
proc sql noprint;
   select sum(ADJWT8*ELCNS8) into :T1 from CBECS_2003 where
   PBA8=16;
   select sum(ADJWT8*SQFT8)  into :T2 from CBECS_2003 where
   PBA8=16;
   select sum(ADJWT8*ELCNS8) into :T3 from CBECS_2003 where
   PBA8=8;
   select sum(ADJWT8*SQFT8)  into :T4 from CBECS_2003 where
   PBA8=8;

   create table lin_subs as
   select STRATUM8,PAIR8,
        sum((1/&T2)*(PBA8=16)*(ADJWT8*ELCNS8
        - (&T1/&T2)*ADJWT8*SQFT8) -
        (1/&T4)*(PBA8=8)*(ADJWT8*ELCNS8
        - (&T3/&T4)*ADJWT8*SQFT8))
              as u_i
     from CBECS_2003
group by STRATUM8,PAIR8;
quit;

* estimate variance of the sum of the linear substitute with
respect to the design;
proc surveymeans data=lin_subs varsum std;
   stratum STRATUM8;
   cluster PAIR8;
   var u_i;
run;
```

Ratio1	Ratio2	Difference	SE_brr	var_brr
27.45229	16.12492	11.32737	2.740111	7.508209

SURVEYMEANS Procedure

Data Summary	
Number of strata	44
Number of clusters	68
Number of observations	88

Statistics		
Variable	Std Dev	Var of Sum
u_i	2.710924	7.349109

Using the figures output by the macro, we find $\hat{R}_1 = 27.4523$ and $\hat{R}_2 = 16.1249$. Thus, the estimated ratio difference is $\hat{R}_1 - \hat{R}_2 = 11.3274$. Moreover, we note $\text{var}_{BRR}(\hat{R}_1 - \hat{R}_2) = 7.5082$. To test whether the difference is significantly different from zero, we can reference the observed test statistic $t = 11.3274/\sqrt{7.5082} = 4.13$ against a reference t distribution with 44 degrees of freedom. The p value is well below 0.05, suggesting the difference is significant, so we can safely conclude that the intensity of inpatient health-care buildings is greater than that of outpatient health-care buildings. The estimated variance using the Woodruff method is 7.3491, very near that acquired from BRR, and so we would have reached the same conclusion either way.

9.9 Degrees of Freedom Adjustments

Examples thus far in the chapter have dealt primarily with utilizing replication to approximate one or more measures of variability. If you wish to have a SURVEY procedure deploy these measures as part of a significance test or to construct confidence intervals, it is important to remain cognizant of SAS' rules for assigning degrees of freedom to the underlying reference distributions. When you specify a replication technique in the VARMETHOD= option in the PROC statement without the REPWEIGHTS statement—but with a STRATA and/or CLUSTER statement, if applicable—SAS uses the same # PSUs – # strata rule of thumb used for VARMETHOD=TAYLOR, which coincides with the recommendation given in Figure 4.20 of Westat (2007). If a REPWEIGHTS statement is present, however, SAS assigns the degrees of freedom to be the number of replicate weight variables. If this is not appropriate, you can reassign the value using the DF=*number* option after the slash in the REPWEIGHTS statement.

Let us revisit the bootstrap replicate weights on the data set TEST_BOOT used for making inferences on the mean of the variable GRADE2.

Program 9.15 builds on syntax shown previously in Program 9.11 by adding the CLM and DF options in the PROC statement to output a 95% confidence interval and the degrees of freedom used for the underlying t distribution. The first SURVEYMEANS run uses the default degrees of freedom, 200, which is vastly overstated. Following Lohr (2012, p. 12), the second run overrides the default with a more appropriate number, DF=4, the # PSUs – # strata in the underlying sample design. Note from the output how the confidence interval is narrower in the first run as compared to the second. Hence, failing to properly adjust for the degrees of freedom would have led us to overstate our precision.

Program 9.15: Adjusting the Degrees of Freedom when Using Replication for Variance Estimation

```
title '1) Bootstrap CI Using Default DF Calculation';
proc surveymeans data=test_boot varmethod=JK
    mean stderr clm df;
  var grade2;
weight weight;
repweights RepWt_1-RepWt_200 / jkcoef=0.005025;
run;

title '2) Bootstrap CI Overriding Default DF Calculation';
proc surveymeans data=test_boot varmethod=JK
    mean stderr clm df;
  var grade2;
weight weight;
repweights RepWt_1-RepWt_200 / jkcoef=0.005025 DF=4;
run;
```

1. Bootstrap CI Using Default DF Calculation

SURVEYMEANS Procedure

Data Summary	
Number of observations	16
Sum of weights	715

Variance Estimation	
Method	Jackknife
Replicate weights	TEST_BOOT
Number of replicates	200

Statistics					
Variable	DF	Mean	Std Error of Mean	95% CL for Mean	
grade2	200	89.317483	1.156067	87.0378387	91.5971263

2. Bootstrap CI Overriding Default DF Calculation

SURVEYMEANS Procedure

Data Summary	
Number of observations	16
Sum of weights	715

Variance Estimation	
Method	Jackknife
Replicate weights	TEST_BOOT
Number of replicates	200

Statistics					
Variable	DF	Mean	Std Error of Mean	95% CL for Mean	
grade2	4	89.317483	1.156067	86.1077265	92.5272386

Similar logic applies if we were to flesh out Programs 9.4, 9.6, and 9.8; in all of those instances, 8 degrees of freedom are assumed, which is still too many. Of course, when restricting analysis to a sparse population domain like we did in Program 9.6, it may be appropriate to reduce the degrees of freedom even further (Rust and Rao, 1996; Chen and Krenzke, 2013). Generally speaking, the same arguments prompting the proposed adjustments discussed in Section 8.7 are pertinent with replication methods.

9.10 Summary

Replication techniques are flexible alternatives that complex survey data analysts can utilize to estimate variances in lieu of TSL, the default method used by SAS and most other software. Instead of a unique formula for each point estimate, replication techniques generally employ a universal formula. For complex estimators, particularly those not readily available in a SURVEY procedure, pursing one of these approaches can save time and effort, and may be the only feasible option. For example, the linearization process hits a snag when the estimator is not differentiable, as is the case for a quantile. Replication techniques can be applied in these instances, although results from Kovar et al. (1988) suggest the jackknife should be avoided. Aside from simplicity in implementation, several other distinct advantages of appending replicate weights to a survey data set were noted, such as there being no need to retain stratum and PSU identifiers and the ability to conduct domain analyses by simply subsetting the data.

We did not consider all replication techniques used in practice. For a more comprehensive treatment of the subject, see Wolter (2007). Replication techniques or subtle variants thereof are sometimes proposed as a way to account for the uncertainty of missing data adjustments. For instance, many researchers (e.g., Valliant, 2004) argue that applying the nonresponse adjustment procedure independently on each replicate weight is needed to capture the uncertainty attributable to techniques compensating for unit nonresponse (see Chapter 10) since TSL variance estimation implicitly assumes a known, fixed weight. We will see a convenient way to do this in Program 10.6. As another example, Efron (1994) proposed a bootstrap approach to account for the uncertainty when imputing missing data for item nonresponse (see Chapter 11).

There may be circumstances in which a particular replication technique as prescribed in this chapter is either difficult or outright impossible to implement. For example, we cannot use BRR or Fay's BRR directly unless there are exactly 2 PSUs in all strata. Moreover, for designs with a large number of PSUs, the number of replicate weights required can become cumbersome, even with modern computing power. The typical workaround is to group PSUs into pseudo-PSUs, collapse strata into pseudo-strata, or perhaps a combination of both. Some general references are Rust (1986), Kott (2001), Appendix D of Westat (2007), and the references cited in the appendix of Mukhopadhyay et al. (2008). These grouping procedures do not bias the variance estimators in any way, but they do sacrifice some degrees of freedom.

10

Weight Adjustment Methods

10.1 Introduction

Nonresponse is a pervasive problem in applied survey research. No matter
how aggressive or ambitious the data collection protocol may be, the reality
is that few surveys are able to collect complete data for the entire sample.
Even surveys with a legal mandate for sampled individuals to participate,
such as the U.S. Decennial Census, are faced with some degree of nonresponse. Common causes for nonresponse include failing to locate or make
contact with the sampling unit, the respondent refusing to answer one or
more sensitive questions, or the respondent refusing to participate in the
survey altogether. Whichever the cause, nonresponse presents a dilemma
during the estimation stage.

Arguably the most frequent "treatment" is to ignore the missing data. That
is, the analyst will utilize only the observed portion of the sample without
making any kind of adjustment. Indeed, the default approach taken by most
software is *list-wise deletion*, maintaining only those records with nonmissing
data on the variables needed for computation. But even a seemingly inconsequential rate of missingness can result in a substantive loss of information
and reduced efficiency. For example, suppose one was fitting a model with 1
outcome variable and 5 predictor variables, each independently missing 3%
of the time. In expectation, list-wise deletion will reduce the sample size by
$1 - (0.97)^6 = 16.7\%$.

Survey sponsors typically conduct nonresponse adjustments prior to
releasing any kind of public-use data set. In fact, this is the case for the commercial buildings energy consumption survey (CBECS), national ambulatory medical care survey (NAMCS), and national survey of family growth
(NSFG) data sets analyzed throughout this book. In this author's view, however, any book on applied survey research would be remiss without some
background on these methods. Over the course of the next two chapters,
we will discuss two of the most frequently used classes of techniques to
compensate for nonresponse. The chapter discusses methods that adjust the
base weights of respondents to more accurately reflect known distributions
of the population (Kalton and Flores-Cervantes, 2003). The second approach

is to impute, or fill in, the missing data (Brick and Kalton, 1996). Imputation methods are covered in Chapter 11.

Four of the most common techniques for conducting weight adjustments are demonstrated in this chapter—namely, the adjustment cell method, the propensity cell method, poststratification, and raking. Syntax examples are motivated by nonresponse in a hypothetical personnel satisfaction survey similar in spirit to the Federal Employee Viewpoint Survey, an annual attitudinal survey of civil servants administered by the U.S. Office of Personnel Management.

10.2 Definitions and Missing Data Assumptions

As we discussed in Chapter 1, it has long been established from survey sampling theory that a randomly selected sample, even of moderate size, can be used to form unbiased (or approximately unbiased) estimates of the attributes of the larger population, a set of units collectively denoted U. Specifically, Horvitz and Thompson (1952) showed that as long as each unit is assigned a fixed, nonzero probability of selection into the sample S, denoted as $\Pr(i \in S)$, unbiased estimation can be achieved by assigning each sampled unit a weight that is the inverse of this probability, or $w_i = 1/\Pr(i \in S)$. This weight goes by several names, including the *base weight*, *sample weight*, or *design weight*, and can be interpreted as the number of population units represented by the sampled unit. The conundrum presented by nonresponse is that, because only a portion of the sample is observed, the unbiasedness properties proved in Horvitz and Thompson (1952) are no longer guaranteed to hold. Analyzing only the observed portion without making any statistical adjustments may introduce *nonresponse error* (Groves, 1989), or a deviation from the quantity that would be computed from the full sample had the complete data been available.

Discussed in Chapter 1 of Groves and Couper (1998), the magnitude of nonresponse error in the sample set S depends on both the statistic at hand and the degree of dissimilarity between S_1, the set of r observed cases and S_0, the set of m missing cases ($r + m = n$ and $S_1 \cup S_0 = S$). For example, suppose that the quantity of interest is a finite population total $Y = \sum_{i \in U} y_i$ of a particular variable taking on strictly positive values. An unbiased estimate of this quantity can be obtained from the sample by finding $\hat{Y}_n = \sum_{i \in S} w_i y_i$. The estimator utilizing only the observed portion of the sample, $\hat{Y}_r = \sum_{i \in S_1} w_i y_i$, is certain to underestimate Y since $\hat{Y}_r < \hat{Y}_n$ ($= \hat{Y}_r + \hat{Y}_m$), where $\hat{Y}_m = \sum_{i \in S_0} w_i y_i$ represents the base-weighted total of the m missing cases. By comparison, suppose that the quantity of interest is a finite population mean $\bar{y} = \sum_{i \in U} y_i / N$, for which an approximately unbiased estimate from the full sample can be computed by finding $\hat{\bar{y}}_n = \sum_{i \in S} w_i y_i / \sum_{i \in S} w_i$. In the presence of nonresponse, if we let

$\hat{\bar{y}}_r = \sum_{i \in S_1} w_i y_i / \sum_{i \in S_1} w_i$ denote the base-weighted mean of the r observed

cases and $\hat{\bar{y}}_m = \sum_{i \in S_0} w_i y_i / \sum_{i \in S_0} w_i$ the like for the m missing cases, then the

nonresponse error can be shown to equal

$$\text{NRerror}(\hat{\bar{y}}_r) = \hat{\bar{y}}_r - \hat{\bar{y}}_n = \frac{\sum_{i \in S_0} w_i}{\sum_{i \in S} w_i} \left(\hat{\bar{y}}_r - \hat{\bar{y}}_m \right) \tag{10.1}$$

In words, nonresponse error is the product of the base-weighted nonresponse rate and the difference in base-weighted means between the observed and missing cases. For the special case of a simple random sample design in which all units share the same base weight, the leading term reduces to the nonresponse rate, or m/n, and the two sample means are calculated without the weights.

In contrast to the negative nonresponse error for $\hat{Y}_r$ when $y_i > 0$ for all $i \in U$, the quantity in Equation 10.1 can be positive or negative. Specifically, if $\hat{\bar{y}}_r > \hat{\bar{y}}_m$, the quantity is positive, but if $\hat{\bar{y}}_r < \hat{\bar{y}}_m$, the quantity is negative. Another important takeaway is that a larger portion of missing data does not necessarily imply a larger nonresponse error, a point that has been demonstrated empirically in the survey methodology literature (Merkle and Edelman, 2002; Groves and Peytcheva, 2008). The reason is that, if $\hat{\bar{y}}_r \approx \hat{\bar{y}}_m$, a base-weighted nonresponse rate of 80% is essentially no more harmful than a rate of 20%. This partially explains why there is no straightforward answer to the following question that frequently arises: What is the minimum response rate necessary to achieve "valid" survey results? The unsatisfying answer is that it depends.

Figure 10.1 is an analogy provided to help understand the concept of nonresponse error for a sample mean. Imagine the outer rectangle represents a three-dimensional water tank (a cube) of which we have a two-dimensional view and that this tank has been partitioned by a separator running perpendicularly to the bottom of the tank, thereby rendering two disjoint water compartments. Let the water level of the left-hand compartment represent the base-weighted respondent mean, and the water level of the right-hand compartment represent the base-weighted mean for nonrespondents. Nonresponse error is the distance between the water level of the left-hand compartment and the resting water level that would be observed if the partition were removed and the two compartments' water commingled. This resting water level is represented by the horizontal dashed line. The relative portion of the tank's length to the left of the separator represents the

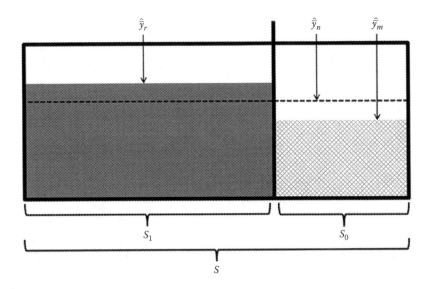

FIGURE 10.1
Visualization of nonresponse error in a sample mean using the analogy of a partitioned water tank.

base-weighted response rate. Regardless of where it falls, if the water levels of both the left- and right-hand side are similar, nonresponse error will be minimal, at least with respect to the sample mean.

There is a colloquial tendency to use the phrases *nonresponse error* and *nonresponse bias* interchangeably, but there is an important distinction to be made between the two. Nonresponse error refers to the deviation of a particular estimate computed from a particular sample, the difference between the estimate computed using only the observed portion of the sample and the like using the entire sample. Nonresponse bias, on the other hand, is most appropriately described as the expected magnitude of nonresponse error over hypothetical repetitions of the sampling process and what is referred to by Little and Rubin (2002) as the *missingness mechanism*.

Before delving into the taxonomy of the three possible missingness mechanisms, this is an apt point to define two general classifications of nonresponse. The first is *unit nonresponse*, which refers to situations when the sampling unit fails to respond to the survey request altogether (i.e., no data are collected). The second is *item nonresponse*, which refers to instances when answers are provided to some, but not all, of the questions in the survey. Figure 10.2 visually contrasts the two for an example survey with four outcome variables labeled y_1, y_2, y_3, and y_4. Ideally, a set of *auxiliary variables* (typically residing on the sampling frame) are known for both observed and missing cases and can be utilized in statistical adjustments to mitigate the nonresponse error. These variables are symbolized by the **X** in the figure.

Unit nonresponse					Item nonresponse				
	Outcome variables					Outcome variables			
X	y_1	y_2	y_3	y_4	**X**	y_1	y_2	y_3	y_4
	?	?	?	?				?	
							?		
	?	?	?	?		?			
									?

FIGURE 10.2
Visual comparison of unit nonresponse versus item nonresponse.

Weight adjustments are the typical remedy for unit nonresponse, whereas imputation is the typical remedy for item nonresponse. These are termed "typical" remedies because there can be some overlap. For example, imputation can be employed to combat unit nonresponse. Weight adjustments are less commonly used to compensate for item nonresponse, but they are feasible. The cumbersome practicality is that separate sets of weights may be needed for separate analyses.

Lessler and Kalsbeek (1992) discuss two traditional perspectives of survey nonresponse at length. The first to emerge in the survey nonresponse literature was the *deterministic* perspective, which asserts that the sampling frame of N units can be partitioned into N_1 units that will always respond and N_0 units that will never respond. This simplistic view served as a decent starting point for thinking through the consequences of nonresponse, but it has since been supplanted by the more realistic *stochastic* perspective, which asserts that each sampling unit has some probability, or *propensity*, of responding to a given survey, which is typically represented by ϕ_i. The terminology and application are most often credited to the ideas appearing in Rosenbaum and Rubin (1983), yet one could argue that the concepts trace back as far as Hartley (1946) and Politz and Simmons (1949).

Little and Rubin (2002) define three missingness mechanisms germane to the stochastic perspective of survey nonresponse. The first assumption is that data are *missing completely at random* (MCAR), which means that the propensities are independent of both the auxiliary variables and outcome variables. This is a strong assumption, essentially positing that the observed cases are a completely random subset of the cases originally sampled. The second assumption is that data are *missing at random* (MAR), which means that the propensities may vary based on the auxiliary variables, but not on the outcome variables. In other words, conditional on the auxiliary variables, data are MCAR. This is the tacit assumption underlying most of the weighting and imputation techniques utilized in practice and demonstrated in this book. The first two missingness mechanisms are collectively referred to by Little and Rubin (2002) as *ignorable missingness mechanisms*. The third assumption is the most perilous, data that are *not missing at random* (NMAR), implying that the propensities depend on the outcome variable beyond what can be explained and adjusted for by the available auxiliary variables. In contrast to the first two, Little and Rubin (2002) refer to this as *ignorable missingness mechanisms*. Naturally, this is a much more difficult situation to handle, one necessitating sophisticated techniques that are beyond the scope of this book. The reader seeking more detail is referred to Chapter 6 of Rubin (1987), Chapter 15 of Little and Rubin (2002), and Andridge and Little (2011).

It should be acknowledged that there are critics who believe nonresponse adjustment methods are futile and should not be performed. As Brick and Kalton (1996) point out, however, foregoing the adjustments is akin to assuming MCAR, which is far more questionable than the MAR assumption implicit in making the adjustments! Despite the difficulty verifying or refuting any one of the three classifications, the plausibility of the MAR assumption increases with a larger number of auxiliary variables.

Given a fixed and known (but not necessarily equal) propensity of responding ϕ_i for all units in the population, Bethlehem (1988) showed that over repeated samples of the same size from a population of N units, the nonresponse bias utilizing $\hat{\bar{y}}_r$, the base-weighted estimate of the sample mean for only the observed portion of the data, is approximately equal to

$$\text{NRbias}(\hat{\bar{y}}_r) \approx \frac{1}{N\bar{\phi}} \sum_{i=1}^{N} (\phi_i - \bar{\phi})(y_i - \bar{y}) \tag{10.2}$$

where $\bar{\phi} = \sum_{i=1}^{N} \phi_i / N$. Brick and Jones (2008) derive bias expressions similarly in spirit for other estimators.

The main takeaway from Equation 10.2 is that the bias is proportional to the covariance of the propensities and the survey variable. For this term to be nonzero, there must be a compelling relationship between the likelihood of responding and the outcome variable. For example, imagine a mail survey charged with estimating the proportion of the electorate that voted in the most recent presidential election. If people who did not vote are less inclined to respond the survey request, then the unadjusted sample proportion could be susceptible to a nontrivial amount of nonresponse error. If there is no relationship between the two, then there are no grounds to anticipate an appreciable amount of nonresponse error.

In effect, all of the weighting methods to be discussed first attempt to partition the sample data set into a mutually exclusive and exhaustive set of groupings referred to as "cells" or "strata." The objective is to construct groups such that the ϕ_is are roughly equal within each. In the voting survey example, perhaps the sampling frame contained auxiliary information regarding the individual's neighborhood, race/ethnicity, and age. These might serve as the candidate variables to be exploited in some form of a grouping scheme. If the assumption of roughly equal propensities holds, the implication is that data are MCAR within any given group, and so the weights of observed cases therein can be inflated to represent also the cases missing data. Because we are arguing that the missingness mechanism occurred more or less at random within the group, it follows that the subset of observed cases can sufficiently represent that group's population.

Of course, we generally do not know the ϕ_is, the true population propensities, but there are a variety of approaches to estimate them, which we distinguish by denoting as $\hat{\phi}_i$s. These different approaches to form these estimates, as well as the format and source of the auxiliary information (whether internal or external to the sampling frame), are the key differentiators of the particular method(s) at one's disposal.

Before continuing on to the specifics of the methods, let us introduce a few key data sets used in examples appearing in this and the following chapter. Syntax examples are derived from a hypothetical employee satisfaction survey motivated in large part by the Federal Employee Viewpoint Survey (www.fedview.opm.gov). Let us assume that the data set POP contains information on 15,000 individuals working for some organization. It is derived from a personnel database (i.e., an employee roster of individuals on board at a specific point in time) rife with auxiliary variables. For tractability, we will consider only the following four:

1. GENDER—M/F indicator of employee gender
2. SUPERVISOR—0/1 indicator of where the employee is a supervisor
3. MINORITY—Y/N indicator of minority status
4. AGE—age of employee at time of survey

GENDER and MINORITY are stored as character variables of length 1, whereas SUPERVISOR and AGE are numeric.

Suppose a single-stage, stratified random sample of 1200 employees was selected using supervisory status as the sole stratification factor. Because they are less prevalent in the population yet constitute a domain of analytic interest, supervisors were sampled at a higher rate than nonsupervisors. Table 10.1 summarizes the population and sample counts for these two strata, as well the corresponding base weights we must use to compensate for the differential sampling rates.

Suppose that the sampled employees were sent a personalized link via email to a web-based survey instrument comprised of a variety of attitudinal questions and certain demographics not readily available in the personnel database. Weekly reminder emails were sent to nonrespondents, but after a few weeks the survey closed with 628 completes, corresponding to a response rate of $628/1200 = 52.3\%$.

TABLE 10.1

Stratification Summary for Hypothetical Personnel Satisfaction Survey

Supervisory Status	Population Size	Sample Size	Base Weight
Nonsupervisor	13,281	1000	13,281/1,000=13.281
Supervisor	1,719	200	1,719/200=8.595

The data set SAMP is the subset of POP consisting of records for the 1200 sampled employees. In addition to the bulleted list of variables just defined, the data set contains a variable called WEIGHT_BASE equaling that shown in the rightmost column of Table 10.1 as well as a 0/1 indicator variable RESPOND flagging those employees who responded. We will assume that there are two key outcome variables in the data set:

1. Q1—a dichotomous 0/1 indicator variable signifying whether an employee replied "Agree" or "Strongly Agree" to the statement "I like the kind of work I do"

2. LOS—a continuous variable quantifying the individual's length of service, or employment duration with the organization

These two variables are missing for the 572 employees who did not respond, the records in SAMP where RESPOND=0. Figure 10.3 is a screenshot of the SAMP data set to visually summarize the mix of variables and their associated coding structures.

Over the span of the next four sections, we will demonstrate a variety of ways to make use of the four auxiliary variables on SAMP to inflate the base weights of respondents to better reflect overall population of the organization.

10.3 Adjustment Cell Method

The first weight adjustment method we will consider is the *adjustment cell method*. The notion behind this technique is to partition the sample into C mutually exclusive groups called *adjustment cells* based on the cross classification of one or more auxiliary variables. Once all sampling units have been allocated into one and only one cell, base weights of nonresponding units are shifted proportionally to the responding units. Operationally, this is accomplished by multiplying each responding unit's base weight by a cell-specific adjustment factor $f_c \geq 1$ defined as the sum of the base weights for all sampling units in that cell divided by the sum of base weights for only the responding units. If we denote the base weight of the ith unit in the cth cell w_{0ci} and let R_{ci}

	GENDER	SUPERVISOR	Minority	AGE	weight_base	Respond	Q1	LOS
1	F	0	N	44.082135524	13.281	0	.	.
2	F	0	N	40.917180014	13.281	1	0	1.4976043806
3	F	0	N	56.750171116	13.281	0	.	.
4	F	0	N	41.664613279	13.281	0	.	.
5	M	0	N	23.082819986	13.281	0	.	.
6	M	1	N	63.915126626	8.595	1	1	34.833675565
7	M	0	N	37.002053388	13.281	1	1	3.0828199863
8	F	1	N	57.497604381	8.595	1	0	24.246406571
9	F	0	N	40.83504449	13.281	0	.	.
10	F	0	N	57.582477755	13.281	1	1	35.748117728
11	M	0	N	30.833675565	13.281	1	1	1.916495551
12	M	1	N	62.168377823	8.595	0	.	.
13	M	0	N	43.748117728	13.281	0	.	.
14	M	0	N	84.082135524	13.281	0	.	.
15	M	0	N	64.583162218	13.281	1	1	42.083504449
16	F	0	Y	54.001368925	13.281	0	.	.
17	M	0	N	30.245037645	13.281	1	1	4.4161533196
18	F	0	N	58.32991102	13.281	1	1	33.245722108
19	M	0	N	64.583162218	13.281	0	.	.
20	M	0	N	57.245722108	13.281	0	.	.

FIGURE 10.3
Partial view of SAMP, a hypothetical survey data set of employees selected for a survey in which unit nonresponse occurred.

symbolize a 0/1 indicator variable for whether the unit responded, then the nonresponse-adjusted weight w_{1ci} can be expressed as

$$w_{1ci} = w_{0ci} \left(\frac{\sum_{i \in S_c} w_{0ci}}{\sum_{i \in S_{1c}} w_{0ci}} \right) R_{ci} = w_{0ci} f_c R_{ci} \tag{10.3}$$

where
  S_c is the set of sampling units in the cth cell
  S_{1c} is the set of responding units in the cth cell

The ultimate effect is that nonresponding units, those for which $R_{ci}=0$, have their weights set to 0, while the set of responding units has their weights multiplied by f_c. Note that there are alternative ways to calculate this inflation factor. Though the form presented here, the inverse of the base-weighted response rate in cell c, is perhaps the most commonly used, Little and Vartivarian (2003) argue it can be inefficient and instead suggest using the inverse of the unweighted cell response rate, or n_c/r_c.

Returning to the employee satisfaction survey, suppose nonresponse adjustment cells are formed by cross classifying the sample by gender and minority status. Program 10.1 compares the base-weighted sums of sampling units and respondents only for each of the four classifications. It is evident that, without any kind of adjustment, the weighted sum of respondents will underestimate the weighted sum of all sampling units. We can also infer that the response rates are not equivalent among the four cells since the two sets of percentages vary somewhat. For example, notice how the sample suggests roughly 10.56% of the population is comprised of minority males, whereas that figure is about 7.33% based on the subset of the sample who responded. Distributional imbalances such as these call the MCAR assumption into question.

Program 10.1: Comparing the Sum of Base Weights for Survey Respondents to the Original Sample within Adjustment Cells

```
title 'Sum of Base Weights for the Original Sample';
proc freq data=samp;
  tables gender*minority / list nocum;
weight weight_base;
run;

title 'Sum of Base Weights for Respondents';
proc freq data=samp;
  where respond=1;
  tables gender*minority / list nocum;
weight weight_base;
run;
```

Sum of Base Weights for Sampling Units

FREQ Procedure

GENDER	Minority	Frequency	Percent
F	N	5101.554	34.01
F	Y	2953.092	19.69
M	N	5361.759	35.75
M	Y	1583.595	10.56

Sum of Base Weights for Respondents

FREQ Procedure

GENDER	Minority	Frequency	Percent
F	N	3028.893	38.41
F	Y	1454.67	18.45
M	N	2824.239	35.81
M	Y	578.124	7.33

Program 10.2 demonstrates syntax to create a new, nonresponse-adjusted weight called WEIGHT_ADJ_CELL derived from WEIGHT_BASE on the data set SAMP. Weights of respondents are inflated by the cell-specific factor f_c named FACTOR_ADJ_CELL in the CELL_FACTORS data set, while the weights for nonrespondents are set to 0. Recall how placing a condition in parentheses as is done with RESPOND=1 in the ultimate DATA step in Program 10.2 returns a 1 if the condition is true and a 0 otherwise. When summed within a cell for only the respondents, WEIGHT_ADJ_CELL should now match the original base-weighted cell sum. It is good practice to verify this holds after completing the weight adjustments to ensure all has gone to plan. Indeed, we observe that the weighted totals output from the ultimate PROC FREQ run in Program 10.2 match those from the first PROC FREQ run in Program 10.1.

Program 10.2: Adjustment Cell Method to Adjust the Weights of Respondents to Compensate for Nonresponse

```
* sum the base weights for sampling units;
proc freq data=samp;
  tables gender*minority / out=counts_s
  (rename=(count=count_s));
weight weight_base;
run;

* sum the base weights for respondents;
proc freq data=samp;
  where respond=1;
  tables gender*minority / out=counts_r
  (rename=(count=count_r));
```

```
weight weight_base;
run;

* create data set of adjustment cell weight inflation factors;
data cell_factors;
  merge counts_s
        counts_r;
  by gender minority;
factor_adj_cell=count_s/count_r;
keep gender minority factor_adj_cell;
run;

* merge these factors into sample data set and assign a new,
  nonresponse-adjusted weight;
proc sort data=samp;          by gender minority; run;
proc sort data=cell_factors; by gender minority; run;
data samp;
  merge samp           (in=a)
        cell_factors (in=b);
  by gender minority;
if a;
weight_adj_cell=weight_base*factor_adj_cell*(respond=1);
run;

* verify sum of adjusted weights for respondents match sum of
  base weights for the full sample;
title 'Sum of Adjustment Cell Weights';
proc freq data=samp;
  where respond=1;
  tables gender*minority / list nocum;
weight weight_adj_cell;
run;
```

Sum of Adjustment Cell Weights

FREQ Procedure

GENDER	Minority	Frequency	Percent
F	N	5101.554	34.01
F	Y	2953.092	19.69
M	N	5361.759	35.75
M	Y	1583.595	10.56

The adjustment cell method is straightforward to carry out once the cell boundaries have been defined. For occasions when the number of auxiliary variables is small, there is little room to debate how best to structure the cells because few permutations are possible. The more formidable task is reducing a large pool of auxiliary variables to only the "best" one(s), since creating cells based on the full cross-classification is prone to produce small or empty cells, yielding unstable weight inflation factors that, in turn, can lead to large variance estimates. A common rule used in practice is to require each cell to contain at least 30 sampling units (Lohr, 2009, p. 342). If that threshold is not met, the cell can be combined with a neighboring cell.

The literature has unequivocally reiterated that the ideal auxiliary variables to employ are those strongly related to the both probability of response *and* the outcome(s) of interest (Kalton, 1983; Bethlehem, 2002; Kalton and Flores-Cervantes, 2003). In a series of simulations, Little and Vartivarian (2005) found that variables characterized as such have the ability to reduce bias and even variances of sample means. In fact, they showed bias can *only* be reduced when the adjustment cells is strongly related to the outcome. Given the multipurpose nature of most surveys, however, it is challenging to craft a single adjustment cell scheme optimal for all outcome variables of interest.

10.4 Propensity Cell Method

A closely related technique to the adjustment cell method is the *propensity cell method*. Rather than forming cells based on a purposively selected subset of auxiliary variables, which some may discount as arbitrary, cells are formed by grouping sampling units with similar estimated response propensities. There are various ways to get the $\hat{\phi}_i$s, but one of the most popular approaches is to use logistic regression. The idea is to fit a logistic regression model on the full sample data set using the response indicator variable as the outcome and auxiliary variables as predictors. After fitting the model, the $\hat{\phi}_i$s are extracted and then ranked to form groups, or *propensity strata*, which play the same role as the nonresponse adjustment cells discussed in Section 10.3 (because of this, no separate formula is given). A common number of propensity strata to form is five, a rule of thumb attributable to Cochran (1968), who concluded from empirical evidence that finer stratification schemes had little effect on estimates and the potential downside of increasing the variance.

The propensity cell approach has several distinct advantages over the adjustment cell method. One is that there are a manageable number of cells, regardless of the number of covariates in the model estimating the propensities. Another is that continuous covariates can be more directly incorporated; in the adjustment cell method, they must be discretized. Additionally, interactions are easy to accommodate.

Program 10.3 estimates the response propensities of the 1200 individuals in the personnel satisfaction survey based on a logistic regression model including the four available covariates as well as all two-way interaction terms. An efficient way to incorporate these interactions is to separate the four variables in the MODEL statement with pipes, and then specify @2 after the last in the list is named. The P= option in the OUTPUT statement of PROC LOGISTIC stores the estimated response propensity in a variable named P_HAT. In the subsequent PROC RANK step, they are sort-ordered and classified into five propensity strata of roughly equal size. The VAR statement points PROC RANK to the variable(s) to be ranked and the RANKS statement names the rank group identifier variable to appear on the output data set. You can modify the number of strata using the GROUPS= option in the PROC RANK statement. Eltinge and Yansaneh (1997) and D'Agostino (1998) discuss diagnostics for determining whether refinements are warranted.

Thereafter, the syntax to adjust the weights is procedurally equivalent to that shown for the adjustment cell method in Program 10.2, except that cells are defined by only one variable, the rank group identifier PROPENSITY_CELL. The propensity cell-adjusted weight is named WEIGHT_ADJ_PROP_CELL.

Program 10.3: Propensity Cell Method to Adjust the Weights of Respondents to Compensate for Nonresponse

```
* estimate response propensities via logistic regression;
proc logistic data=samp descending;
  class gender minority / param=ref;
  model respond (event='1')=age|gender|supervisor|minority @2;
output out=samp p=p_hat;
run;

* stratify the propensities into 5 cells;
proc rank data=samp out=samp groups=5;
  var p_hat;
  ranks propensity_cell;
run;

* sum base weights of all sampling units in a cell;
proc freq data=samp;
  tables propensity_cell / out=counts_s
  (rename=(count=count_s));
weight weight_base;
run;

* sum base weights of responding sampling units in a cell;
proc freq data=samp;
  where respond=1;
  tables propensity_cell / out=counts_r
  (rename=(count=count_r));
weight weight_base;
run;

* create data set of adjustment cell weight inflation factors;
data cell_factors;
  merge counts_s
        counts_r;
  by propensity_cell;
factor_prop_cell=count_s/count_r;
```

```
keep propensity_cell factor_prop_cell;
run;

* merge these into sample data set and assign new,
nonresponse-adjusted weight;
proc sort data=samp;          by propensity_cell; run;
proc sort data=cell_factors; by propensity_cell; run;
data samp;
  merge samp          (in=a)
        cell_factors (in=b);
  by propensity_cell;
if a;
weight_adj_prop_cell=weight_base*factor_prop_cell*(respond=1);
run;
```

10.5 Poststratification

Poststratification (Holt and Smith, 1979) was originally proposed as a method to balance a sample's covariate distribution for the case of full response (i.e., in the absence of unit or item nonresponse), but nowadays it is used to mitigate nonresponse or coverage imparities between the sampling frame and target population. The method closely resembles the adjustment cell method, with the principle difference being that control totals are not formed from the sample; rather, they are assumed known for the entire population and often emanate from an external source.

The benchmark source need not be external, however. In the employee satisfaction survey, for example, the data set POP contains a record for all employees in the population. Using that data set, covariates can be cross classified to obtain N_c, the population total for the cth cell. Note the subtle distinction between this approach and the adjustment cell method, which uses the base-weighted cell total $\sum_{i \in S_c} w_{0ci} = \hat{N}_c$. Granted, $\hat{N}_c$ is an unbiased estimate of N_c.

We can formally express the poststratified weight of the ith unit in the cth cell as

$$w_{1ci} = w_{0ci} \left(\frac{N_C}{\sum_{i \in S_{1c}} w_{0ci}} \right) R_{ci} = w_{0ci} f_c R_{ci} \qquad (10.4)$$

where
S_{1c} is the set of responding units in the cth cell
w_{0ci} is the base weight of the ith unit in the cth cell
R_{ci} denotes a 0/1 indicator variable of whether that unit responded

We can easily adapt the syntax from Program 10.2 to poststratify the base weights according to the cross-classification of gender and minority status. The only change necessary is to redefine the numerator of the cell-specific weight inflation factor f_c. We can do this by substituting an unweighted PROC FREQ run on the data set POP for the base-weighted PROC FREQ run on the data set SAMP. Thanks to the advent of the POSTSTRATA statement in PROC SURVEYMEANS in Version 9.4, however, there is more syntactically efficient method we can use, which is illustrated in Program 10.4.

The POSTSTRATA statement assumes that the cross-classification of any categorical variables listed form the poststrata within which weights are to be adjusted following Equation 10.4. A preliminary step is to create a secondary data set containing the poststratification variables and their control totals, or N_cs. This is done in the PROC FREQ step at the top of Program 10.4. We point PROC SURVEYMEANS to this data set with the PSTOTAL= option after the slash in the POSTSTRATA statement. In addition to the poststratification variables, SAS will be looking for a key variable named _PSTOTAL_. The output data set named in the OUT= option stores the poststratified base weight in a column named _PSWT_. Lastly, since we are only interested in using the POSTSTRATA statement to generate poststratified weights, the VAR statement is omitted and the NOPRINT option is specified in the PROC SURVEYMEANS statement to suppress all output.

Program 10.4: Poststratification to Adjust the Weights of Respondents to Compensate for Nonresponse Using the POSTSTRATA Statement in PROC SURVEYMEANS

```
* create supplemental data set with control totals;
proc freq data=pop;
  tables gender*minority / list out=counts_p
  (rename=(count=_PSTOTAL_));
run;

* set nonrespondent weights to missing;
data samp2;
  set samp;
if respond=0 then weight_base=.;
run;
proc surveymeans data=samp2 noprint;
  weight weight_base;
poststrata gender minority / pstotal=counts_p
out=samp_weight_ps;
run;
```

In fact, the POSTSTRATA statement can be employed to generate nonresponse-adjusted weights for any cell-based procedure. For example, the syntax in Program 10.5 is functionally equivalent to that from Program 10.2. That is, the variable _PSTWT_ on the output data set SAMP_WEIGHT_ADJ_CELL is exactly the same as the variable WEIGHT_ADJ_CELL produced in Program 10.2.

Interestingly, as we saw from the Program 10.1 output, the control totals in this particular example are not integers, but include decimals. This does not pose a problem when calculating the f_cs. Along these same lines, an alternative benchmark data set orientation is to store a sequence of proportions (or percentages) in a key variable named _PSPCT_ that sum to 1 (or 100), and point SAS to it using the PSPCT= option after the slash in the POSTSTRATA statement.

Program 10.5: Demonstrating the POSTSTRATA Statement in PROC SURVEYMEANS to Adjust the Weights of Respondents Using the Adjustment Cell Technique

```
* create supplemental data set with adjustment cell base-
weighted totals;
proc freq data=samp;
  tables gender*minority / out=counts_s
  (rename=(count=_PSTOTAL_));
weight weight_base;
run;

* set nonrespondent weights to missing;
data samp2;
  set samp;
if respond=0 then weight_base=.;
run;

proc surveymeans data=samp2 noprint;
  weight weight_base;
poststrata gender minority / pstotal=counts_s
out=samp_weight_adj_cell;
run;
```

Another utile feature of the POSTSTRATA statement is the ability to conduct the weight adjustment method on a set of replicate weights. As was mentioned in Chapter 9, many researchers believe this is imperative to properly reflect the uncertainty associated with the nonresponse adjustment process. To motivate an example, recall in Program 9.3 when we appended a set of BRR replicate weights to the data set TEST, consisting of mathematical examination scores for a sample of high school students. It was implicitly assumed that the survey achieved full response, and so the base weight variable WEIGHT was assigned only to compensate for unequal selection probabilities. Suppose now that there was also unit nonresponse, and we sought to poststratify each replicate weight independently such that the weights for students in grades 9 through 12 summed to 400, 360, 395, and 405, respectively. Program 10.6 demonstrates how to store this information into a

	Class	Homeroom	Weight	Poststratification Weight	Replicate Weight	Replicate Weight
1	12	1	49.5	102.80769231	202.5	202.5
2	12	1	49.5	102.80769231	202.5	202.5
3	12	2	48	99.692307692	0	0
4	12	2	48	99.692307692	0	0
5	11	1	47.5	98.232984293	197.5	0
6	11	1	47.5	98.232984293	197.5	0
7	11	2	48	99.267015707	0	197.5
8	11	2	48	99.267015707	0	197.5
9	10	1	39	91.764705882	180	180
10	10	1	39	91.764705882	180	180
11	10	2	37.5	88.235294118	0	0
12	10	2	37.5	88.235294118	0	0
13	9	1	40	90.909090909	200	0
14	9	1	40	90.909090909	200	0
15	9	2	48	109.09090909	0	200
16	9	2	48	109.09090909	0	200

FIGURE 10.4
Partial view of TEST_BRR_PS, the data set TEST appended with a set of poststratified, BRR replicate weights.

supplemental data set and utilize the POSTSTRATA statement to accomplish this task.

When the OUTWEIGHTS= option is specified for a replication technique requested via the VARMETHOD= option in the PROC SURVEYMEANS statement, the OUT= option in the POSTSTRATA statement is ignored. Instead, the data set identified in the OUTWEIGHTS= option is appended with the _PSWT_ variable, the poststratified weight, and the set of post-stratified replicate weights numbered sequentially beginning with REP_WT1. Figure 10.4 is a partial view of the data set TEST_BRR_PS provided to help visualize these new weights and referenced against those shown in Figure 9.1 as generated from Program 9.3. Notice how only weights for observations in a selected PSU are adjusted, whereas the other observations' weights remain set to 0.

Program 10.6: Demonstrating the POSTSTRATA Statement in PROC SURVEYMEANS to Independently Poststratify a Set of Replicate Weights

```
* create supplemental data set with class totals;
data class_totals;
  input class _PSTOTAL_;
datalines;
```

```
 9  400
10  360
11  395
12  405
;
run;

proc surveymeans data=test varmethod=BRR
    (outweights=test_BRR_ps) noprint;
  stratum class;
  cluster homeroom;
weight weight;
poststrata class / pstotal=class_totals;
run;
```

10.6 Raking

There are occasions when benchmark totals are available, just not in the cross-classified nature necessary to conduct poststratification. For instance, the marginal distributions of gender and minority status for an employee population may be available, but not the joint distribution of the two. Even when the desired joint distributions are available, cross-classifying even a moderate amount of covariates can result in the small cell size dilemma alluded to earlier. In these situations, you can use the algorithm known as *raking* to calibrate the respondents' weights such that they simultaneously match the marginal totals of two or more covariates.

The origins of raking can be traced to Deming and Stephan (1940), who first conceived of the technique to ensure certain tabulations from the 1940 U.S. Decennial Census and samples taken from it agreed with one another. The process begins by making a poststratification-type adjustment using marginal totals of the first covariate, or *raking dimension*. Next, a similar adjustment is made for the second covariate, although this may disrupt the totals with respect to the first covariate. Nonetheless, all covariates are calibrated sequentially. After adjusting the last covariate, the algorithm returns the first and iterates through the covariate set anew. The process continuous until the adjustment factors and/or weights change less than some prespecified tolerance.

There is no need to write code from scratch to conduct raking because the %RAKING macro developed by Izrael et al. (2000) is free and straightforward to use. With respect to the employee satisfaction survey, suppose that we wanted to use all four available covariates as raking dimensions. The variables MINORITY, SUPERVS, and GENDER are all dichotomous, but AGE is continuous. Because raking variables must be categorical, a new variable called AGECAT has been created as a 0/1 indicator of whether the employee is over the age of 40.

The first step is to calculate and store the marginal totals for the population. This is accomplished by the PROC FREQ step in Program 10.7. The data sets output by PROC FREQ are given the same name as the covariates themselves, with the default COUNT variable renamed to MRGTOTAL, a key variable the %RAKING macro will be looking for in the supplemental data sets. To compile the macro, syntax from the Izrael et al. (2000) paper can be copied and pasted into one's SAS program. It is assumed this has been done prior to the macro execution step in Program 10.7. (The code is too lengthy to include as part of this text.)

The macro parameter INDS= specifies the input data set. Notice how the data set RESPONDENTS is created and passed to the macro instead of the full SAMP data set as was done in previous examples. There is no parameter to identify a respondent, so the sample data must be subset accordingly beforehand. The OUTDS=parameter names the output data set, to contain all variables in the input data set plus the raked version of the input weight identified in the INWT= parameter. The raked weight is named what the user provides in the OUTWT= parameter. The FREQLIST= parameter requires a list of the marginal total data sets separated by spaces, those defined in the PROC FREQ step in the present example. The VARLIST= parameter identifies the corresponding variables appearing in the input data set. The macro also requires that the count of variables (or marginal total data sets) be specified in the NUMVAR= parameter. The CNTOTAL= parameter is an optional way to specify the population size, only necessary when marginal information is specified as percentages rather than totals—in essence, the macro will convert the percentages to totals by multiplying each by the value supplied in the CNTOTAL= parameter.

The parameter TRMPREC= allows the user to control the precision of raked weight totals relative to the controls. Setting TRMPREC=1 requires the sum of each covariate category's raked weight to be within 1 unit of the control total. The final parameter is NUMITER=, which assigns a maximum number of iterations to be performed if the raked weights' totals have not yet converged to less than the tolerable amount. Typically, convergence occurs within a few iterations, but not always. As discussed in Brick et al. (2003) and Battaglia et al. (2009), two of the leading causes of convergence failure are (1) a small (or zero) respondent count in one or more raking dimension cells or (2) the grand totals of two raking dimensions not agreeing with one another.

An abridged version of the output is provided following Program 10.7. At the first iteration, the macro outputs each covariate's base-weighted sum in the "Calculated Margin" column alongside the marginal control total input by the user. The column "Difference" quantifies how far away the two are from one another. As we have already seen, the summed base weights for respondents underestimate the controls. The macro then makes a sequence of poststratification-type adjustments until the convergence criteria are met for all covariates. Intermediate output is suppressed, but we can verify from what is reported following the fourth iteration that the "Difference" column is less than 1 for all covariate categories, forcing the macro to terminate.

Program 10.7: Raking to Adjust the Weights of Respondents to Compensate for Nonresponse

```
* create the supplemental data sets of marginal totals;
proc freq data=pop;
  tables gender      / out=gender     (drop=percent
  rename=(count=MRGTOTAL));
  tables agecat      / out=agecat     (drop=percent
  rename=(count=MRGTOTAL));
  tables minority    / out=minority   (drop=percent
  rename=(count=MRGTOTAL));
  tables supervisor / out=supervisor (drop=percent
  rename=(count=MRGTOTAL));
run;

* filter sample data set for only respondents;
data respondents;
  set samp;
where respond=1;
run;

%RAKING(
inds=respondents,
outds=respondents_raked,
inwt=weight_base,
freqlist=gender agecat minority supervisor,
outwt=weight_raked,
varlist=gender agecat minority supervisor,
numvar=4,
cntotal=,
trmprec=1,
numiter=50
);
```

Raking by GENDER, Iteration –1

Obs	GENDER	Calculated Margin	Margin Control Total	Difference
1	F	4483.56	7,563	3079.44
2	M	3402.36	7,437	4034.64
		7885.93	15,000	

Raking by AGECAT, Iteration –1

Obs	agecat	Margin Control Total	Calculated Margin	Difference
1	0	11,859	11,878.42	–19.4170
2	1	3,141	3,121.58	19.4170
		15,000	15,000.00	

Raking by MINORITY, Iteration –1

Obs	Minority	Margin Control Total	Calculated Margin	Difference
1	N	10,833	11,282.15	449.150
2	Y	4,167	3,717.85	449.150
		15,000	15,000.00	

Raking by SUPERVISOR, Iteration –1

Obs	SUPERVISOR	Margin Control Total	Calculated Margin	Difference
1	0	13,281	13,371.27	–90.2670
2	1	1,719	1,628.73	90.2670
		15,000	15,000.00	

Raking by GENDER, Iteration –4

Obs	GENDER	Margin Control Total	Calculated Margin	Difference
1	F	7,563	7,563.21	–0.21370
2	M	7,437	7,436.79	0.21370
		15,000	15,000.00	

Raking by AGECAT, Iteration –4

Obs	agecat	Margin Control Total	Calculated Margin	Difference
1	0	11,859	11,858.90	0.10292
2	1	3,141	3,141.10	–0.10292
		15,000	15,000.00	

Raking by MINORITY, Iteration –4

Obs	Minority	Margin Control Total	Calculated Margin	Difference
1	N	10,833	10,832.36	0.037238
2	Y	4,167	4,167.04	–0.037238
		15,000	15,000.00	

Raking by SUPERVISOR, Iteration −4

Obs	SUPERVISOR	Margin Control Total	Calculated margin	Difference
1	0	13,281	13,280.97	0.030828
2	1	1,719	1,719.03	−0.030828
		15,000	15,000.00	

**** Program terminated at iteration 4 because all calculated margins differ from marginal control totals by less than 1.

Enhancements to the original %RAKING macro were made in Izrael et al. (2004) and Izrael et al. (2009). In the former, diagnostics about the convergence process were added. The latter introduced capabilities to trim the raked weights, which some researchers prefer to do in an effort to control variance increases caused by highly variable weights (Elliott, 2008). The core algorithm embedded in the macro is essentially the same, however.

10.7 Summary

This chapter has explored some of the common weight adjustment procedures used in practice to compensate for unit nonresponse in surveys when certain sampling units do not respond to the survey request. The methods discussed here are not all-inclusive. For instance, Deville and Särndal (1992) comment on how poststratification and raking are but two special cases of a larger class of calibration estimators. Additionally, alternative methods to define adjustment cells are available using various classification or search algorithms such as the ones discussed in Sonquist et al. (1974), Kass (1980), and Breiman et al. (1993). These alternatives, however, operate under the same central objective: to use auxiliary variables known for the entire sample to group units into cells with similar estimated response propensities. Regardless of how cells are created, the ultimate assumption is that data are MCAR within each.

A slight twist on the nonresponse adjustment cell approach is to construct cells with units having similar y_is. Thinking back to the nonresponse bias formula specified in Equation 10.2, if there is little variability amongst the y_is, then the covariance of the y_is and the ϕ_is will be minimal. So a variant of the propensity cell approach, in which we estimate and rank the $\hat{\phi}_i$s, is to use auxiliary information to estimate the predicted values of the outcome variable, or $\hat{y}_i$s, and then rank those to form cells. With respect to the syntax shown in Program 10.3, one could simply substitute Q1 for RESPOND as the dependent variable in the MODEL statement of PROC LOGISTIC—a comparable process could be followed using PROC REG, say, for a continuous outcome variable. (Recall that SAS modeling procedures will generally output a predicted value even if the outcome variable is missing, as long as all independent variables are nonmissing.) Little (1986) terms this approach "predictive mean stratification." Its primary disadvantage is that the stratification scheme is optimal only for a single variable.

Finally, although a separate section was devoted to each method, practitioners often use two or more methods sequentially. For instance, nonresponse adjustments might be implemented via the propensity cell method, with the resulting set of weights poststratified or raked to known totals, possibly with a different set of covariates used in each step. Indeed, a multistage procedure similar in spirit has been adopted for the Federal Employee Viewpoint Survey (U.S. Office of Personnel Management, 2013).

11

Imputation Methods

11.1 Introduction

Whereas Chapter 10 described methods to adjust the base weights of the responding sampling units to compensate for the nonresponding units' missing data, this chapter discusses techniques to *impute*, or substitute a plausible value for, the missing data. Imputation is most often utilized to handle item nonresponse, which occurs when the respondent provides an answer to some, but not all, of the outcome variables (see Figure 10.2). The motivation behind imputing for item nonresponse is to salvage the partially complete data and produce a completed, "rectangular" data set upon which traditional statistical techniques can be applied.

Comparatively speaking, the theory and application of imputation methods are more complex than weight adjustment methods. The benefits and legitimacy of imputation also tend to be tougher to convince an unfamiliar audience. The most frequently voiced concern is that imputation appears to involve fabricating data. While this is an understandable initial reaction, the goal of imputation is not exactly to locate the value "closest" to a nonrespondent's unobserved, true value. More precisely, the goal is to exploit relationships between the auxiliary variables and the observed data to recapture a portion of the uncertainty caused by the missing data. The amount of information one can recover is a function of how strong these relationships are. In fact, several ingenious methods have been proposed to estimate the unrecoverable portion of uncertainty. In Section 11.6, we will discuss one such quantification, Rubin's (1987) *fraction of missing information* (FMI) statistic.

This chapter begins by outlining a typology of imputation methods and modeling approaches used in practice. They are demonstrated in a univariate missingness setting in Section 11.4 via syntax examples drawing upon the same personnel satisfaction survey data introduced in Chapter 10. These ideas are then extended to multivariate missingness scenarios in Section 11.5, and we conclude the chapter with some remarks about recent literature developments to incorporate features of the complex survey data into the imputation process.

11.2 Definitions and a Brief Taxonomy of Imputation Techniques

The purpose of this section is to classify the various imputation methods used in practice and briefly comment on their respective advantages and disadvantages. It is assumed that the reader has a firm grasp on the terminology and fundamental missingness assumptions discussed in Section 10.2 as that material will be referenced in this chapter without review. A quick refresher on the SAMP data set variables and their associated coding structure (also discussed in Section 10.2) would also be helpful.

One of the earliest approaches proposed to address item nonresponse, and one used frequently to this day, is to fill in data for the missing cases with the mean of the observed cases. This is also known as "unconditional mean imputation" and is an example of a *deterministic* imputation model. The approach does not alter certain descriptive statistics such as the sample mean—the sample means prior to and after this form of imputation are equivalent—but it can unjustifiably attenuate measures of variability. To see why, suppose that a simple random sample of size n has been selected in which r units provided a response to some outcome variable Y and m units did not ($r + m = n$). The variance for $\hat{\bar{y}}_r$, the unadjusted sample mean of the r observed cases, is given by $\text{var}(\hat{\bar{y}}_r) = \sum_{i=1}^{r}(y_i - \hat{\bar{y}}_r)^2/r(r-1)$. Notice how substituting $\hat{\bar{y}}_r$ for the m missing cases has no effect on the numerator. In other words, despite the summation now running to n instead of r, the m added values each contribute a squared deviation of 0. At the same time, the denominator increases from $r(r-1)$ to $n(n-1)$. Taken together, we can reason that the variance decreases, which seems ill advised given the absence of any new information. A preferred alternative, at least from a variance perspective, is to employ a *stochastic* imputation model in which we begin with the expected value and then add on some kind of residual.

To compare and contrast concepts of deterministic versus stochastic imputation models, let us consider the task of imputing length of service (LOS) for the 572 nonrespondents in the SAMP data set using the auxiliary variable AGE. Figure 11.1 shows the relationship of these two variables for a subset of the 628 observed cases. If we let y_i denote the LOS for the ith individual and x_i denote the age of the ith individual, an intuitive approach might be to suppose a simple linear regression model such as $y_i = \beta_0 + \beta_1 x_i + \varepsilon_i$ governs the data, where $\varepsilon_i \sim N(0, \sigma_\varepsilon^2)$. The idea is to estimate the parameters of this model using the observed data (e.g., using methods discussed in Section 5.2) and then derive a plausible value of y_i given x_i for the m nonrespondents. So the imputed value for a 40-year-old nonrespondent, say, would be assigned as $y_i = \hat{\beta}_0 + \hat{\beta}_1(40) + z_i\sqrt{\hat{\sigma}_\varepsilon^2}$, where z_i denotes a random normal variate. On the other hand, a deterministic imputation model approach might assign the

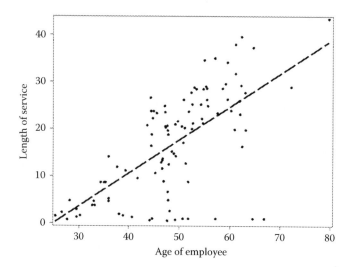

FIGURE 11.1

Scatterplot illustrating the relationship between a continuous auxiliary variable and a continuous outcome variable, overlaid with a least squares regression line.

missing LOS value as $y_i = \hat{\beta}_0 + \hat{\beta}_1(40)$ (i.e., substituting the value from the regression line).

The stochastic imputation model discussed earlier is an example of an *explicit* or a *parametric* imputation model. In contrast, an example of an *implicit* or a *nonparametric* imputation model would be to categorize sampling units into cells based on some grouping of ages (e.g., <40, 40–60, and >60). Within a cell, the missing values for *beggars* are populated by randomly selecting values from *donors* or the set of cases with the key outcome variable observed. Andridge and Little (2010) review this widely used approach, typically referred to in the nonresponse literature as "hot-deck imputation," and remark on the origins of its phraseology as follows (p. 41):

> Historically, the term "hot deck" comes from the use of computer punch cards for data storage, and refers to the deck of cards for donors available for a non-respondent. The deck was "hot" since it was currently being processed, as opposed to the "cold deck" which refers to using pre-processed data as the donors, i.e. data from a previous data collection or a different data set.

Guidance on how best to form cells essentially follows what was discussed in Chapter 10. Thinking back to Equation 10.2, the nonresponse bias formula from Bethlehem (1988), the goal is to create cells in which either the ϕ_is or y_is (or both) are similar. However, when the cells are formed, it is important to keep in mind the missing data assumption is the same as that of an explicit model: data are assumed MAR or MCAR given the auxiliary variables **X**. The only nuance is that **X** can be perceived as a series of cell membership indicator variables under the implicit model specification.

There are advantages and disadvantages to using either explicit and implicit imputation models. One potential downside to using an explicit model such as the one described earlier is that the imputed value returned could be nonsensical—for instance, a negative LOS. This is less of an issue in the cell-based approach, which at least returns a value actually observed in the data. A related advantage is that the implicit approach does not require error terms to conform to any prespecified, "well-behaved" distribution. For instance, one could call into question the distribution of residuals in the simple linear regression model portrayed in Figure 11.1. Indeed, there is some evidence of heteroscedasticity as residuals tend to increase in magnitude for older employees. Of course, one or more variable transformations could be performed to rectify a situation like this.

A key advantage of the explicit model approach, however, is that it can accommodate a larger number of covariates. Many researchers assert that the imputation model should include as many covariates as possible (Rubin, 1996; Schafer, 1997; Van Buuren et al., 1999) to bolster the plausibility of the MAR assumption. By comparison, defining cells based on the joint distribution of a large number of covariates is prone to yield small or empty cells. We commented in Chapter 10 on the potentially undesirable variance inflation that can result from performing weight adjustments within small cells. Schenker and Taylor (1996, p. 430) issue similar warnings when performing hot-deck imputation, stating it is undesirable to draw from too small a pool of observed values or to reuse the same observed value too frequently.

There are, however, techniques effectively bridging the gap between strictly explicit and strictly implicit imputation models. These are referred to as "partially parametric" or "semiparametric" techniques. One example implementation we will discuss in Section 11.4.4 calls for fitting an explicit regression model but affixing "real" residuals by randomly drawing from neighboring observations.

The aforementioned ideas translate to categorical outcome variables, but the visualization and modeling approaches differ somewhat. As an example, Figure 11.2 shows the relationship between employee gender and the proportion of respondents reacting positively to the statement "I like the kind of work I do," found by taking the mean of the 0/1 indicator variable Q1 on the data set SAMP. We can gather from the observed data that females are more likely to respond positively, so this auxiliary variable is at least somewhat useful for recapturing a portion of the missing data uncertainty. As opposed to a linear regression model, however, deriving imputations using a logistic regression model would be more appropriate in this setting.

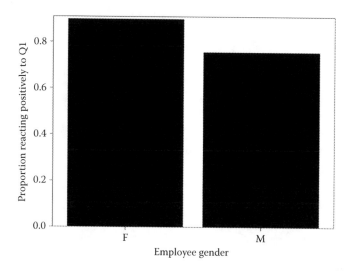

FIGURE 11.2
Histogram demonstrating the relationship between a categorical auxiliary variable and a categorical outcome variable.

11.3 Multiple Imputation as a Way to Incorporate Missing Data Uncertainty

While it was argued in the previous section that stochastic imputation models lead to a much less pronounced underestimation of variance than do deterministic imputation models, there is still a residual component of uncertainty ignored by treating the imputed values in the completed data set as truth. A range of techniques have been proposed in the survey methodology literature to inflate measures of variability to account for this uncertainty (e.g., Rao and Shao, 1992; Särndal, 1992; Efron, 1994; Fay, 1996; Kim and Fuller, 2004), but the most popular is *multiple imputation* (MI) (Rubin, 1987). Indeed, the SAS/STAT procedures PROC MI and PROC MIANALYZE were developed exclusively for conducting MI, and Berglund and Heeringa (2014) is an excellent reference text highlighting the capabilities of those two procedures.

The notion behind MI is to impute the missing data independently M times ($M \geq 2$), thereby producing M completed data sets, after which traditional statistical techniques can be independently applied. Results from the M independent analyses are then coalesced according to a few straightforward rules to be defined shortly. Figure 11.3 provides a visualization of the MI process. Note that, even when single imputation is the objective, it is most efficient to make use of the wide variety of tools built into PROC MI. As with

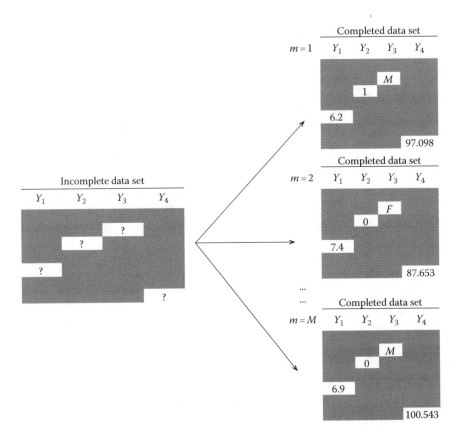

FIGURE 11.3
Visual depiction of the multiple imputation process.

other comparable routines designed to multiply impute missing data, there is an option to output only a single copy of one of the M completed data sets (or you could simply maintain one selected at random).

Let us define the statistical details of the MI approach prior to illustrating any syntax examples. If we symbolize the mth completed data set estimate by $\hat{\theta}_m$, the overall MI estimate is found by averaging the M completed data set estimates or

$$\hat{\theta}_M = \frac{1}{M}\sum_{m=1}^{M}\hat{\theta}_m \qquad (11.1)$$

General "theta" notation is used here to emphasize that this and all subsequent MI formulas apply for any estimator, such as a mean, quantile, and correlation coefficient. Naturally, this is a very appealing feature of the approach.

The overall MI variance is the sum of two measures of variability. The first is the M average completed data set estimated variances or

$$W_M = \frac{1}{M} \sum_{m=1}^{M} \text{var}(\hat{\theta}_m) \tag{11.2}$$

which is referred to as the "within-imputation variance," and the second is

$$B_M = \frac{1}{M-1} \sum_{m=1}^{M} (\hat{\theta}_m - \hat{\theta}_M)^2 \tag{11.3}$$

which is referred to as the "between-imputation variance." Observe how the between-imputation term functions as a means to capture the variability of the M completed data set point estimates themselves.

Taken together, the estimated variance of the overall MI estimate is

$$\text{var}(\hat{\theta}_M) = W_M + \left(1 + \frac{1}{M}\right) B_M \tag{11.4}$$

where the term $(1+(1/M))$ represents a finite imputation correction factor that converges to 1 as $M \to \infty$.

We will walk through a variety of syntax examples over the course of this chapter, but to briefly summarize the MI process, the following sequence of steps is taken:

1. Use PROC MI or an alternative routine to independently create M completed data sets (see Figure 11.3).
2. Compute the M point estimate(s) and associated measure(s) of variability using standard complete data methods.
3. Supply the results from step 2 to PROC MIANALYZE, which will carry out the calculations defined in Equations 11.1 through 11.4.

A tenet of the MI technique advocated by Rubin (1987) is that one must account for the imputation *model* uncertainty when deriving imputations. Failing to do so is *improper* (pp. 112–128) in his terminology. This is a complex task often overlooked or otherwise not well understood. It is most readily comprehendible from the Bayesian perspective of MI discussed in more detail in Rubin (1978) and Chapter 3 of Rubin (1987). A few simple examples are given next to help illuminate the concept and further justify how MI variances properly reflect the missing data uncertainty.

As before, suppose that a simple random sample of n units has been drawn, but that only r provided data while m did not. Let us denote the set of r observed cases as S_1 and the set or m missing cases as S_0. Perhaps we are willing to assume data are MCAR, yet still wish to compensate for the non-response by multiply imputing the missing values. We will now define two proper MI strategies. The first is essentially a hot-deck routine treating the sample set S ($S = S_1 \cup S_0$) as a single cell. It begins by selecting r cases with replacement from S_1. Symbolizing this set as S_1^*, the second step is to select m missing cases with replacement from S_1^* to create a new set we can symbolize by S_0^*. The m values in S_0^* serve as substitutes for each of the m missing values. This two-stage process is then independently repeated M times to create M completed data sets. Rubin and Schenker (1986) refer to this technique as the *approximate Bayesian bootstrap* (ABB).

The second proper MI strategy is a parametric version of the first. Suppose that the sample mean of interest was actually a proportion or the mean of a 0/1 indicator variable. From set S_1, we compute the proportion $\hat{p}_r$ and its variance $\text{var}(\hat{p}_r) = (\hat{p}_r(1-\hat{p}_r))/(r-1)$. To derive an imputed value for the ith case in S_0, we first compute $\hat{p}_r^* = \hat{p}_r + z\sqrt{\text{var}(\hat{p}_r)}$, where z is a random normal deviate. From there, we draw r_i, a random uniform variable between 0 and 1, and impute a 1 for if $r_i < \hat{p}_r^*$ and 0 otherwise. This process is conducted independently M times.

The remarkable property of either approach is that, in expectation as $M \to \infty$, the MI variance estimate approximately matches the variance estimate from only the observed portion of the data. The intrigued reader is encouraged to verify this by carrying out a simple simulation. Note that without the respective first steps (i.e., creating r^* or $\hat{p}_r^*$), some degree of a variance underestimation will remain even if the MI procedure is otherwise followed as prescribed.

11.4 Univariate Missingness

11.4.1 Introduction

We begin the illustration of SAS' imputation tools by addressing *univariate missingness*, the scenario in which one or more fully observed variables is used to impute missing values of a single variable. Univariate missingness is actually a special case of the more general *monotone missingness pattern*, which will be formally defined in Section 11.5 when it is contrasted with an *arbitrary missingness pattern* (see Figure 11.5). But at this point, suffice it to say this explains why the handiest built-in imputation algorithms for this setting reside within the MONOTONE statement of PROC MI. Examples appearing over the course of the next two sections focus solely on illustrating methods to create one or more completed data sets; examples of how to actually analyze multiply imputed data are deferred until Section 11.6.

11.4.2 Methods Based on Explicit Models

Suppose we wanted to multiply impute the missing values of LOS in the data set SAMP $M = 5$ times using a model that includes all four available auxiliary variables: GENDER, SUPERVISOR, AGE, and MINORITY. The PROC MI syntax in Program 11.1 illustrates one way to accomplish this task. Since LOS is a continuous variable, we could opt for a linear regression imputation approach, which can be implemented with the REG option of the MONOTONE statement. Although there is output generated in the listing, the key "output" is the $M = 5$ stacked completed data sets stored in the data set named SAMP_MI_5 specified in the OUT= option of the PROC MI statement. For the present case, then, the output data set consists of $5 \times 1200 = 6000$ observations. A numeric variable named _IMPUTATION_ with values 1 through 5 is appended and can be used to identify the mth completed data set. In the event your objective is to create only one completed data set, you can specify NIMPUTE=1 or retain one of the completed data sets at random.

Categorical variables must be listed in the CLASS statement and must also appear in the VAR statement alongside continuous variables. Since the variable SUPERVISOR is stored numerically as a 0/1 indicator, it does not need to be listed in the CLASS statement. In general, variables in the VAR statement must be listed in ascending order with respect to their rates of missingness, but in the present univariate missingness situation, we only need to ensure LOS is listed last. The imputation model is specified in parentheses after the REG option in the MONOTONE statement with syntax abiding by traditional MODEL statement rules.

Program 11.1: Multiple Imputation of a Continuous Outcome Variable Using a Linear Regression Model

```
proc mi data=samp out=samp_MI_5 nimpute=5 seed=943222;
  class gender minority;
  var gender minority age supervisor los;
  monotone reg (los=age supervisor gender minority / details);
run;
```

Missing Data Patterns									
							Group Means		
Group	GENDER	Minority	AGE	SUPERVISOR	LOS	Freq	Percent	AGE	SUPERVISOR	LOS
1	X	X	X	X	X	628	52.33	48.700175	0.154459	16.272640
2	X	X	X	X	.	572	47.67	49.844704	0.180070	.

Regression Models for Monotone Method									
Imputed Variable	Effect	GENDER	Minority	Obs-Data	Imputation				
					1	2	3	4	5
LOS	Intercept			0.03733	−0.048952	0.081051	0.122400	0.017561	−0.049626
LOS	AGE			0.55099	0.543273	0.569269	0.522167	0.615283	0.532826
LOS	SUPERVISOR			0.04699	0.036015	0.023554	0.039521	0.002420	0.079313
LOS	GENDER	F		0.22018	0.251815	0.166224	0.198130	0.174576	0.260237
LOS	Minority		N	−0.06712	−0.004828	−0.110064	−0.145315	−0.116411	−0.011814

Only the portion of the output generated from Program 11.1 relevant to the present discussion is shown here. The "Missing Data Patterns" section provides a concise visualization of the missingness pattern: an X denotes observed data and a period denotes missing data. The count and relative percent of each pattern are also reported, as are the respective means for any continuous variables. The DETAILS option after the slash in the imputation model specification step of the syntax requests that the model coefficients for the observed portion of the data to be output as well as the perturbed values. For example, the supervisory status model parameter is approximately 0.047 for the model fit to the observed data, but is 0.036 for the purposes of generating the first set of imputations and 0.024 for generating the second set of imputations, and so on. Perturbing the coefficients is needed to account for the imputation model uncertainty or to make the imputation process *proper*, as explained in Section 11.3.

The multistep perturbation process is succinctly summarized for a variety of models in the appendix of Raghunathan et al. (2001) with a few references from the Bayesian statistics literature justifying the various approaches. It is enlightening to walk through the steps for a simple linear regression model, if nothing else to foster an appreciation for the complexity of the process and the convenience of being able to call upon a tool such as PROC MI to grind through the computational legwork on our behalf.

Following the notation and terminology from Chapter 5, let us denote the fully observed covariates for the r respondents $\mathbf{X}$, which we can think of as an $r \times p$ design matrix, where p is the number of unique parameters in the model, which may include an intercept. Using this matrix and $\mathbf{Y}$, the $r \times 1$ outcome vector, we estimate the model $\mathbf{Y} = \mathbf{X}\hat{\boldsymbol{\beta}} + \mathbf{e}$ for the r observed cases, where $\mathbf{e}$ is the $r \times 1$ vector of residuals. Symbolizing the mean squared error of this model by $\hat{\sigma}_\varepsilon^2 = \text{SSE}/df$, where SSE is the model's sum of squares for error and $df = r - p$ degrees of freedom, the following process is followed to derive a set of imputed values:

1. Define $\hat{\sigma}_\varepsilon^{2*} = \text{SSE}/\chi_{df}^2$, where χ_{df}^2 is a random chi-square deviate with df degrees of freedom.

2. Define $\hat{\boldsymbol{\beta}}^* = \hat{\boldsymbol{\beta}} + \hat{\sigma}_\varepsilon^{2*}\mathbf{V}\mathbf{z}$, where $\mathbf{V}$ is the upper triangular matrix in the Cholesky decomposition, such that $\mathbf{V}^{\mathrm{T}} = (\mathbf{X}^{\mathrm{T}}\mathbf{X})^{-1}$ (a

multivariate analog to the variance term used in Section 11.3), and z is a p-dimensional vector of random normal deviates.

3. For the ith unit in the set of missing cases, assign $y_i = x_i\hat{\beta}^* + z_i\sqrt{\hat{\sigma}_\varepsilon^{2*}}$, where x_i represents the unit's covariate vector (i.e., its row in the design matrix) and z_i represents an independently generated random normal variate.

A similar sequence of steps is followed for other explicit models, such as the logistic regression imputation model approach demonstrated in the next syntax example, but for brevity we will not detail the process for every possible model. The reader seeking more detail is referred to the appendix of Raghunathan et al. (2001), Rubin (1987), and the PROC MI documentation.

We next consider techniques for a dichotomous outcome variable plagued by missing values. The LOGISTIC option in the MONOTONE statement of PROC MI is equipped to handle variables of this type. As the name suggests, it does so by fitting a logistic regression imputation model to the observed data. The syntax in Program 11.2 demonstrates how to impute the 0/1 indicator variable Q1 in the data set SAMP exploiting the same four auxiliary variables used in Program 11.1. As before, since we have specified NIMPUTE=5, the output data set SAMP_MI_5 consists of five times the number of observations appearing in the input data set. Note that the fully observed data and imputation-specific perturbed model coefficients are still given, but here they correspond to logistic regression model coefficients, not linear regression model coefficients as generated in Program 11.1. The perturbed parameters are used to derive $\hat{p}_i^*$, the ith sampling unit's predicted probability of whichever outcome was treated as the "event." From there, a random uniform variate between 0 and 1 is drawn. Representing this variate as r_i, an "event" is imputed if $r_i < \hat{p}_r^*$ and a "nonevent" imputed otherwise.

Program 11.2: Multiple Imputation of a Dichotomous Outcome Variable Using a Logistic Regression Model

```
proc mi data=samp nimpute=5 seed=7000726 out=samp_MI_5;
   class gender minority Q1;
   var gender minority age supervisor Q1;
   monotone logistic (Q1=gender minority age supervisor /
   details);
run;
```

Logistic Models for Monotone Method									
Imputed Variable	Effect	GENDER	Minority	Obs-Data	Imputation				
					1	2	3	4	5
Q1	Intercept			−1.43774	−1.494506	−1.447076	−1.509665	−1.608827	−1.444244
Q1	GENDER	F		−0.28083	−0.205150	−0.428327	−0.327285	−0.412485	−0.143868

(Continued)

Logistic Models for Monotone Method									
Imputed Variable	Effect	GENDER	Minority	Obs-Data	Imputation				
					1	2	3	4	5
Q1	Minority		N	0.24397	0.416990	0.295714	0.215823	0.388679	0.199811
Q1	AGE			0.30238	0.429127	0.316463	0.252488	0.322257	0.243506
Q1	SUPERVISOR			0.12546	0.158775	0.176997	0.178259	0.052431	0.107608

The LOGISTIC option can actually be used to impute variables with three or more distinct values, but be aware that PROC MI defaults to the cumulative logit model (see Section 6.7.3), which is only appropriate if you are willing to assume an ordinal scale. This may be plausible for the variable Q1_FULL on the data set SAMP, the precollapsed version of Q1, a numeric variable coded 1–5 representing the respondent's selection on a five-point Likert-type scale ranging from "Completely Disagree" to "Completely Agree." Program 11.3 demonstrates syntax to impute this variable by way of the cumulative logit model. Relative to Program 11.2, the only difference is that Q1_FULL has been substituted for Q1. No additional options are necessary. Notice how the coefficient structure resembles that from Program 6.10.

Program 11.3: Multiple Imputation of an Ordinal Outcome Variable Using a Cumulative Logistic Regression Model

```
proc mi data=samp nimpute=5 seed=655231 out=samp_MI_5;
   class gender minority Q1_full;
   var gender minority age supervisor Q1_full;
   monotone logistic (Q1_full=gender minority age supervisor /
   details);
run;
```

Logistic Models for Monotone Method									
Imputed Variable	Effect	Q1_full	GENDER	Minority	Obs-Data	Imputation			
						1	2	3	4	5
Q1_full	Intercept	1.000000			−2.93758	−2.836662	−2.853194	−2.895202	−2.875302	−2.715477
Q1_full	Intercept	2.000000			−1.70689	−1.517590	−1.440681	−1.693735	−1.690207	−1.813471
Q1_full	Intercept	3.000000			−1.36532	−1.265700	−1.100017	−1.447711	−1.255164	−1.432451
Q1_full	Intercept	4.000000			0.50225	0.592986	0.765561	0.426195	0.595230	0.471246
Q1_full	GENDER		F		−0.11895	−0.205543	−0.103870	−0.078735	−0.092025	−0.180083
Q1_full	Minority			N	0.16933	0.119564	0.034361	0.182713	0.185098	0.046157
Q1_full	AGE				0.16048	0.058989	0.115903	0.177908	0.191523	0.098895
Q1_full	SUPERVISOR				0.13373	0.120930	0.054343	0.117971	0.161478	−0.032329

For nominally scaled variables, you can override this default in PROC MI by specifying the LINK=GLOGIT option in parentheses. This will invoke the multinomial (i.e., generalized) logit model discussed in Section 6.7.2. Program 11.4 illustrates how to fit this alternative model. We can observe how the coefficient structure now follows the same pattern seen previously in Program 6.8.

Program 11.4: Multiple Imputation of a Nominal Outcome Variable Using a Multinomial Logistic Regression Model

```
proc mi data=samp nimpute=5 seed=609210 out=samp_MI_5;
  class gender minority Q1_full;
  var gender minority age supervisor Q1_full;
  monotone logistic (Q1_full=gender minority age supervisor /
                     link=glogit details);
run;
```

						Logistic Models for Monotone Method					
Imputed Variable	Effect	Q1_full	GENDER	Minority	Obs-Data	Imputation				
						1	2	3	4	5
Q1_full	Intercept	1.000000			−2.10348	−2.116840	−2.327672	−1.581006	−1.812303	−1.966763
Q1_full	Intercept	2.000000			−1.39183	−1.392740	−1.153268	−1.142631	−1.439214	−1.150987
Q1_full	Intercept	3.000000			−2.27848	−2.592118	−2.333159	−2.107670	−1.676443	−2.353367
Q1_full	Intercept	4.000000			0.06941	0.173900	0.007018	0.332447	0.265175	0.235939
Q1_full	GENDER	1.000000	F		0.06758	0.599094	0.108255	−0.156474	−0.273615	−0.063272
Q1_full	GENDER	2.000000	F		−0.28355	−0.041634	−0.215016	−0.177080	−0.602895	−0.202969
Q1_full	GENDER	3.000000	F		−0.48667	−0.721612	−0.440727	−0.211410	−0.357030	−0.802269
Q1_full	GENDER	4.000000	F		0.07483	0.037953	0.135534	0.174650	−0.115692	−0.035540
Q1_full	Minority	1.000000		N	0.20316	0.351092	0.352006	−0.064361	−0.219998	−0.125090
Q1_full	Minority	2.000000		N	0.28241	0.315237	−0.171594	0.237992	0.160560	0.204405
Q1_full	Minority	3.000000		N	0.41792	0.718260	0.050195	0.328764	0.022040	0.201854
Q1_full	Minority	4.000000		N	0.08150	0.039222	0.054641	−0.127627	−0.186791	−0.123649
Q1_full	AGE	1.000000			0.63896	0.509675	0.455469	0.297686	0.659578	0.163565
Q1_full	AGE	2.000000			0.19993	0.395424	0.292218	0.168453	0.302754	0.180081
Q1_full	AGE	3.000000			0.10600	−0.028067	0.353483	−0.234919	0.318478	0.109974
Q1_full	AGE	4.000000			−0.03600	0.127718	.023389	−0.009764	0.059752	−0.182849
Q1_full	SUPERVISOR	1.000000			0.10460	0.204834	−.036334	0.193534	−0.105980	0.130918
Q1_full	SUPERVISOR	2.000000			0.28178	0.318493	0.469111	0.322705	0.113759	0.199853
Q1_full	SUPERVISOR	3.000000			0.10444	0.055629	0.311096	0.027678	0.404330	−0.163029
Q1_full	SUPERVISOR	4.000000			0.12787	0.149117	0.105398	0.135395	0.066444	0.038027

Another modeling technique available for nominal variables is the *discriminant method* (Brand, 1999). This can be implemented in PROC MI via the DISCRIM option in the MONOTONE statement. The approach is bit restrictive, however, because it requires multivariate normality of the auxiliary variables, effectively nullifying the possibility of categorical predictor variables. A more widely applicable alternative is to utilize an implicit, or cell-based, imputation model, which is the topic of the next section.

11.4.3 Methods Based on Implicit Models

The first implicit imputation model technique we will illustrate is *hot-deck* imputation, the imputation analog of the adjustment cell approach discussed in Section 10.3. Instead of inflating the weights of the observed cases, however, we randomly select observed values within a cell and designate those as replacements for the missing values. We can do this singly or multiply; if we opt for the latter, as alluded previously in this chapter, we must follow the ABB approach described in Rubin and Schenker (1986).

To further cement the connection between hot-deck imputation to the adjustment cell approach to weighting for nonresponse, let us revisit the objectives underlying Program 10.2. Recall the assumption was that data were MCAR conditional upon the cross classification of gender and minority status. Program 11.5 illustrates syntax to independently carry out the ABB $M = 5$ times to multiply impute Q1_FULL within each of the four cells defined by crossing GENDER and MINORITY on the data set SAMP. There are no options in PROC MI to conduct hot-deck imputation, and most user-written macros are difficult to generalize because they are often developed with a particular survey's missing data scenarios in mind, are rarely amenable to performing proper MI, and tend to be based on antiquated code that is now inefficient and unnecessarily complex given some of the features recently incorporated into PROC SURVEYSELECT. Hence, this author generally finds it necessary to customize one's own syntax as we do here.

The initial PROC SQL step in Program 11.5 stores the counts of observed and missing cases within each of the four cells, which will later be passed to PROC SURVEYSELECT as part of the ABB's two stages of sampling. Recall that the first stage involves selecting a with-replacement sample of observed values within each cell of equal size to the number of observed values in the cell. Several features of PROC SURVEYSELECT highlighted in Chapter 2 are exploited to accomplish this task. Specifically, cells are defined by placing GENDER and MINORITY in the STRATA statement, with-replacement sampling is requested by the METHOD=URS (unrestricted sampling) option, and the $M = 5$ independent iterations of sampling is facilitated efficiently with the REPS=5 option.

The output data set ABB_PART1 is composed of five independent samples drawn according to our specifications and distinguishable by the numeric variable REPLICATE coded 1–5. This variable is then used as an additional stratification factor in the second stage of sampling, which calls for selecting a with-replacement sample from the observed values selected in the first stage, only this time the sample size is dictated by the number of missing values within a cell. The COUNTS_MISS data set created in the initial PROC SQL step maintains these counts, but in a subsequent DATA step the data set is concatenated with itself five times and a counter variable named REPLICATE is appended such that it can serve as the supplemental data set specified in the SAMPSIZE= option. Note that using REPLICATE as an additional stratification factor precludes the need to use the REPS= option in the second PROC SURVEYSELECT step.

Ultimately, the $M = 5$ sets of observed LOS values designated to replace the missing LOS values are stored in the data set named ABB_PART2. They are randomly sorted within each cell and then merged back into the data set SAMP_MI_5, which is the SAMP data set (with missing values) stacked on top of itself five times. Note that the PROC SORT step immediately preceding the ultimate DATA step forces the missing values of LOS to appear first within each cell of each copy of the incomplete data set. Because the ABB_PART2 data set is listed second in the MERGE statement, these missing values are overwritten with the donor values. Lastly, if we wanted to modify this code to produce a single completed data set, we could simply replace REPS=5 with REPS=1 and change both instances of DO REPLICATE=1 TO 5 to DO REPLICATE=1 TO 1.

Program 11.5: Multiple Imputation in a Hot-Deck (Cell-Based) Setting

```
* get counts of observed and missing cases within cells;
proc sql;
   create table counts_obs as
   select gender,minority,count(*) as _NSIZE_
     from samp
     where Q1_full ne .
     group by gender,minority;

   create table counts_miss as
   select gender,minority,count(*) as _NSIZE_
     from samp
     where Q1_full = .
     group by gender,minority;
quit;

* select donors within cells;
proc sort data=samp;
   by gender minority;
run;

* first step of ABB is to select WR sample of observed cases;
proc surveyselect data=samp (where=(Q1_full ne .))
     sampsize=counts_obs method=URS seed=5441254 reps=5
     outhits out=ABB_part1 (keep=replicate gender minority
     Q1_full);
     strata gender minority;
run;

* second step is to select WR sample of missing cases;
* create supplemental sample size file with the M=5 replicate
codes;
data counts_miss;
   set counts_miss;
do replicate=1 to 5;
   output;
end;
run;
```

```
proc sort data=counts_miss;
  by replicate gender minority;
run;
proc sort data=ABB_part1;
  by replicate gender minority;
run;
* select the sample of donors;
proc surveyselect data=ABB_part1
   sampsize=counts_miss method=URS seed=8744063
   outhits out=ABB_part2 (keep=replicate gender minority
   Q1_full);
   strata replicate gender minority;
run;

* randomly sort donor values within a cell;
data ABB_part2;
  set ABB_part2;
random=ranuni(34234433);
run;
proc sort data=ABB_part2 out=ABB_part2 (drop=random);
  by replicate gender minority random;
run;

* merge donor values into SAMP to create M=5 completed data
sets;
data samp_MI_5;
  set samp;
do replicate=1 to 5;
  output;
end;
run;

* sort stacked completed data sets such that missing values
appear first;
proc sort data=samp_MI_5;
  by replicate gender minority Q1_full;
run;
data samp_MI_5;
  merge samp_MI_5
        ABB_part2 (in=b);
  by replicate gender minority;
run;
```

Another implicit imputation model technique frequently employed in practice is the *propensity cell* approach, which shares its name with the weighting method counterpart discussed in Section 10.4. The two are functionally related much in the same way hot-deck imputation and adjustment cell weighting are. The setup is the same in either case. The first step is to estimate the propensity, or probability, that the given outcome variable is missing as a function of the observed covariates. In Chapter 10, we illustrated doing so via a logistic regression model in which the response indicator was the dependent variable. PROC MI follows the same approach when the PROPENSITY option is specified in the MONOTONE statement.

Once estimated for the entire sample, all units' propensities are sort-ordered and allocated into a mutually exclusive and exhaustive set of cells. Since all units within a cell should have approximately equivalent propensities, the assumption is that data are MCAR. As with the propensity cell weighting approach, allocating the sample into five cells is typically deemed sufficient. Indeed, this is the default number of cells created by PROC MI, but note that the NGROUPS= option can be used to declare an alternative number. A random sample of observed cases is selected to fill in values for the missing cases. To select these donors while accounting for the (implicit) imputation model uncertainty, PROC MI utilizes the two-stage ABB algorithm illustrated in Program 11.5.

In Program 10.3, we created five cells from a propensity model that included the four auxiliary variables as predictors as well as all two-way interaction terms. Program 11.6 demonstrates a comparable approach using the PROPENSITY option in the MONOTONE statement of PROC MI. In expectation, estimates formed after applying either nonresponse compensation technique are equivalent.

Program 11.6: Multiple Imputation within Propensity Cells

```
proc mi data=samp nimpute=5 seed=655231 out=samp_MI_5;
   class gender minority;
   var gender minority age supervisor Q1_full;
   monotone propensity (Q1_full=age|gender|supervisor|minority
   @2);
run;
```

One potential obstacle to applying the propensity cell imputation method in PROC MI at the time of this writing is that only numeric outcome variables are accommodated. This is a limitation of PROC MI, not the approach, which is scale indifferent. In all likelihood, PROC MI will be updated shortly to accommodate categorical outcome variables, but a general work-around in the meantime is to exploit a user-defined format to create a numeric "shadow" variable for purposes of imputation and then back-transform to the original character variable in the completed data set. Lewis (2013b) shows syntax to do this for a simple example.

Of course, another work-around is to conduct the approach "by hand," fitting a logistic regression model and then ranking the propensities into cells, within which the ABB is implemented. Essentially, you would combine elements of Programs 10.3 and 11.5. Yet another possibility would be to use the auxiliary variables as part of a model to predict the outcome itself, forming cells based on the rank-ordered predicted values. Little (1986) terms this approach *predictive mean stratification* within the realm of weight adjustment methods, but the approach is valid just the same for imputation.

11.4.4 A Semiparametric Method

We close this section on methods addressing univariate missingness by discussing a semiparametric approach available for continuous outcome variables called "predictive mean matching." The technique is labeled semiparametric because it blends elements of both explicit and implicit imputation models. Based on ideas in Heitjan and Little (1991) and Schenker and Taylor (1996), the notion is to fit an explicit linear regression model relating the auxiliary variables to the outcome variable, but then select a donor case from a pool of neighboring observed cases. This pool is sometimes referred to as the "neighborhood" and is defined by cases sharing similar predicted values of the outcome variable. The method can be implemented in PROC MI in the "proper" sense (i.e., accounting for the imputation model uncertainty by independently perturbing model coefficients) by specifying the REGPMM option in the MONOTONE statement.

Figure 11.4 is provided to help visualize the mechanics of the method. The figure depicts the same simple linear regression model from Figure 11.1 relating employee age to LOS for a subset of cases in the data set SAMP. As before, let us consider a nonrespondent who is 40 years old. Instead of starting at the regression line and adding a random residual term in proportion to the

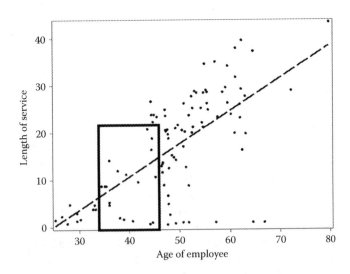

FIGURE 11.4
Visual representation of a neighborhood of similar predicted values within which a donor is selected as part of a semiparametric imputation model.

estimated root-mean-squared error model, we define a pool of responding individuals whose predicted LOS resides in the neighborhood of the predicted LOS of the missing case. The rectangle frames one possible neighborhood definition for the 40-year-old lacking data for the LOS item. From it, a donor value is selected randomly. In fact, we can control the size of the neighborhood by specifying an integer in the K= parameter in the REGPMM option of the MONOTONE statement, which defines the number of nearest cases to draw from for purposes of assigning that case's imputed value. The default is K=5.

A few advantages of this semiparametric approach relative to a strictly explicit imputation model approach are that (1) it guarantees imputed values are "real," in the sense that they are actually observed in the data, and (2) it is less sensitive to parametric assumptions about the residuals. Both of these are advantages already associated with the implicit imputation model approach, but the hybrid method's added benefit is being able to handle more auxiliary variables in the prediction model since small or empty cell sizes are no longer a cause for concern. One disadvantage is that, at the time of this writing, there is no capability in PROC MI to adapt the approach to a categorical variable using, say, a logistic regression model as the underlying explicit imputation model.

Program 11.7 illustrates an application of the predictive mean matching approach, multiply imputing LOS $M = 5$ times based on an explicit regression model including all four available auxiliary variables. Structurally, the syntax is very similar to that of the regression method shown in Program 11.1. The K=15 option specified after the slash increases the number of nearest neighbors considered for imputation from 5 (the default) to 15.

Program 11.7: Semiparametric Method for Multiple Imputation of a Continuous Outcome Variable

```
proc mi data=samp nimpute=5 seed=2311109 out=samp_MI_5;
  class gender minority;
  var gender minority age supervisor LOS;
  monotone regpmm (LOS=age gender supervisor minority / k=15);
run;
```

11.5 Multivariate Missingness

11.5.1 Introduction

The goal of this section is to extend ideas presented in the preceding sections to multivariate missingness problems, those in which missingness is not confined to a single outcome variable. We begin by defining the two fundamental missing data patterns one can encounter, which determines the particular statements and options within PROC MI at one's disposal.

11.5.2 Methods for Monotone Missingness Patterns

The only time multiple MONOTONE statements can be specified in a single PROC MI step for two or more outcome variables subject to item nonresponse is when the data set is characterized by a *monotone missingness pattern*. Figure 11.5 offers a side-by-side comparison of this pattern and its counterpart, an *arbitrary missingness pattern*, for an example survey with four outcome variables labeled y_1, y_2, y_2, and y_4. Methods to handle arbitrary missingness patterns are deferred to Section 11.5.3. Note that those methods can also be used to handle monotone patterns of nonresponse, but the methods discussed in this present section cannot be used directly to handle arbitrary patterns of nonresponse.

More formally, a survey data set with $k = 1, ..., K$ outcome variables subject to item nonresponse is characterized by a monotone missingness pattern if the variables can be arranged such that all sampling units with the kth variable missing also have all variables thereafter missing, those indexed from $k + 1$ to K. A complementary definition is that the variables can be arranged such that, for all sampling units, a particular variable being observed implies all previous variables are observed. Note that univariate missingness, when $K = 1$, is a special case of a monotone missingness pattern.

Van Buuren (2007) reviews the statistical intricacies behind imputation of more than one outcome variable and explains how a monotone pattern

Monotone missingness	Arbitrary missingness
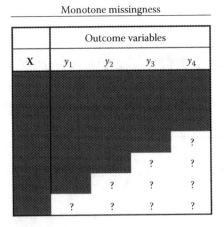	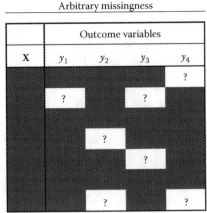

FIGURE 11.5
Visualization of a monotone versus arbitrary pattern of missingness.

of missingness is much easier to handle than an arbitrary pattern. In fact, the degree of simplification is so desirable that some researchers have suggested singly imputing for the purpose of achieving a monotone pattern of missingness, at which point the remaining missing data are multiply imputed (Schafer, 1997; Berglund, 2010). The reason things simplify is that a monotone pattern of missingness allows one to factor the joint distribution of expected values of $y_1, ..., y_K$ into a sequence of univariate distributions. If we represent all parameters associated with the kth variable's imputation model by θ_k (e.g., the MSE and beta terms in the case of a linear regression model), the idea is to begin by imputing values for the first variable by drawing from $f(y_1|\mathbf{X},\theta_1)$, the conditional distribution of y_1 given $\mathbf{X}$ and θ_1. Note that y_1 is the least missing variable or the one with the lowest item non-response rate. If we denote the completed (observed plus imputed) version of y_1 by y_1^*, then the next step is to impute values for the second variable by drawing from $f(y_2|\mathbf{X},y_1^*,\theta_2)$. The sequence continues until y_K is reached, the variable with the highest item nonresponse rate, which gets imputed by drawing from the conditional distribution $f(y_K|\mathbf{X},y_1^*,...,y_{K-1}^*,\theta_K)$, and the process is independently conducted anew to generate completed data set 2, 3, ..., M.

In the data set SAMP, Q1 and LOS are concomitantly missing or observed, which is to say unit nonresponse and item nonresponse are one and the same. Therefore, their order in the VAR statement is arbitrary as long as they both appear at the end of the list. Generally speaking, however, it is advisable to listed variables in the VAR statement in order of their missingness rates, least missing variables first. Without any imputation model covariates specified for a given outcome variable, PROC MI uses all aforementioned variables appearing in the VAR statement, which may include variables previously imputed.

Program 11.8 illustrates PROC MI syntax to impute LOS and, subsequently, Q1 using this default approach. Notice from the output how the imputation model for LOS consists of the four fully observed auxiliary variables, whereas the imputation model for Q1 also includes the completed version of LOS. If desired, however, you can tailor the set of predictor variables utilized for each variable.

Program 11.8: Multivariate Imputation to Address Multivariate Missingness in a Data Set Characterized by a Monotone Pattern of Nonresponse

```
proc mi data=samp nimpute=5 seed=2311002 out=samp_MI_5;
   class gender minority Q1;
   var age supervisor gender minority LOS Q1;
   monotone reg       (LOS / details);
   monotone logistic (Q1  / details);
run;
```

Regression Models for Monotone Method									
Imputed Variable	Effect	GENDER	Minority	Obs-Data	Imputation				
					1	2	3	4	5
LOS	Intercept			0.03733	0.078320	0.031989	0.018884	0.072501	0.134158
LOS	AGE			0.55095	0.545726	0.552659	0.481884	0.545204	0.546268
LOS	SUPERVISOR			0.04699	−0.018410	0.109854	0.113643	0.041053	0.080477
LOS	GENDER	F		0.22018	0.211729	0.259209	0.203366	0.201167	0.225245
LOS	Minority		N	−0.06712	−0.086724	−0.046748	−0.044605	0.069757	−0.112897

Logistic Models for Monotone Method									
Imputed Variable	Effect	GENDER	Minority	Obs-Data	Imputation				
					1	2	3	4	5
Q1	Intercept			−1.52646	−1.558413	−1.544903	−1.542054	−1.545839	−1.162852
Q1	AGE			0.74298	0.760925	1.125417	0.648479	0.427119	0.501718
Q1	SUPERVISOR			0.18687	0.326286	0.118693	0.226960	0.236919	0.141469
Q1	GENDER	F		−0.06099	−0.056570	−0.084886	−0.142290	−0.195548	−0.097480
Q1	Minority		N	0.19038	0.266162	0.199569	0.173322	0.097349	0.131788
Q1	LOS			−0.93964	−1.101397	−0.964325	−1.007601	−0.800782	−0.812402

11.5.3 Methods for Arbitrary Missingness Patterns

Some of the earliest proposed methods to combat arbitrary patterns of nonresponse called for specifying a parametric multivariate density for the entire battery of auxiliary and outcome variables. This leads to a tractable solution only in certain circumstances, such as when all variables are jointly distributed as multivariate normal. These approaches are not well suited for arbitrary missingness patterns in data sets consisting of mixed variable types, such as commingled continuous, semicontinuous, count, and categorical variables. In response, techniques have been proposed to conduct variable-by-variable imputations that effectively bypass the need to formally specify the multivariate density. They have been ascribed different names by their inventors, including *sequential regression* (Raghunathan et al., 2001), *iterated univariate imputation* (Gelman, 2004), and *chained equations* (Van Buuren and Groothuis-Oudshoorn, 2011), but all operate similarly in spirit. Van Buuren (2007) terms this class of approaches *fully conditional specification* (FCS), which is the terminology adopted by SAS in PROC MI.

The FCS technique proceeds as follows. As before, let $\mathbf{X}$ represent the design matrix of fully observed auxiliary variables and let $y_1, \ldots, y_K$ represent the sequence of outcome variables subject to missingness ordered according to their item nonresponse rates, smallest to largest. The first step is to impute values of y_1 conditional on $\mathbf{X}$ using an appropriate model (e.g., linear regression for a continuous outcome, logistic regression for a binary outcome). Next, values of y_2 are imputed conditional upon $\mathbf{X}$ and y_1^*, where y_1^* reflects the completed version of y_1 (including both observed and imputed values), and then the process continues to impute y_3 conditional upon $\mathbf{X}$, y_1^*, and y_2^*, and so on.

At first glance, there is nothing new relative to what was described for the monotone missingness case, but the critical difference is that once the first completed data set has been created, the approach calls for cycling back through all K variables and reimputing the originally missing values a predetermined number of times. These are synonymously referred to as "cycles," "rounds," or "burn-in iterations." The iterative process builds interdependence and allows the technique to stabilize, or converge, to the multivariate joint density it is implicitly approximating. The default number of iterations used in software to conduct the FCS technique is sometimes as few as 5 or 10, but there is generally an option for the user to specify an alternative. Only after these iterations are finished is the mth completed data set "released." Note that this iterative step is not necessary when the missingness pattern is monotone.

Critics of the FCS approach to multivariate imputation are apprehensive about the lack of theory justifying its use for certain missing data problems and the fact that there is no convergence guarantee. Despite these potential weaknesses, the method has been shown to work quite well in a broad range of settings (Brand, 1999; Horton and Lipsitz, 2001; Van Buuren et al., 2006), and in situations where the multivariate density *can* be explicitly specified (e.g., when missing data are multivariate normal), it is possible to use formal statistical theory to prove convergence to the stated joint distribution.

Program 11.9 demonstrates the FCS statement in PROC MI. Because the missingness pattern in the data set SAMP is monotone, for illustrative purposes the program begins by imparting an arbitrary pattern of missingness by randomly deleting the observed values of LOS and Q1 at rates of 25% and 10%, respectively, storing the result in a new data set named SAMP_ARB. The Missing Data Patterns table of the PROC MI output reflects this new pattern of missing data.

The FCS statement syntax is patterned after the MONOTONE statement, with the REG, LOGISTIC, REGPMM, and DISCRIM options available. Multiple FCS statements are allowed, but one difference relative to the MONOTONE statement is that variables listed in the VAR statement need not be ordered in terms of increasing rates of missingness. Note, however, that PROC MI imputes variables according to their order in the VAR statement. This is evident from the output of Program 11.9 as we notice Q1 was imputed first even though it appears in the second FCS statement. Note, also, that the default number of iterations PROC MI uses is 20, which can be modified by the NBITER= option in the FCS statement.

Program 11.9: Multiple Imputation to Address Multivariate Missingness in a Data Set Characterized by an Arbitrary Pattern of Nonresponse

```
data samp_arb;
  set samp;
if LOS ne . and ranuni(882377) < .25 then LOS=.;
if Q1  ne . and ranuni(243113) < .1  then Q1=.;
```

```
run;
proc mi data=samp_arb nimpute=5 seed=873772 out=samp_MI_5;
  class gender minority Q1;
  var age supervisor gender minority Q1 LOS;
  fcs reg (LOS / details);
  fcs logistic (Q1  / details);
run;
```

									Group Means		
Group	AGE	SUPERVISOR	GENDER	Minority	Q1	LOS	Freq	Percent	AGE	SUPERVISOR	LOS
1	X	X	X	X	X	X	418	34.33	48.999582	0.167464	16.820217
2	X	X	X	X	X	.	144	12.00	48.115864	0.131944	.
3	X	X	X	X	.	X	50	4.17	49.870445	0.140000	17.735469
4	X	X	X	X	.	.	588	49.00	49.644228	0.175871	.

Missing Data Patterns

Imputed Variable	Effect	GENDER	Minority	Imputation				
				1	2	3	4	5
Q1	Intercept			1.596296	−1.730205	−1.703117	−1.376551	−1.592933
Q1	AGE			0.969982	0.718085	0.884634	0.798718	0.823566
Q1	SUPERVISOR			0.292778	0.486419	0.178292	0.150075	0.258213
Q1	GENDER	F		0.032520	−0.076563	−0.001279	−0.172497	−0.098352
Q1	Minority		N	0.277019	0.233725	0.402920	0.146328	0.351572
Q1	LOS			−1.220286	−1.039853	−1.081966	−0.865587	−1.035971

Logistic Models for FCS Method

Imputed Variable	Effect	GENDER	Minority	Q1	Imputation				
					1	2	3	4	5
LOS	Intercept				−0.194696	−0.157448	−0.113303	−0.119262	−0.178684
LOS	AGE				0.571511	0.569955	0.589343	0.623881	0.601827
LOS	SUPERVISOR				0.016753	−0.044060	0.020574	−0.013147	−0.040495
LOS	GENDER	F			0.158178	0.112932	0.166012	0.210020	0.092271
LOS	Minority		N		−0.078002	−0.140693	−0.054859	−0.089310	−0.052694
LOS	Q1			0	−0.366181	−0.387676	−0.361469	−0.341433	−0.459718

Regression Models for FCS Method

11.6 Inferences from Multiply Imputed Data

11.6.1 Introduction

Over the course of this chapter, we have discussed various methods to multiply impute missing data. In each example, the objective was to create $M = 5$ completed data sets, a number commonly used in practice (e.g., Schenker et al., 2006; Lewis et al., 2014). In this section, we revisit

Rubin's (1987) fundamental combination rules we defined in Equations 11.1 through 11.4 to illustrate how the M point estimates and corresponding variance approximations can be consolidated into a single MI point estimate and measure of variability. PROC MIANALYZE is a handy SAS/STAT procedure you can employ to carry out these computations, but independent estimation of each completed data sets is a necessary antecedent. The setup is the same regardless of the analysis at hand, with the exception that the input data sets and syntax for univariate inferences differ somewhat from multivariate inferences. An example illustrating the former is discussed in Section 11.6.2, and an example illustrating the latter is given in Section 11.6.3.

11.6.2 Univariate Inferences

Suppose the goal is to estimate the overall MI mean and standard error for the multiply imputed LOS variable produced from, say, Program 11.7. Once the M completed data sets have been created, the first step is to conduct the standard complete data analysis independently on each. Recall that PROC MI stacks the $M = 5$ completed data sets on top of one another, which can be distinguished by the automatically appended numeric variable _IMPUTATION_. Therefore, we simply place that variable in the BY statement of the analysis procedure we would employ in the absence of missing data, which is PROC SURVEYMEANS in this case. The ODS statement in Program 11.10 captures the key portion of the output and stores the result in a data set named STATS_MI_5. Figure 11.6 previews the contents of this summary data set, which consists of a point estimate and standard error associated with each completed data set analyzed. This is what gets input to PROC MIANALYZE. We specify the column in this data set housing the point estimate (MEAN) in the MODELEFFECTS statement, the column housing the standard error (STDERR) in the STDERR statement, and PROC MIANALYZE infers M from the number of rows.

	Imputation Number	Variable Name	Mean	StdErr
1	1	LOS	16.523988	0.328866
2	2	LOS	16.523568	0.327185
3	3	LOS	16.581456	0.331456
4	4	LOS	16.407314	0.322938
5	5	LOS	16.089813	0.326479

FIGURE 11.6
Partial view of data set STATS_MI_5 generated by Program 11.10, the five sets of sample means estimated from the multiply imputed survey data set SAMP.

The overall MI estimate can be found in the Parameter Estimates table. Following Equation 11.1, the value is 16.425, the average of the five completed data set estimates $(16.524 + \cdots + 16.090)/5$. The overall MI standard error is the square root of the overall MI variance (see Equation 11.4), which is the sum of two terms: the average within-imputation variance and a term reflecting the between-imputation uncertainty. These two terms are reported in the Variance Information table. For example, the value 0.1072 reported in the column labeled "Within" is found by averaging the five completed data set variance estimates $[(0.3289)^2 + \cdots + (0.3265)^2]/5$.

While several other quantities are provided in the Variance Information table, one particularly useful diagnostic of the two variance terms' respective sizes and the efficacy of the imputation model is the FMI. This is discussed in depth in Section 3.3 of Rubin (1987). Following the notation defined earlier, if we symbolize the within-imputation variance term W_M and the between-imputation variance term B_M, the FMI of the overall MI point estimate $\hat{\theta}_M$ is approximately equal to

$$FMI(\hat{\theta}_M) \approx \frac{(1+(1/M))B_M}{W_M + (1+(1/M))B_M} \tag{11.5}$$

where, again, the term $(1 + (1/M))$ represents a finite imputation correction factor that tends to 1 as $M \to \infty$. In words, the FMI is the between-imputation variance component divided by the total MI variance. In an uninformative imputation model, such as a single-cell, hot-deck routine, the FMI approximates the item nonresponse rate. To the extent the FMI is smaller than the item nonresponse rate, it provides evidence that the covariates employed in the imputation model serve to recapture a portion of the missing data uncertainty. In this particular analysis, the FMI is approximately 33.5%, which is a little less than the item nonresponse rate of 47.7%, suggesting that the four covariates exhibit a moderate amount of explanatory power. The FMI is estimate specific, however, so the same conclusion may not prevail for all statistics produced.

Program 11.10: Univariate Variance Estimation following Multiple Imputation

```
* replicating the predictive mean matching imputation model
approach from Program 11.7;
proc mi data=samp nimpute=5 seed=2311109 out=samp_MI_5;
  class gender minority;
  var gender minority age supervisor LOS;
  monotone regpmm (LOS=age gender supervisor minority / k=15);
run;
```

```
* compute and store the M estimated means and standard errors
of LOS;
proc sort data=samp_MI_5;
  by _IMPUTATION_;
run;

ods output statistics=stats_MI_5;
proc surveymeans data=samp_MI_5 mean stderr;
  by _IMPUTATION_;
  var LOS;
weight weight_base;
run;

* get the overall MI estimate and standard error;
proc mianalyze data=stats_MI_5;
  modeleffects mean;
  stderr stderr;
run;
```

Model Information	
Data Set	WORK_STATS_MI_5
Number of imputations	5

Variance Information							
	Variance				Relative Increase in	Fraction Missing	Relative
Parameter	Between	Within	Total	DF	Variance	Information	Efficiency
Mean	0.039164	0.107189	0.154185	43.054	0.438448	0.334997	0.937208

Parameter Estimates										
		Std	95% Confidence						t for H0: Parameter	
Parameter	Estimate	Error	Limits		DF	Minimum	Maximum	Theta0	= Theta0	Pr > |t|
Mean	16.425228	0.392665	15.63337	17.21708	43.054	16.089813	16.581456	0	41.83	<.0001

11.6.3 Multivariate Inferences

To motivate a multivariate MI inference problem, suppose that we wanted to exploit the four auxiliary variables as part of a logistic regression model to jointly estimate their impact on the likelihood that the employee agrees with the statement "I like the kind of work I do." Recall this is stored in the 0/1 indicator variable Q1. Without loss of generality, we will use the $M = 5$ completed data sets created using the logistic regression explicit imputation model demonstrated in Program 11.2. For quick reference, that PROC MI syntax is reproduced at the top of Program 11.11.

For multivariate inferences on a set of p distinct point estimates, we need to provide two input data sets to PROC MIANALYZE. The first should contain the M point estimates, and the second should contain M copies of the corresponding variance–covariance matrices. With respect to the logistic regression model fitted in Program 11.11, $p = 5$, since the model contains an intercept and four other coefficients. The ODS statement immediately preceding PROC SURVEYLOGISTIC stores the five sets of estimated model coefficients in a data set named BETAS and the five sets of variance–covariance matrices in a data set named COV_MATRIX. Note that the COVB option after the slash in the MODEL statement must be specified to have these matrices sent to the output. Figure 11.7 is a screenshot of the BETAS data set, and Figure 11.8 is the like for the COV_MATRIX data set (only the first two sets of each are shown).

The point estimate data set must be identified in the PARMS= option of the PROC MIANALYZE statement. SAS will recognize any variable named PARAMETER, EFFECT, VARIABLE, or PARM as the one housing the point estimates. In this case, PROC SURVEYLOGISTIC names the column of logistic regression parameters VARIABLE by default, so no additional recoding is necessary.

Similar rules apply to the variance–covariance matrices, which must be specified in the COVB= option of the PROC MIANALYZE statement. The key variable PARAMETER will be recognized, but note how the effect naming convention differs slightly relative to the VARIABLE column in the BETAS data set. This will cause PROC MIANALYZE to stop executing and issue an error to the log. The IF/THEN logic and the RENAME statement in the DATA step overwriting the COV_MATRIX data set ensures compatible effect names in the two input data sets. These five effects (i.e.,

	Imputation Number	Variable	Level of CLASS Variable 1 for Variable	DF	Estimate	Standard Error	Wald Chi-Square	Pr> Chi-Square
1	1	Intercept		1	3.4547	0.4120	70.3274	<.0001
2	1	GENDER	F	1	0.3342	0.0747	20.0427	<.0001
3	1	Minority	N	1	−0.2578	0.0903	8.1473	0.0043
4	1	AGE		1	−0.0373	0.00776	23.1316	<.0001
5	1	SUPERVISOR		1	−0.4345	0.1764	6.0660	0.0138
6	2	Intercept		1	2.7696	0.3665	57.0955	<.0001
7	2	GENDER	F	1	0.3616	0.0724	24.9635	<.0001
8	2	Minority	N	1	−0.2375	0.0859	7.6419	0.0057
9	2	AGE		1	−0.0266	0.00709	14.0955	0.0002
10	2	SUPERVISOR		1	−0.3065	0.1790	2.9333	0.0868

FIGURE 11.7
Partial view of data set BETAS generated by Program 11.11, the five sets of logistic regression parameters estimated from the multiply imputed survey data set SAMP.

	Imputation Number	Parameter	Intercept	GENDER	Minority	AGE	SUPERVISOR
1	1	Intercept	0.169704	-0.00096	-0.00539	-0.00311	0.000607
2	1	GENDER	-0.00096	0.005574	0.000092	0.000037	0.0004
3	1	Minority	-0.00539	0.000092	0.008155	0.000019	-0.00145
4	1	AGE	-0.00311	0.000037	0.000019	0.00006	-0.00013
5	1	SUPERVISOR	0.000607	0.0004	-0.00145	-0.00013	0.031123
6	2	Intercept	0.134353	-0.00119	-0.00342	-0.00252	0.00364
7	2	GENDER	-0.00119	0.005237	0.00002	0.00004	0.000556
8	2	Minority	-0.00342	0.00002	0.007384	-7.86E-6	-0.00148
9	2	AGE	-0.00252	0.00004	-7.86E-6	0.00005	-0.00018
10	2	SUPERVISOR	0.00364	0.000556	-0.00148	-0.00018	0.032036

FIGURE 11.8
Partial view of data set COV_MATRIX generated by Program 11.11, the five sets of logistic regression parameter covariance matrices estimated from the multiply imputed survey data set SAMP.

model parameters) are listed in the MODELEFFECTS statement of PROC MIANALYZE.

The combination rules are multivariate analogs to those from the univariate setting. For example, the overall MI estimates reported in the Parameter Estimates table are found by averaging each entry in the vector of model parameters, and the overall MI variance is the sum of a within- and between-imputation component.

If we denote the mth completed data set's vector of estimates $\hat{\theta}_m$, the overall MI estimate is

$$\hat{\theta}_M = \frac{1}{M} \sum_{m=1}^{M} \hat{\theta}_m \tag{11.6}$$

The within-imputation variance component is found by averaging all entries of the M variance–covariance matrices. If we denote the matrix for the mth completed data set as $\text{cov}(\hat{\theta}_m)$, this is defined as

$$\mathbf{W}_M = \frac{1}{M} \sum_{m=1}^{M} \text{cov}(\hat{\theta}_m) \tag{11.7}$$

while the between-imputation variance component is given by

$$\mathbf{B}_M = \frac{1}{M-1} \sum_{m=1}^{M} (\hat{\theta}_m - \hat{\theta}_M)(\hat{\theta}_m - \hat{\theta}_M)^{\mathsf{T}} \tag{11.8}$$

and so the total MI variance is

$$\mathrm{var}(\hat{\boldsymbol{\theta}}_M) = \mathbf{W}_M + \left(1 + \frac{1}{M}\right)\mathbf{B}_M \tag{11.9}$$

Program 11.11: Multivariate Variance Estimation following Multiple Imputation

```
* replicating the logistic regression imputation model
approach from Program 11.2;
proc mi data=samp nimpute=5 seed=7000726 out=samp_MI_5;
  class gender minority Q1;
  var gender minority age supervisor Q1;
  monotone logistic (Q1=gender minority age supervisor /
  details);
run;

ods output ParameterEstimates=betas
           covb=cov_matrix;
proc surveylogistic data=samp_MI_5;
  by _IMPUTATION_;
  class gender minority;
  model Q1(event='1') = gender minority age supervisor / covb;
run;

* rename the variables in cov_matrix to match those in betas;
data cov_matrix;
  set cov_matrix;
if Parameter='GENDERF' then Parameter='GENDER';
if Parameter='MinorityN' then Parameter='Minority';
rename GENDERF   = GENDER
       MinorityN = Minority;
run;

proc mianalyze parms=betas covb=cov_matrix;
  modeleffects intercept gender minority age supervisor;
run;
```

Model Information	
PARMS data set	WORK.BETAS
COVB data set	WORK.COV_MAT
Number of imputations	5

Variance Information							
	Variance				Relative Increase in Variance	Fraction Missing Information	Relative Efficiency
Parameter	Between	Within	Total	DF			
Intercept	0.095704	0.150768	0.265613	21.396	0.761734	0.478911	0.912590
Gender	0.006710	0.005318	0.013370	11.029	1.514049	0.658942	0.883557
Minority	0.000237	0.007592	0.007877	3060.5	0.037508	0.036781	0.992697
Age	0.000040807	0.000055161	0.000104	18.088	0.887724	0.520503	0.905715
Supervisor	0.020032	0.031028	0.055066	20.991	0.774719	0.483504	0.911826

Parameter Estimates										
Parameter	Estimate	Std Error	95% Confidence Limits		DF	Minimum	Maximum	Theta0	t for H0: Parameter= Theta0	Pr> $\lvert t \rvert$
Intercept	2.991897	0.515377	1.92132	4.06247	21.396	2.761432	3.454689	0	5.81	<.0001
Gender	0.294057	0.115629	0.03964	0.54847	11.029	0.156403	0.361562	0	2.54	0.0273
Minority	−0.246250	0.088754	−0.42027	−0.07223	3060.5	−0.258586	−0.223282	0	−2.77	0.0056
Age	−0.029932	0.010204	−0.05136	−0.00850	18.088	−0.037321	−0.024454	0	−2.93	0.0089
Supervisor	−0.399888	0.234662	−0.88791	0.08813	20.991	−0.577348	−0.214672	0	−1.70	0.1031

11.7 Accounting for Features of the Complex Survey Data during the Imputation Modeling and Analysis Stages

Despite missing data problems in complex surveys being the catalyst for much of the literature on imputation methods over the past 40 or so years, historically, most underlying theory assumes a simple random sample design. In recent years, however, researchers have begun to study in earnest the considerations of the complex survey features, and a clearer picture regarding best practices is beginning to emerge. To be sure, much more research is needed, but the purpose of this section is to briefly summarize recent advances in this burgeoning area of the literature.

Based on guidance provided in Rubin (1996), the working rule for accommodating complex survey features into the imputation modeling process has been to include the weight as a continuous predictor variable and include indicator variables to represent the effects of strata and PSUs. A series of simulations by Reiter et al. (2006) exposed how egregious biases can occur if one fails to account for the effects of strata and PSUs when the outcome variable's expected value differs amongst them. Berglund and Heeringa (2014, p. 52) suggest that the most parsimonious way to account for both simultaneously is to create a unique PSU identification variable and treat as a CLASS variable in the imputation model. Some minor recoding may be necessary if the codes identifying PSUs are "reused" across strata. For example, considering the TEST data set from Chapter 9, we would not want to use the HOMEROOM variable as is, but instead create a new variable in a DATA step with syntax like PSU_UNIQUE=GRADE|HOMEROOM, which concatenates the stratum and PSU codes into a set of unique PSU identifiers. Reiter et al. (2006, p. 148) acknowledge this approach may be cumbersome in the presence of dozens or hundreds of PSUs, but suggest "imputers can simplify the model for the design variables, for example, collapsing cluster categories or including proxy variables (e.g., cluster size) that are related to the outcome of interest."

In retrospect, recall that the employee satisfaction survey used for examples in this chapter was a single-stage, stratified design in which $n = 1200$ employees were sampled from $H = 2$ strata, one stratum of nonsupervisors and one stratum of supervisors. Because each employee represents a cluster of size 1, we did not need to concern ourselves with that feature of complex survey data. We did, however, account for the stratification by including the SUPERVS variable in the imputation model. Hence, adding the base weights into the imputation model as a continuous predictor would have been redundant.

Consideration of weights has cropped up repeatedly in applications of cell-based (e.g., hot-deck) imputation. In effect, ignoring the weights when randomly selecting donors assumes that there is no relationship between the weights, the propensity to respond, and the expected value of the outcome variable within a cell. Several alternatives have been proposed to account for the weights during the selection process (Cox, 1980; Platek and Gray, 1983; Rao and Shao, 1992) by, for example, selecting a PPS sample (see Section 2.2.3) of donors where their weights serve as the measure of size. Recent research by Andridge and Little (2009), however, suggests that these alternatives can be inefficient. Instead, they argue one should focus on creating cells that are homogeneous with respect to response propensities and the outcome variable, but utilize an unweighted selection process therein (e.g., as was done in Programs 11.5 and 11.6).

In Chapter 4 of Berglund and Heeringa (2014), a few practical recommendations are given with respect to the analysis stage of multiply imputed complex survey data. One is to remain cognizant of the complex survey degrees of freedom, defined as the number of distinct PSUs minus the number of strata under the default TSL variance estimation approach. By default, PROC MIANALYZE assumes the sample size is large enough to effectively consider infinite degrees of freedom. While this may have been acceptable in the $n = 1200$ employee satisfaction survey used in this chapter, it may not be for other sample designs. The EDF=*number* option in the PROC MIANALYZE statement can be used to assign a more appropriate degrees of freedom.

Another issue raised by Berglund and Heeringa (2014) is that one may occasionally restrict imputation to occur for only a portion of the survey data set. For example, imagine the variable YRS_SMOKE reflects the number of years an individual has been smoking, which is only relevant for individuals who currently smoke. A logical approach for imputing missing values of YRS_SMOKE might be to utilize a WHERE statement or otherwise subset the survey data set prior to invoking PROC MI; however, this may not correspond to a proper subset (see definition in Section 8.2), and so subsequent analyses of the completed data sets output by PROC MI could result in erroneous inferences. One work-around would be to concatenate the excluded records (e.g., data for nonsmokers) to the survey data set in each of the M data sets output and then perform a domain analysis as prescribed in Chapter 8. If you pursue this option, however, be advised that you should fill in the missing data in the analysis variable for records excluded from the imputation process (e.g., substitute in YRS_SMOKE=0 for the nonsmokers). It does not matter what the value is since it falls outside the domain of interest, but any record with missing data for the analysis variable is dropped by the given SURVEY procedure at the outset, defeating the whole purpose. The other work-around, as suggested by Berglund and Heeringa (2014), is to use a SURVEY procedure to append replicate weights (see Chapter 9) to the full complex survey data set prior to subsetting and executing the PROC MI step. So long as the replicate weights are maintained on the data sets output by PROC MI, they can be used in subsequent analyses—recall that, because the replicate weights wholly reflect the features of the complex survey design, subsetting is technically permitted.

In closing, we make mention of a recent proposal by Zhou (2014) that eliminates many of the complexities cited in this section. She calls for a two-step MI procedure in which the first step effectively "uncomplexes" the sample design by creating synthetic populations, each of which is susceptible to the nonresponse patterns exhibited in the sample data. This is as if one were simulating hypothetical iterations of a census of the finite population. For each synthetic population, the second step is to implement traditional imputation methods, those assuming a simple random sample design. This means there is no longer a need to account for the complex survey features during the imputation modeling stage. Moreover, there is no longer a need to account for these features during the analysis stage of the M completed data sets, and so traditional, non-SURVEY procedures can be employed. Preliminary results are promising. It will be interesting to see whether the method, or a future adaptation of it, is adopted into mainstream nonresponse adjustment techniques in a complex survey setting.

11.8 Summary

Virtually all real-world survey efforts are affected by some degree of item nonresponse, which occurs when a sampling unit provides an incomplete or invalid response to a survey question. This chapter dealt with a popular approach to compensate for it: imputing, or filling in, the missing data. Imputation is challenging on many fronts. It is challenging to formulate a model with covariates that are both predictive of the outcome variable susceptible to missingness itself *and* the probability that the outcome variable is missing, two conditions needed for the MAR assumption to hold. Many surveys simply lack the necessary auxiliary variables to do so. Another challenge is properly reflecting the uncertainty associated with substituting a plausible value in place of the unobserved true value. Although a variety of methods have been proposed in the literature, MI (Rubin, 1987) has gained the most traction. Indeed, many of the tools built into SAS and demonstrated in this chapter reside within the SAS/STAT procedure PROC MI. Because of the intricate nature of the imputation process, it is wise to make use of these capabilities if possible even if the goal is to produce a single completed data set.

There are numerous other utile features of PROC MI that were not demonstrated in this chapter. For example, the MIN= and MAX= options in the PROC MI statement can be used to bound the imputed values for continuous variables, which can reduce or eliminate the chance that implausible values are returned. Moreover, the EM and MCMC statements are capable of applying the maximum likelihood (Dempster et al., 1977) and Markov chain Monte Carlo (Schafer, 1997) methods, respectively. These methods are bit restrictive, however, in that they are only applicable to multivariate normal data. Although the TRANSFORM statement can be a handy on-the-fly patch to achieve normality in certain circumstances, this author would argue that categorical variables are more appropriately handled using one of the alternatives demonstrated as part of the MONOTONE or FCS statements, depending upon whether the missing data pattern is monotone or arbitrary (see Figure 11.5).

References

Agresti, A. (2013). *Categorical Data Analysis*, 3rd edn. New York: Wiley.

Allison, P. (1982). Discrete-time methods for the analysis of event histories, *Sociological Methods*, **13**, 61–98.

Allison, P. (2008). Convergence failures in logistic regression, *Proceedings of the SAS Global Forum*. Available online at: http://www2.sas.com/proceedings/forum2008/360-2008.pdf. Accessed on 31 January, 2016.

Allison, P. (2010). *Survival Analysis Using SAS®: A Practical Guide*, 2nd edn. Cary, NC: SAS Institute.

Allison, P. (2012). *Logistic Regression Using SAS®: Theory and Application*, 2nd edn. Cary, NC: SAS Institute.

Andridge, R. and Little, R. (2009). The use of sample weights in hot deck imputation, *Journal of Official Statistics*, **25**, 21–36.

Andridge, R. and Little, R. (2010). A review of hot deck imputation for survey non-response, *International Statistical Review*, **78**, 40–64.

Andridge, R. and Little, R. (2011). Proxy pattern-mixture analysis for survey nonresponse, *Journal of Official Statistics*, **27**, 153–180.

Archer, K. (2001). Goodness-of-fit tests for logistic regression models developed using data collected from a complex sampling design, PhD thesis, Ohio State University, Columbus, OH.

Archer, K. and Lemeshow, S. (2006). Goodness-of-fit test for a logistic regression model fitted using survey sample data, *The Stata Journal*, **6**, 97–105.

Battaglia, M., Hoaglin, D., and Frankel, M. (2009). Practical considerations in raking survey data, *Survey Practice*, **2**. Available online at: http://www.surveypractice.org/index.php/SurveyPractice/article/view/176/html. Accessed on 31 January, 2016.

Berglund, P. (2004). Analysis of complex sample survey data using the SURVEY PROCEDURES and macro coding, *Proceedings of the Annual Conference of the MidWest SAS Users Group*. Available online at: http://www.lexjansen.com/mwsug/2004/Statistics/S7_Berglund.pdf. Accessed on 31 January, 2016.

Berglund, P. (2010). An introduction to multiple imputation of complex sample data using SAS® v9.2, *Proceedings of the SAS Global Forum*. Available online at: http://support.sas.com/resources/papers/proceedings10/265-2010.pdf.

Berglund, P. (2011). An overview of survival analysis using complex sample data, *Proceedings of the SAS Global Forum*. Available online at: http://support.sas.com/resources/papers/proceedings11/338-2011.pdf. Accessed on 31 January, 2016.

Berglund, P. and Heeringa, S. (2014). *Multiple Imputation of Missing Data Using SAS®*. Cary, NC: SAS Institute.

Bethlehem, J. (1988). Reduction of nonresponse bias through regression estimation, *Journal of Official Statistics*, **4**, 251–260.

Bethlehem, J. (2002). Weighting nonresponse adjustments based on auxiliary information, in *Survey Nonresponse*, Groves, R., Dillman, D., Eltinge, J., and Little, R. (eds). New York: Wiley.

Bienias, J. (2001). Replicate-based variance estimation in a SAS® macro, *Proceedings of the Annual Conference of the NorthEast SAS Users Group (NESUG)*. Available online at: http://www.lexjansen.com/nesug/nesug01/st/st9001.pdf. Accessed on 31 January, 2016.

Binder, D. (1981). On the variances of asymptotically normal estimators from complex surveys, *Survey Methodology*, **7**, 157–170.

Binder, D. (1983). On the variances of asymptotically normal estimators from complex surveys, *International Statistical Review*, **51**, 279–292.

Binder, D. (1990). Fitting Cox's proportional hazards models from survey data, *Proceedings of the Joint Statistical Meetings of the American Statistical Association*, American Statistical Association, Alexandria, VA.

Binder, D. and Roberts, G. (2009). Design- and model-based inference for model parameters, in *Handbook of Statistics Volume 29B: Sample Surveys: Inference and Analysis*, Pfeffermann, D. and Rao, C.R. (eds.). Amsterdam, the Netherlands: Elsevier.

Bowley, A. (1906). Address to the economic science and statistics section of the British Association for the Advancement of Science, *Journal of the Royal Statistical Society*, **69**, 548–557.

Brand, J. (1999). Development, implementation and evaluation of multiple imputation strategies for the statistical analysis of incomplete data sets, PhD dissertation, Erasmus University, Rotterdam, the Netherlands.

Breiman, L., Friedman, J., Olshen, R., and Stone, C. (1993). *Classification and Regression Trees*. New York: Chapman & Hall.

Breslow, N. (1974). Covariance analysis of censored survival data, *Biometrics*, **30**, 89–99.

Brewer, K. (1963). A model of systematic sampling with unequal probabilities, *Australian Journal of Statistics*, **5**, 93–105.

Brick, J.M. and Jones, M. (2008). Propensity to respond and nonresponse bias, *METRON-International Journal of Statistics*, **66**, 51–73.

Brick, J.M. and Kalton, G. (1996). Handling missing data in survey research, *Statistical Methods in Medical Research*, **5**, 215–238.

Brick, J.M., Montaquila, J., and Roth, S. (2003). Identifying problems with raking estimators, *Proceedings of the Survey Research Methods of the American Statistical Association*, American Statistical Association, Alexandria, VA.

Brown, L., Cai, T., and DasGupta, A. (2001). Interval estimation for a binomial proportion, *Statistical Science*, **16**, 101–133.

Cassell, D. (2007). Don't be loopy: Re-sampling and simulation the SAS® Way, *Proceedings of the SAS Global Forum*. Available online at: http://www2.sas.com/proceedings/forum2007/183-2007.pdf. Accessed on 31 January, 2016.

Chen, S. and Krenzke, T. (2013). Inference of domain parameters through an automatic adjustment of degrees of freedom, *Proceedings of the 2013 Federal Committee on Statistical Methodology (FCSM) Research Conference*. Available online at: https://fcsm.sites.usa.gov/files/2014/05/D1_Chen_2013FCSM.pdf. Accessed on 31 January, 2016.

Chromy, J. (1979). Sequential sample selection methods, *Proceedings of the Joint Statistical Meetings of the American Statistical Association*, American Statistical Association, Alexandria, VA.

Clopper, C. and Pearson, E. (1934). The use of confidence or fiducial limits illustrated in the case of the binomial, *Biometrika*, **26**, 404–413.

Cochran, W. (1968). The effectiveness of adjustment by subclassification in removing bias in observational studies, *Biometrics*, **24**, 295–313.

Cochran, W. (1977). *Sampling Techniques*, 3rd edn. New York: Wiley.

Converse, J. (1987). *Survey Research in the United States: Roots and Emergence*. Berkeley, CA: University of California Press.

Copen, C., Daniels, K., Vespa, J., and Mosher, W. (2012). First marriages in the United States: Data from the 2006–2010 National Survey of Family Growth, *National Health Statistics Reports*, **49**. Available online at: www.cdc.gov/nchs/data/nhsr/nhsr049.pdf.

Couper, M. (2008). *Designing Effective Web Surveys*. New York: Cambridge University Press.

Cox, B. (1980). The weighted sequential hot deck imputation procedure, *Proceedings of the Joint Statistical Meetings of the American Statistical Association*, American Statistical Association, Alexandria, VA.

Cox, D. (1972). Regression models and life tables, *Journal of the Royal Statistical Society, Series B*, **20**, 187–220 (with discussion).

Cox, D. (1975). Partial likelihood, *Biometrika*, **62**, 269–276.

Cox, D. and Snell, E. (1989). *The Analysis of Binary Data*, 2nd edn. London, U.K.: Chapman & Hall.

D'Agostino, R. (1998). Propensity score methods for bias reduction in the comparison of a treatment to a non-randomized control group, *Statistics in Medicine*, **17**, 2265–2281.

de Leeuw, E. (2005). To mix or not to mix data collection modes in surveys, *Journal of Official Statistics*, **21**, 233–255.

Deming, W. and Stephan, F. (1940). On a least squares adjustment of a sampled frequency table when the expected marginal totals are known, *The Annals of Mathematical Statistics*, **11**, 427–444.

Dempster, A., Laird, N., and Rubin, D. (1977). Maximum likelihood from incomplete data via the EM algorithm, *Journal of the Royal Statistical Society, Series B*, **39**, 1–38.

Deville, C. and Särndal, C. (1992). Calibration estimators in survey sampling, *Journal of the American Statistical Association*, **87**, 376–382.

Dillman, D., Smythe, J., and Christian, L. (2009). *Internet, Mail, and Mixed-Mode Surveys: The Tailored Design Method*. Hoboken, NJ: Wiley.

Dorfman, A. and Valliant, R. (1993). Quantile variance estimators in complex surveys, *Proceedings of the Survey Research Methods Section of the American Statistical Association*, American Statistical Association, Alexandria, VA, pp. 866–871.

Durbin, J. (1959). A note on the application of Quenouille's method of bias reduction to the estimation of ratios, *Biometrika*, **46**, 477–480.

Efron, B. (1977). The efficiency of Cox's likelihood function for censored data, *Journal of the American Statistical Association*, **72**, 557–565.

Efron, B. (1994). Missing data, imputation, and the bootstrap, *Journal of the American Statistical Association*, **89**, 463–479.

Efron, B. and Tibshirani, R. (1993). *An Introduction to the Bootstrap*. New York: Chapman & Hall.

Elliott, M. (2008). Model averaging methods for weight trimming, *Journal of Official Statistics*, **24**, 517–540.

Eltinge, J. and Yansaneh, I. (1997). Diagnostics of formation of nonresponse adjustment cells, with an application to income nonresponse in the U.S. consumer expenditure survey, *Survey Methodology*, **23**, 33–40.

Fay, R. (1996). Alternative paradigms for the analysis of imputed survey data, *Journal of the American Statistical Association*, **91**, 490–498.

Fuller, W. (1975). Regression analysis for sample survey, *Sankhyā*, **37**, Series C, Pt. 3, 117–132.

Gelman, A. (2004). Parameterization and Bayesian modeling, *Journal of the American Statistical Association*, **99**, 537–545.

Graubard, B. and Korn, E. (1996). Survey inference for subpopulations, *American Journal of Epidemiology*, **144**, 102–106.

Graubard, B., Korn, E., and Midthune, D. (1997). Testing goodness-of-fit for logistic regression with survey data, *Proceedings from the Joint Statistical Meetings of the American Statistical Association*, American Statistical Association, Alexandria, VA.

Groves, R. (1989). *Survey Errors and Survey Costs*. New York: Wiley.

Groves, R. and Couper, M. (1998). *Nonresponse in Household Interview Surveys*. New York: Wiley.

Groves, R., Fowler, F., Couper, M., Lepkowski, J., Singer, E., and Tourangeau, R. (2009). *Survey Methodology*. Hoboken, NJ: Wiley.

Groves, R. and Peytcheva, E. (2008). The impact of nonresponse rates on nonresponse bias: A meta-analysis, *Public Opinion Quarterly*, **72**, 167–189.

Hansen, M. (1987). Some history and reminiscences on survey sampling, *Statistical Science*, **2**, 180–190.

Hansen, M., Hurwitz, W., and Madow, W. (1953). *Sample Survey Methods and Theory*, Vols. I and II. New York: Wiley.

Hartley, H. (1946). Discussion of "A review of recent statistical developments in sampling and sampling surveys" by F. Yates, *Journal of the Royal Statistical Society, Series A*, **109**, 37–38.

Hawkes, R. (1997). Implementing balanced replicated subsampling designs in SAS® Software, *Proceedings of the SAS Users Group International (SUGI) Conference*. Available online at: http://www2.sas.com/proceedings/sugi22/STATS/PAPER279.PDF. Accessed on 31 January, 2016.

Heeringa, S., West, B., and Berglund, P. (2010). *Applied Survey Data Analysis*. Boca Raton, FL: Chapman & Hall/CRC Press.

Heitjan, F. and Little, R. (1991). Multiple imputation for the fatal accident reporting system, *Applied Statistics*, **40**, 13–29.

Hidiroglou, M., Fuller, W., and Hickman, R. (1980). *Super Carp*. Ames, IA: Statistical Laboratory, Iowa State University.

Hing, E., Gousen, S., Shimizu, I., and Burt, C. (2003). Guide to using masked design variables to estimate standard errors in public use files of the National Ambulatory Medical Care Survey and the National Hospital Ambulatory Medical Care Survey, *Inquiry*, **40**, 401–415.

Holt, D. and Smith, T. (1979). Poststratification, *Journal of the Royal Statistical Society, Series A*, **142**, 33–46.

Horton, N. and Lipsitz, S. (2001). Multiple imputation in practice: Comparison of software packages for regression models with missing variables, *The American Statistician*, **55**, 244–254.

Horvitz, D. and Thompson, D. (1952). A generalization of sampling without replacement from a finite universe, *Journal of the American Statistical Association*, **47**, 663–685.

Hosmer, D. and Lemeshow, S. (1980). Goodness-of-fit tests for the multiple logistic regression model, *Communications in Statistics: Theory and Methods, Part A*, **9**, 1043–1069.

Hosmer, D., Lemeshow, S., and Sturdivant, R. (2013). *Applied Logistic Regression*, 3rd edn. New York: Wiley.

Izrael, D., Battaglia, M., and Frankel, M. (2009). Extreme survey weight adjustment as a component of sample balancing (a.k.a. Raking), *Proceedings of the SAS Global Forum*. Available online at: http://support.sas.com/resources/papers/proceedings09/247-2009.pdf. Accessed on 31 January, 2016.

Izrael, D., Hoaglin, D., and Battaglia, M. (2000). A SAS macro for balancing a weighted sample, *Proceedings of the SAS Users Group International (SUGI) Conference*. Available online at: http://www2.sas.com/proceedings/sugi25/25/st/25p258.pdf. Accessed on 31 January, 2016.

Izrael, D., Hoaglin, D., and Battaglia, M. (2004). To rake or not to rake is not the question anymore with the enhanced raking macro, *Proceedings of the SAS Users Group International (SUGI) Conference*. Available online at: http://www2.sas.com/proceedings/sugi29/207-29.pdf. Accessed on 31 January, 2016.

Judkins, D. (1990). Fay's method for variance estimation, *Journal of Official Statistics*, **6**, 223–240.

Kalbfleisch, J. and Prentice, R. (2002). *The Statistical Analysis of Failure Time Data*, 2nd edn. New York: Wiley.

Kalton, G. (1983a). *Compensating for Missing Survey Data*. Ann Arbor, MI: Institute for Social Research, University of Michigan.

Kalton, G. (1983b). *Introduction to Survey Sampling*, Sage University Paper Series on Quantitative Applications in the Social Sciences, 07-035. Newbury Park, CA: Sage.

Kalton, G. and Flores-Cervantes, I. (2003). Weighting methods, *Journal of Official Statistics*, **19**, 81–97.

Kass, G. (1980). An exploratory technique for investigating large quantities of categorical data, *Applied Statistics*, **29**, 119–127.

Kiaer, A. (1895). Observations et Experiences Concernat des Denombrements Representatives, *Bulletin of the International Statistical Institute*, **9**, Liv. 2, 176–183.

Kim, J. and Fuller, W. (2004). Fractional hot deck imputation, *Biometrika*, **91**, 559–578.

Kish, L. (1965). *Survey Sampling*. New York: Wiley.

Kish, L. (1992). Weighting for unequal P_i, *Journal of Official Statistics*, **8**, 183–200.

Korn, E. and Graubard, B. (1990). Simultaneous testing of regression coefficients with complex survey data: Use of Bonferroni T statistics, *The American Statistician*, **44**, 270–276.

Korn, E. and Graubard, B. (1998). Confidence intervals for proportions with small expected number of positive counts estimated from survey data, *Survey Methodology*, **24**, 193–201.

Korn, E. and Graubard, B. (1999). *Analysis of Health Surveys*. New York: Wiley.

Kott, P. (1991). A model-based look at linear regression with survey data, *The American Statistician*, **45**, 107–112.

Kott, P. (2001). The delete-a-group jackknife, *Journal of Official Statistics*, **17**, 521–526.

Kovar, J., Rao, J.N.K., and Wu, C. (1988). Bootstrap and other methods to measure errors in survey estimates, *Canadian Journal of Statistics*, **16**, 25–45.

Lahiri, P. (2003). On the impact of bootstrap in survey sampling and small-area estimation, *Statistical Science*, **18**, 199–210.

Lee, E. and Forthofer, R. (2006). *Analyzing Complex Survey Data*, 2nd edn. Thousand Oaks, CA: Sage.

Lepkowski, J., Mosher, W., Davis, K., Groves, R., and Van Hoewyk, J. (2010). *The 2006–2010 National Survey of Family Growth: Sample Design and Analysis of a Continuous Survey. Vital Health Statistics*, Series 2, Vol. 150. Hyattsville, MD: National Center for Health Statistics.

Lessler, J. and Kalsbeek, W. (1992). *Nonsampling Error in Surveys*. New York: Wiley.

Leung, K.-M., Elashoff, R., and Afifi, A. (1997). Censoring issues in survival analysis, *Annual Review of Public Health*, **18**, 83–104.

Lewis, T. (2007). PROC LOGISTIC: The logistics behind interpreting categorical variable effects, *Proceedings of the Annual Conference of the Northeast SAS Users Group (NESUG)*. Available online at: http://www.lexjansen.com/nesug/nesug07/sa/sa11.pdf. Accessed on 31 January, 2016.

Lewis, T. (2013a). PROC SURVEYSELECT as a tool for drawing random samples, *Proceedings of the Annual Conference of the Southeast SAS Users Group (SESUG)*. Available online at: http://analytics.ncsu.edu/sesug/2013/SD-01.pdf. Accessed on 31 January, 2016.

Lewis, T. (2013b). Tools for imputing missing data, *Proceedings of the Annual Conference of the Western Users of SAS Software (WUSS)*. Available online at: http://www.lexjansen.com/wuss/2013/27_Paper.pdf. Accessed on 31 January, 2016.

Lewis, T., Goldberg, E., Schenker, N., Beresovsky, V., Schappert, S., Decker, S., Sonnenfeld, N., and Shimizu, I. (2014). The relative impacts of design effects and multiple imputation on variance estimates: A case study with the 2008 National Ambulatory Medical Care Survey, *Journal of Official Statistics*, **30**, 147–161.

Lewis, T., Hess, K., Ezoua, S., and Marcu, M. (2014). The challenge in forecasting federal employee retirements, *Presentation given at the Federal Forecasters Conference (FFC)*, Bureau of Labor Statistics, Washington, DC.

Li, J. and Valliant, R. (2009). Survey weighted hat matrix and leverages, *Survey Methodology*, **35**, 15–24.

Li, J. and Valliant, R. (2011a). Linear regression influence diagnostics for unclustered survey data, *Journal of Official Statistics*, **27**, 99–119.

Li, J. and Valliant, R. (2011b). Detecting groups of influential observations in linear regression using survey data—Adapting the forward search method, *Pakistan Journal of Statistics*, **27**, 507–528.

Little, R. (1986). Survey nonresponse adjustments for estimates of means, *International Statistical Review*, **54**, 139–157.

Little, R. and Rubin, D. (2002). *Statistical Analysis with Missing Data*, 2nd edn. New York: Wiley.

Little, R. and Vartivarian, S. (2003). On weighting the rates in non-response weights, *Statistics in Medicine*, **22**, 1589–1599.

Little, R. and Vartivarian, S. (2005). Does weighting for nonresponse increase the variance of survey means? *Survey Methodology*, **31**, 161–168.

Lohr, S. (2009). *Sampling: Design and Analysis*, 2nd edn. Boston, MA: Brooks/Cole.

Lohr, S. (2012). Using SAS® for the design, analysis, and visualization of complex surveys, *Proceedings of the SAS Global Forum*. Available online at: http://support.sas.com/resources/papers/proceedings12/343-2012.pdf. Accessed on 31 January, 2016.

Lohr, S. and Rao, J.N.K. (2000). Inference from dual frame surveys, *Journal of the American Statistical Association*, **95**, 271–290.

McCarthy, P. (1966). Replication: An approach to the analysis of data from complex surveys, *Vital and Health Statistics*, Series 2, **14**. Washington, DC: U.S. Government Printing Office.

McCarthy, P. and Snowden, C. (1985). The bootstrap and finite population sampling, *Vital and Health Statistics, Series 2*, **95**. Washington, DC: U.S. Government Printing Office.

McCullagh, P. (1980). Regression models for ordinal data, *Journal of the Royal Statistical Society, Series B*, **42**, 109–142 (with discussion).

Merkle, D. and Edelman, M. (2002). Nonresponse in exit pools: A comprehensive analysis, in *Survey Nonresponse*, Groves, R., Dillman, D., Eltinge, J., and Little, R. (eds.). New York: Wiley.

Milhøj, A. (2013). *Practical Time Series Analysis Using SAS®*. Cary, NC: SAS Institute.

Moore, D. and McCabe, G. (2006). *Introduction to the Practice of Statistics*, 5th edn. New York: W.H. Freeman and Company.

Morel, J. (1989). Logistic regression under complex survey designs, *Survey Methodology*, **15**, 203–223.

Mukhopadhyay, P. (2010). Not hazardous to your health: Proportional hazards modeling for survey data with the SURVEYPHREG procedure, *Proceedings of the SAS Global Forum*. Available online at: https://support.sas.com/resources/papers/proceedings10/254-2010.pdf. Accessed on 31 January, 2016.

Mukhopadhyay, P., An, A., Tobias, R., and Watts, D. (2008). Try, try again: Replication-based variance estimation methods for survey data analysis in SAS® 9.2, *Proceedings of the SAS Global Forum*. Available online at: http://www2.sas.com/proceedings/forum2008/367-2008.pdf. Accessed on 31 January, 2016.

Murthy, M. (1957). Ordered and unordered estimators in sampling without replacement, *Sankhyā*, **18**, 379–390.

Nagelkerke, N. (1991). A note on a general definition of the coefficient of determination, *Biometrika*, **78**, 691–692.

Neyman, J. (1934). On the two different aspects of the representative method: The method of stratified sampling and the method of purposive selection, *Journal of the Royal Statistical Society*, **97**, 558–606.

Pfeffermann, D. (1996). The use of sampling weights for survey data analysis, *Statistical Methods in Medical Research*, **5**, 239–261.

Platek, R. and Gray, G. (1983). Imputation methodology: Total survey error, in *Incomplete Data in Sample Surveys*, Vol. 2, Madow, W., Olkin, I., and Rubin, D. (eds), pp. 249–333. New York: Academic Press.

Politz, A. and Simmons, W. (1949). An attempt to get the not-at-homes into the sample without callbacks, *Journal of the American Statistical Association*, **44**, 9–31.

Quenouille, M., (1949). Approximate tests of correlation in time series, *Journal of the Royal Statistical Society, Series B*, **11**, 68–84.

Raghunathan, T., Lepkowski, J., Van Hoewyk, J., and Solenberger, P. (2001). A multivariate technique for multiply imputing missing values using a sequence of regression models, *Survey Methodology*, **27**, 85–95.

Raghunathan, T., Solenberger, P., and Van Hoewyk. J. (2002). *IVEware: Imputation and Variance Estimation Software: User Guide*. Ann Arbor, MI: Institute for Social Research, University of Michigan.

Rao, J.N.K. and Scott, A. (1981). The analysis of categorical data from complex sample surveys: Chi-squared tests for goodness of fit and independence in two-way tables, *Journal of the American Statistical Association*, **76**, 221–230.

Rao, J.N.K. and Scott, A. (1984). On chi-squared tests for multiway contingency tables with cell proportions estimated from survey data, *Annals of Statistics*, **12**, 46–60.

Rao, J.N.K. and Shao, J. (1992). Jackknife variance estimation with survey data under hot deck imputation, *Biometrika*, **79**, 811–822.

Rao, J.N.K. and Wu, C. (1988). Resampling inference with complex survey data, *Journal of the American Statistical Association*, **83**, 231–241.

Reiter, J., Raghunathan, T., and Kinney, S. (2006). The importance of modeling the sampling design in multiple imputation for missing data, *Survey Methodology*, **32**, 143–149.

Roberts, G., Rao, J.N.K., and Kumar, S. (1987). Logistic regression analysis of sample survey data, *Biometrika*, **74**, 1–12.

Rosenbaum, P. and Rubin, D. (1983). The central role of the propensity score in observational studies for causal effects, *Biometrika*, **70**, 41–55.

Rubin, D. (1978). Multiple imputations in sample surveys—A phenomenological Bayesian approach to nonresponse, *Proceedings of the Joint Statistical Meetings of the American Statistical Association*, American Statistical Association, Alexandria, VA.

Rubin, D. (1987). *Multiple Imputation for Nonresponse in Surveys*. New York: Wiley.

Rubin, D. (1996). Multiple imputation after 18+ years, *Journal of the American Statistical Association*, **91**, 473–489.

Rubin, D. and Schenker, N. (1986). Multiple imputation for interval estimation from simple random samples with ignorable nonresponse, *Journal of the American Statistical Association*, **81**, 366–374.

Rust, K. (1985). Variance estimation for complex estimators in sample surveys, *Journal of Official Statistics*, **1**, 381–397.

Rust, K. (1986). Efficient formation of replicates for replicated variance estimation, *Proceedings of the Survey Research Methods Section of the American Statistical Association*, American Statistical Association, Alexandria, VA.

Rust, K. and Hsu, V. (2007). Confidence intervals for statistics for categorical variables from complex samples, *Proceedings of the Survey Research Methods Section of the American Statistical Association*, American Statistical Association, Alexandria, VA.

Rust, K. and Rao, J.N.K. (1996). Replication methods for analyzing complex survey data, *Statistical Methods in Medical Research: Special Issue on the Analysis of Complex Surveys*, **5**, 283–310.

Sampford, M. (1967). On sampling without replacement with unequal probabilities of selection, *Biometrika*, **54**, 499–513.

Särndal, C. (1992). Methods for estimating the precision of survey estimates when imputation has been used, *Survey Methodology*, **18**, 241–252.

Schafer, J. (1997). *Analysis of Incomplete Multivariate Data*. New York: Chapman & Hall.

Scheaffer, R., Mendenhall, W., and Ott, R. (1996). *Elementary Survey Sampling*, 5th edn. Belmont, CA: Duxbury.

Schenker, N., Raghunathan, T., Chiu, P.L., Makuc, D., Zhang, G., and Cohen, A. (2006). Multiple imputation of missing income data in the National Health Interview Survey, *Journal of the American Statistical Association*, **101**, 924–933.

Schenker, N. and Taylor, J. (1996). Partially parametric techniques for multiple imputation, *Computational Statistics and Data Analysis*, **22**, 425–446.

Shah, B., Holt, M., and Folsom, F. (1977). Inference about regression models from sample survey data, *Bulletin of the International Statistical Institute*, **41**, 43–57.

Singer, J. and Willett, J. (2003). *Applied Longitudinal Data Analysis: Modeling Change and Event Occurrence*. New York: Oxford University Press.

Skinner, C., Holt, D., and Smith, T. (1989). *Analysis of Complex Surveys*. New York: Wiley.

Sonquist, J., Baker, E., and Morgan, J. (1974). *Searching for Structure*. Ann Arbor, MI: Institute for Social Research, University of Michigan.

Thomas, D. and Rao, J.N.K. (1987). Small-sample comparisons of level and power for simple goodness-of-fit statistics under cluster sampling, *Journal of the American Statistical Association*, **82**, 630–636.

Tourangeau, R., Rips, L., and Rasinski, K. (2000). *The Psychology of Survey Response*. New York: Cambridge University Press.

Tukey, J. (1958). Bias and confidence in not-quite large samples (abstract), *Annals of Mathematical Statistics*, **29**, 614.

U.S. Office of Personnel Management. (2013). 2013 Federal Employee Viewpoint Survey Results: Technical report. Available online at: http://www.fedview.opm.gov/2013/Published/.

Valliant, R. (2004). The effect of multiple weight adjustments on variance estimation, *Journal of Official Statistics*, **20**, 1–18.

Valliant, R., Brick, J.M., and Dever, J. (2008). Weight adjustments for the grouped jackknife variance estimator, *Journal of Official Statistics*, **24**, 469–488.

Valliant, R., Dever, J., and Kreuter, F. (2013). *Practical Tools for Designing and Weighting Survey Samples*. New York: Springer.

Valliant, R., and Rust, K. (2010). Degrees of freedom approximations and rules-of-thumb, *Journal of Official Statistics*, **26**, 585–602.

Van Buuren, S. (2007). Multiple imputation of discrete and continuous data by fully conditional specification, *Statistical Methods in Medical Research*, **16**, 219–242.

Van Buuren, S., Boshuizen, H., and Knook, D. (1999). Multiple imputation of missing blood pressure covariates in survival analysis, *Statistics in Medicine*, **18**, 681–694.

Van Buuren, S., Brand, J., Groothuis-Oudshoorn, K., and Rubin, D. (2006). Fully conditional specification in multivariate imputation, *Journal of Statistical Computation and Simulation*, **76**, 1049–1064.

Van Buuren, S. and Groothuis-Oudshoorn, K. (2011). MICE: Multivariate Imputation by Chained Equations in R, *Journal of Statistical Software*, **45**, 1–67.

Vijayan, K. (1968). An exact π_{ps} sampling scheme: Generalization of a method of Hanurav, *Journal of the Royal Statistical Society, Series B*, **30**, 556–566.

Vollset, S. (1993). Confidence intervals for a binomial proportion, *Statistics in Medicine*, **12**, 809–824.

Weisberg, S. (2013). *Applied Linear Regression*, 4th edn. Hoboken, NJ: Wiley.

West, B. (2010). Accounting for multi-stage sample designs in complex sample variance estimation, Institute for Social Research Technical Report Prepared for National Survey of Family Growth (NSFG) User Documentation. Available online at: http://www.isr.umich.edu/src/smp/asda/first_stage_ve_new.pdf. Accessed on 31 January, 2016.

Westat. (2007). *WesVar® User's Guide*. Available online at: https://www.westat.com/our-work/information-systems/wesvar%C2%AE-support/wesvar-documentation. Accessed on 31 January, 2016.

Wicklin, R. (2010). *Statistical Programming with SAS/IML Software*. Cary, NC: SAS Institute.

Wilson, E. (1927). Probable inference, the law of succession, and statistical inference, *Journal of the American Statistical Association*, **22**, 209–212.

Wolter, K. (1984). An investigation of some estimators of variance for systematic sampling, *Journal of the American Statistical Association*, **79**, 781–790.

Wolter, K. (2007). *Introduction to Variance Estimation*, 2nd edn. New York: Springer.

Woodruff, R. (1952). Confidence intervals for medians and other position measures, *Journal of the American Statistical Association*, **47**, 635–646.

Woodruff, R. (1971). A simple method for approximating the variance of a complicated estimate, *Journal of the American Statistical Association*, **66**, 411–414.

Zhou, H. (2014). Accounting for complex sample designs in multiple imputation using the finite population Bayesian bootstrap, PhD dissertation, University of Michigan, Ann Arbor, MI.

Index

Printed in the United States
by Baker & Taylor Publisher Services